HANDLOADING for HANDGUNNERS

By Maj. George C. Nonte, Jr.

DBI BOOKS, Inc., Northfield, Ill.

Handloading for Handgunners
Editorial Staff

Editor
Robert S.L. Anderson

Production Manager
Pamela J. Johnson

Cover Photography
John Hanusin

Associate Publisher
Sheldon L. Factor

ISBN 0-695-81199-1 Library of Congress Catalog Card Number 78-64340

CONTENTS

Introduction

"**ANOTHER** handloading book? Good Lord, what for?"

One friend exclaimed those very words when on a visit he asked what my next book would be and I told him, HANDLOADING FOR HANDGUNNERS. A few others voiced similar though less emphatic comments, but a few responded, "Good idea! None of the others I've read get into the real nitty-gritty. They all talk mostly about rifles."

Actually, over the past 20 years I've received thousands of letters from readers inquiring about handloading. One thing I've noted about them is that the writers seem to have had the most trouble locating information on loading for handguns. Loading data manuals traditionally contain lots of handgun loads but very little detail on assembling them or adapting them to various guns and uses. On the other hand, rifle loading is covered in excruciating detail (and I don't mean that's bad; I've written a lot of it). But the data extends to factors which are of interest to only a minute fraction of today's handloaders.

Handgun handloaders on the other hand, haven't been that fortunate. Many books ignore the loading of handgun calibers entirely, while some give it treatment ranging from fair to good. All this seems to be based on the assumption that rifle loading is a highly developed art (it is), but that pistoleros are so inaccurate and unfeeling that information on merely refilling the fired case with primer, propellant and bullet will give them all they can use.

Tain't true, brethren of the short gun, because producing the best six-gun and auto loads for *your* particular purposes can be just as exacting an art and just as involved as assembling bench-rest rifle ammunition. And, it appears to be little known that the techniques for doing so differ much from rifle techniques. Even the basic tools and operations are often different for handgun loading. The differences are multitudinous as a result of new components, techniques, and even guns and cartridges developed over the past couple of decades. Think back over the many fine magazine articles published on the subject since about 1960, and the uniqueness of *modern* handgun loading will become clearly evident. Handgun hunting has given rise to special techniques, as has development of high-performance factory ammunition, *truly* effective expanding bullets, and new understandings of the factors pertinent to accuracy and reliability.

In short, proper loading of handgun cartridges is no longer just a quick and dirty duplication of basic factory loads dating from before 1900. The process is fast developing, paralleling the earlier and slower growth of rifle techniques begun in the 1930's and the quick revolution of shotshell loading that came in the late 1950's and early 1960's.

But, new developments alone wouldn't necessarily justify a book devoted purely to handgun loading. Forty years ago handgun handloaders were few and far between, and handguns themselves were hardly as plentiful as now. Today, over *2½ million new* handguns are sold every year, and at least as many used specimens change hands as well. Conservative estimates place in excess of *50 million* handguns in the possession of private citizens. Not all of them are shot often, but millions are. No one knows how many millions of handgun *shooters* exist in this country, but we do know they constitute a vast multitude. There are at least *seven* different forms of organized competition with modern handguns, and each is growing by leaps and bounds—and there is hardly a competitive shooter alive who doesn't handload to improve his performance *and* to save money.

All of these factors combine to make a separate book on handloading not only desirable, but even necessary if today's shooters are to reap the benefits of work by dozens of experimenters during the past two decades. Actually much of the information in this book has probably been published *somewhere*. To find it all, though, a fellow would have to dig through dozens of books and thousands of magazines published since the 1920's. What we hope here is that we have managed to put the best and most useful information of both the new and the old between a single pair of covers to form a single-source reference second to none in the field. We don't pretend this book will answer *all* questions and please *all* readers—but it will come a lot closer to doing so than any *three* other books in existence.

Whether you handload for fun or money and whether you shoot six-guns or autos for meat, trophies, or to save your own life, this book is intended to enable you to do it all better. If it does so, even in a small measure, then the time and money has been well spent and I shall forever be pleased that in some small way I have contributed to the growth of the finest sport I know—SHOOTING.

George C. Nonte, Jr.
Maj. ORDC, Ret. USA
Peoria, Illinois
June, 1978

CHAPTER 1

Why Handload?

WHY INDEED? Why under the sun would any reasonably intelligent human being and handgunner want to burden himself down with the purchase of tools and equipment, the building of a bench, and the storage and use of a wide variety of primers, propellants, cartridge cases, bullets, etc.? Handloading requires space, tools, investment, time and effort. Obviously, then, a person isn't going to become involved in something like that unless he can see significant advantages to be gained, or unless he simply has a consuming passion for handloading. We'll give you a handful of them here, and you'll discover more with every chapter you read through this book.

Economy is the reason most often cited for most handloading, and personally I believe that this has been overdone. Certainly, economy is a major underlying reason, and the savings that can be achieved by some types of handloading are little short of phenomenal. Yet, he who loads for economy alone is missing many of the advantages of this game. Sure, it's nice to be able to say that for 4¢ you can produce at your bench a cartridge equal in performance to one costing seven or eight times that much at a store. But I feel it is equally valuable to be able to say that you can *produce* a cartridge, possessing unique capabilities to suit a particular purpose, when no such cartridge is available from the factories. In some instances, you might spend several times the factory price in assembling a special-purpose load, and that is certainly no economy when measured in pure dollars and cents—however, it demonstrates that handloading gives you the capability of producing a cartridge that the factories either *cannot* or *will not* make and sell. So, while economy may be the most *common* reason for handloading, at least in the beginning, to a great many people it is by no means the most *important*.

In any event, since economy is probably the first word that came into your mind when you read the title of this chapter, let's go into it in a bit more detail. In loading for handguns, there are roughly three levels of economy to be obtained. Maximum savings are achieved in the assembly of light, target-type, lead-bullet loads (including the ultra-light types described in a later chapter) which require minimum amounts of lead and propellant and which place the least stress and strain on cases and therefore afford the longest case life. And, of course, this maximum

FACTORY COSTS *VS.* RELOADING COSTS

Cost of 500 Rounds

	Factory	Reload (50 factory loads 9 times)
.45 ACP Semi-Wadcutler Mid-Range Target Load	$ 14.75 (Box of 50) Total for 500 $147.50 Cost per Shot 29.5¢	$14.75 (Box 50 factory bullets) 3.83 (450 primers) 2.45 (1/4-pound Hercules Bullseye powder) 1.95 (13 pounds lead) .55 (lubricant) Total for 500 $23.53 Cost per Shot 4.7¢
.44 Magnum	$ 17.50 (Box of 50) Total for 500 $175 Cost per Shot 35¢	$17.50 (Box 50 factory bullets) 3.83 (450 primers) 13.00 (powder) 31.50 (Factory jacketed expanding bullets) .55 (lubricant) Total for 500 $66.38 Cost per Shot 13.2¢

economy is obtained only when one makes (casts, sizes, and lubricates) his own bullets from low-cost (or free) scrap lead and utilizes the most efficient and therefore the smallest amount of propellant. Let's examine the savings that can be obtained in this fashion in two of our most popular target-shooting calibers, the .45 ACP and .38 Special. We'll do it on the basis of 500 rounds purchased at regular retail price as opposed to purchasing *50 rounds* and reloading those 50 fired cases nine times to produce a total of 500 rounds.

At $14.75 per box of 50, those 500 rounds of .45 ACP semi-wadcutter, mid-range target loads will cost $147.50; the same amount of .38 Special mid-range wadcutter cartridges will cost $103 at $10.30 per box of 50. Incidentally, those are rather impressive costs, considering the fact that one could easily burn up 500 rounds in a couple of pleasant weekends of shooting.

In handloading the .45 ACP, we start with $14.75 for the initial box of 50 factory cartridges which we fire and produce a like number of fired cases (assuming you don't lose some of them in the grass and weeds); then we add the cost of 450 large pistol primers (standard type), which is $3.83 at the current retail price of $8.50 per 1,000. To that we must add the cost of propellant, slightly less than 1/4-pound of Hercules Bullseye (the most efficient and therefore most economical powder), costing about $2.45 at the current price of $9.81 per 1-pound cannister. All we need now is 450 lead bullets suitably lubricated and sized; a typical weight for the type is 200 grains, which means we'll use about 13 pounds of lead if waste is held to a minimum. Automobile wheel weights which serve the purpose well can usually be picked up from your most favored service station for around 15¢ a pound if not for the asking. Bullet metal cost is then about $1.95, and a suitable first-class lubricant might add as much as another 55¢ to bullet cost, so let's call the total $2.50. Add all that together, materials and components plus the cost of the first box of cartridges, and we come up with a total cost of $23.53 for 500 shots. Viewed on an individual basis, that figures out to 4.7¢ per shot for your home-brewed product. Factory loads cost 29.5¢ each. If just the simple comparison of those figures isn't adequate to impress you with the savings, look at it this way—your handloads cost

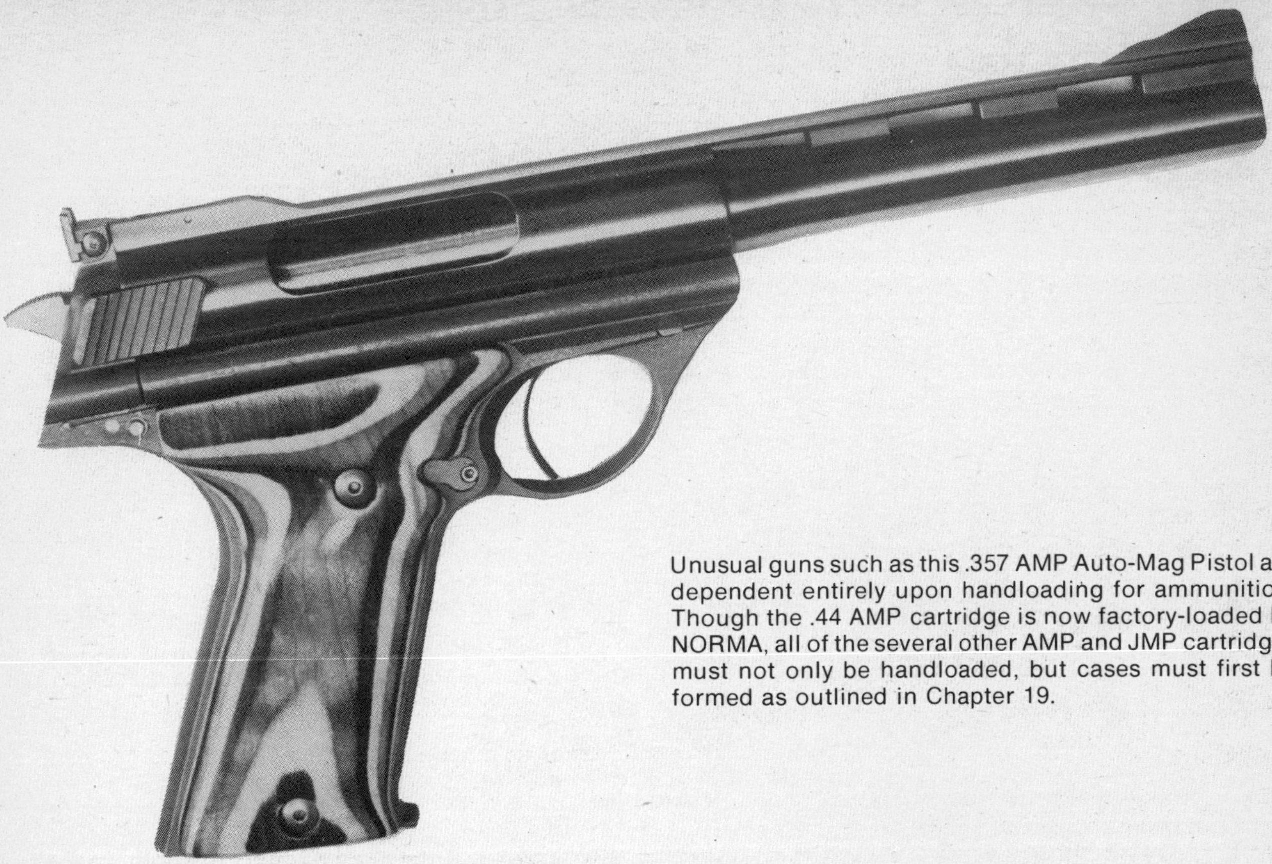

Unusual guns such as this .357 AMP Auto-Mag Pistol are dependent entirely upon handloading for ammunition. Though the .44 AMP cartridge is now factory-loaded by NORMA, all of the several other AMP and JMP cartridges must not only be handloaded, but cases must first be formed as outlined in Chapter 19.

only 15.9 percent of the factory price or, you can have 6 I handloaded shots for the cost of one factory shot. Pretty impressive, I'd say.

The .38 Special factory load costs substantially less and uses 20 percent less propellant (Bullseye again) and bullet costs will be about 25 percent less. When that is all added up, your home-loaded .38 Specials will cost you about 3.8¢ each as opposed to 20.6¢ for the factory loads. The saving isn't quite as great as with the .45, but I think you can look all over town and not find a bargain that even closely approaches this.

Savings for full-charge, lead-bullet loads (those approximating factory ballistic performance) cost a bit more to assemble for three reasons: first, they use more powder; second, they usually require heavier bullets and therefore more lead; and last, case life is a wee bit less. Actually, in anything less than full-charge magnum calibers, case life differs so little as to be insignificant; far more cases are lost on the range or damaged in loading and handling than will ever be "worn out." In any event, using the same basis for comparison as above, that first 500 full-charge 45 ACP reload will cost you 4.7¢ per shot as opposed to the 29.5¢ cost of factory fodder—a savings of 24¢ per round. In as much as the factory service-type reload costs a bit more than the target reload, the cost of your service-type handload is therefore slightly higher, reducing the savings over the lighter loads. In .38 Special, the full-charge factory load costs 20.6¢ per round and your first 500 handloads cost 3.5¢ each—a 17-cent cost-per-round reduction. Not quite as much money as you saved with the target loads, but still one hell of a deal.

High-performance loads with jacketed bullets offer the least savings of all. This is due to the simple fact that case life is much less, much larger quantities of propellant are required, and jacketed, expanding bullets are quite costly as factory products; and even if you are equipped to make your own, the materials cost is several times that for lead bullets. Still, quite substantial savings can be achieved.

Again using our 500-round factory/handload comparison, .44 Magnum cartridges (the most expensive of all production handgun ammunition) cost an astonishing 35¢ each—$175 per 500! Comparable handloads will require over $13 for

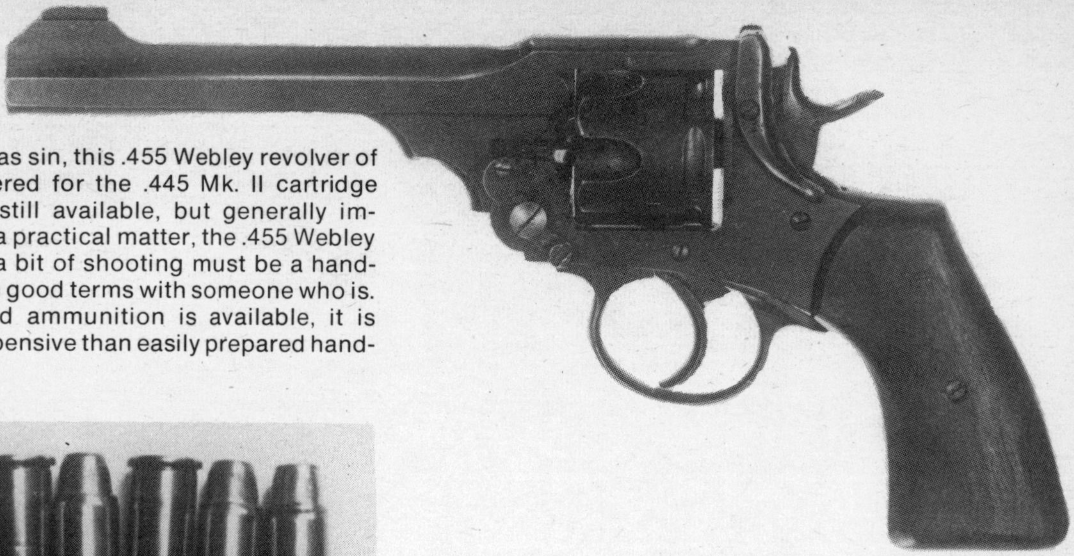

Big, clumsy, and ugly as sin, this .455 Webley revolver of WWI fame is chambered for the .445 Mk. II cartridge which is technically still available, but generally impossible to obtain. As a practical matter, the .455 Webley owner wishing to do a bit of shooting must be a handloader or at least be on good terms with someone who is. Even if factory-loaded ammunition is available, it is much, much, more expensive than easily prepared handloads.

This may help you visualize how much money handloading can save; at left is a single round of factory-loaded .38 Special ammunition, and at right are the half-dozen identical rounds that you can load for its cost; about six times as much shooting for the same money.

propellant and about $31.50 for factory jacketed expanding bullets; and, of course, the usual primer cost which is the same for all types of loads. When these figures are all run through the calculator, the per-shot cost of those high-performance handloads comes out to 13.2¢ per shot—that is 38 percent of factory-load cost.

It isn't always quite that bad, for as already indicated, if you're willing to invest in the equipment, you can make your own jacketed expanding bullets. With care, they can be assembled for a materials cost sometimes as little as one-half the price of comparable commercial bullets, thus reducing the per-shot cost to about 10¢—that's 25¢ less than the factory load.

Naturally, in discussing economy as a reason for handloading, we cannot overlook the cost of tools and equipment. This subject is discussed in detail further along in this volume, but a bare-bones minimum outfit for loading fired cases and casting and sizing lead bullets will cost from $50 to $75—and can go to many, many times that. But, using the minimum outfit for comparison, you can see that with light target loads (remember maximum savings) the entire cost of a $75 starting outfit can be amortized over the loading of only 300 rounds of .45 ACP or 450 rounds of .38 Special. If, on the other hand, you choose to amortize the tool and equipment cost over a period of several years, or several thousand rounds, it raises the per-shot cost of your handloads such a small amount as to be virtually insignificant.

But, if you shoot much, or if you're a professional gun handler, you'll soon discover that economy is by no means everything. If you want to get into handgun metallic silhouette shooting (at ranges of 50 to 200 meters) you'll very quickly discover that no factory load exists that will produce optimum performance at the longer ranges on those heavy steel-plate targets which *must* be knocked over for a hit to score. The best commercial .44 Magnum (and to date this cartridge is the one most commonly used) does not transfer its energy efficiently to the steel targets, even at its 35¢ per round cost. To obtain the best results possible, one requires a fully-jacketed "hardball" type bullet of very strong construction. Such bullets may be purchased from independent makers, or if you possess the equipment, they can be made at home, or you may alter (by squeezing them down) .45 ACP ball bullets. Then, when these more efficient bullets are assembled with the other .44 components, you have a load that is substantially superior to the factory product for this particular purpose—and,

Above, left — Often we need, or at least want, a load which the factories simply do not supply in a given caliber. Here is a good example in that on the left is the *only* factory loading available in the excellent little .32 S&W Long cartridge. If you shoot this number, especially in one of those fine S&W or Colt target revolvers of years gone by, you must make do with this loading, import an expensive foreign load, or handload a full-wadcutter target round such as the one at the right. It not only shoots with superb accuracy in .32 revolvers, but functions perfectly in those very expensive foreign .32 S&W Long target autoloaders. Above, right — At left is a single round of factory-loaded .45 ACP target ammunition, and to its right is the 25¢ to be saved for *each shot fired* by handloading the lead-bullet load of equal performance at the extreme right.

just coincidently, it costs less; about 65 percent less than the factory fodder. I say the monetary savings is *coincidental* simply because those metallic silhouette shooters, who demand the best in ammunition, would handload to get it, even if their product costs several times *more* than factory ammunition.

There are other special-purpose loads—a good many of them, in fact—that are simply not available from the factories. While one might have occasion to be a bit irritated with the factories because of this situation, it must be kept in mind that they are in business to make a profit, and unless the sales potential of a particular load is in the many millions of rounds, they simply cannot afford the development, tooling, and equipment costs to produce it. Since the handloader doesn't have this problem, he can concoct all manner of nonstandard or special-purpose loads with relatively little cost and difficulty. For example, truly target-grade ammunition capable of fine accuracy is offered only in .45 ACP and .38 Special, and even in those calibers, one is limited to relatively lightweight bullets and low velocities. If a pistolero desires a target-grade .357 Magnum load (or perhaps, in .32 S&W Long, 9mm Parabellum, etc.), he'll either handload it or do without. This isn't to say that the existing factory loads in those calibers are *inaccurate,* but generally speaking their standard of accuracy is below

Two .44 Magnum cartridges loaded to the hilt with 180-grain JSP bullets — can you tell the difference? Performance-wise there isn't any difference, but the round on the left is fresh out of the factory box at over 35¢ each, while the one on the right is a meticulously prepared handload costing approximately one-quarter as much.

Common, garden-variety lead wheel weights scrounged from your local service station will, when melted, cleaned and cast into ingots, make superb bullets of whatever shape, size, and weight you wish to cast and load. The raw material is often free or ridiculously cheap — the bulk of your bullet cost is in your own labor.

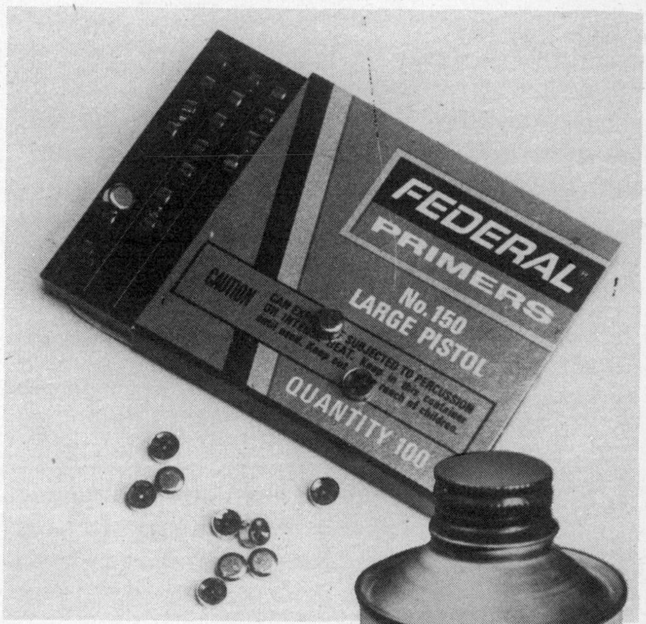

Primers are the one component for which the cost won't change, whether you load a large caliber or small caliber, a pip-squeak or a real loud loudenboomer. Though several makes and sizes are available, retail price is generally the same, and every cartridge requires just one.

that required for fine target work. In some calibers it is not always possible to handload for really top-grade accuracy simply because of inadequacies in the gun; nevertheless, it is usually possible to develop a lead-bullet handload which will shoot more accurately than the standard factory service load. A couple of examples are .44-40 caliber revolvers and .380 ACP autoloaders. The .44 is usually loaded with bullets which are substantially undersized for the chamber throats and bores of most revolvers, and in addition, the bullets are of sufficiently strong construction that they don't upset well upon firing. By matching a lead bullet to the gun, accuracy can be improved considerably. To a lesser extent, the same situation often exists with .380 pistols and ammunition, with the result that careful selection and sizing of a lead bullet will produce a significant increase in the accuracy of defensive loads. This, in turn, enhances the first-shot hit probability. One may also assemble multiple-projectile loads which have never been offered commercially by any of our major domestic ammunition makers.

With the large number of military-surplus and other obsolete handguns that came into this country some years ago, handloading offers the

This 12-ounce can of Winchester-Western ball propellant costs barely $6 as this is written and yet will provide the propulsive force for over 1,600 .45 ACP handloads.

tremendous advantage of loading ammunition in calibers which simply are no longer available commercially. For example, the Swedish Browning/Husqvarna M1907 service pistols were chambered for the 9mm Browning Long cartridge. Though once loaded in this country, this number has not been produced since before World War II, and while imported cartridges may occasionally be found, there is no dependable source and they're quite expensive. Yet, by simply trimming the .38 ACP or .38 Super Auto cases to the proper length and assembling them with the correct bullet and propellant powder charge, one may produce first-class 9mm Browning Long ammunition quite cheaply. With sometimes less, sometimes more effort, most other obsolete or unavailable handgun cartridges can be made up by the astute and knowledgeable handloader using, as a basis, some readily available cartridge case. Among them are the .455 Webley/Enfield, .45 S&W, .450 Revolver, 9mm Steyr, 9mm Bergmann-Bayard, etc. This aspect of handloading offers sufficient potential to the enthusiast who is willing to learn and experiment. We have devoted a separate chapter to the subject of case forming and wildcat cartridges. Cases for virtually every cartridge for which handguns have been chambered can be made from some readily-available modern case.

A good many handgun cartridges are loaded by the factories with only one type and weight of bullet at one standard performance level. The non-handloader must either shoot that or nothing. A typical example might be the .32 S&W Long (or Short) with its 98-grain, round-nose lead bullet at 705 fps. In solid-frame guns of 20th century design, this little cartridge may be loaded to levels of velocity and energy that double or triple its effectiveness for either hunting or defense use. Yet, to take advantage of that potential, one *must* handload. Simply substituting a semi-wadcutter bullet and increasing the propellant charge to give a velocity in the vicinity of 1000 fps turns this pip-squeak round into a most excellent small-game load which also has considerable potential for defensive use. At the other end of the scale, substituting a 92-grain wadcutter bullet and reducing the propellant charge slightly produces a delightful target load for ranges out to 50 yards. Substituting an even lighter bullet and propellant charge gives the .32 S&W Long such a quiet and dignified demeanor (combined yet with

An example of handloading versatility, the .44 Special is available factory-loaded only with the 246-grain, round-nose, lead bullet driven at barely over 700 fps shown at left. Without handloading, that's all you can shoot. To the right, we have first a superbly-accurate, mid-range, wadcutter target load; a light-bullet/low velocity plinking load; a hollow-point/semi-wadcutter/lead-bullet defense load cranking up over 1200 fps for maximum effectiveness, a 250-grain, semi-wadcutter, hard lead bullet at about 1050 fps for long-range shooting of any sort; and, last, a fine all-around load employing a 240-grain, semi-wadcutter, lead bullet at about 900 fps. Without handloading, the shooter could not use any of the last five loads.

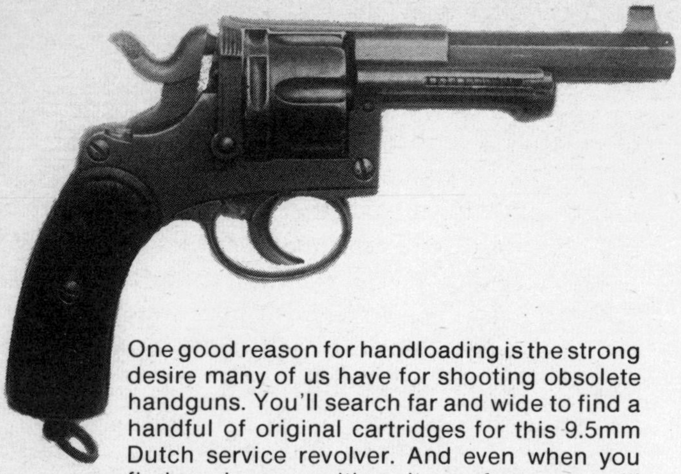

One good reason for handloading is the strong desire many of us have for shooting obsolete handguns. You'll search far and wide to find a handful of original cartridges for this 9.5mm Dutch service revolver. And even when you find such ammunition, its performance will most likely be sub-standard, and it may be impossible to reload the fired cases. However, as outlined in Chapter 19, you can *make* excellent-quality new cases to fit this gun, and with a modest supply in hand, can handload to shoot as much as you like.

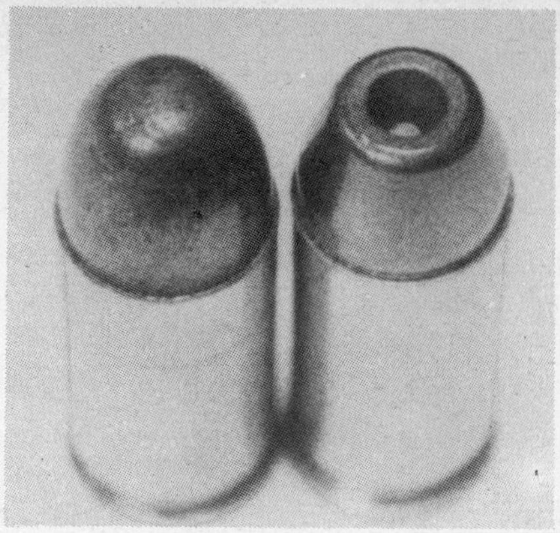

Until just the past few years, autoloading pistol cartridges were offered by the factories assembled only with the full-jacketed, round-nose bullet at left. This bullet depended purely upon its cross-sectional area and velocity for effectiveness against any type of live target. By handloading, though, bullets of the jacketed, hollowpoint (JHP) type at right may be used at varying velocities to achieve an effect upon live targets that is vastly more effective than the full-jacketed, "hardball" factory bullet.

superb accuracy in some guns) that it can be shot in your basement or garage. Even something so offbeat as the 7.65mm French military cartridge can be vastly improved for a number of purposes by substituting a lighter, home-swaged, jacketed, expanding bullet and a larger propellant charge to give it higher velocity, flatter trajectory, and greater energy. In fact, there is hardly any handgun cartridge whatever which cannot be improved for at least some shooting purposes by increasing or decreasing bullet weight and propellant charge. And so it goes.

The last reason for handloading we are going to give here is by no means the least—in fact, once one gets a bit deeper into the game, it often becomes more important than any of the others. That is simply the self-satisfaction and personal pride and sense of accomplishment that one obtains from, first, making his own ammunition and, second, making it to suit his particular needs better than can be done by the factories. This is no small matter in these days when our technological society has taken over so many functions that man used to perform for himself. Man has always taken pleasure in making things and has always felt a genuine *need* to produce worthwhile objects. This basic need has never been more evident than it has in the tremendous upsurge of "do-it yourself" activities in recent years. Certainly a fellow feels productive when he's assembled an end table or piece of lawn furniture, or perhaps has plastered a ceiling, but these are more or less passive accomplishments. Once completed, such projects don't really accomplish anything except to be seen or used in a rather prosaic manner. On the other hand, the making of ammunition—handloading (or reloading, if you wish)—produces much more active functionality. To brew up a hunting handload, then take it into the woods and use it—to the accompaniment of thunder, lightning, and felt recoil—to down 100 pounds of tasty venison is a hell of a lot different than simply sitting in a lawn chair you've made. Action is the key to accomplishment here, and few things we may own and use today can be considered more active than a loaded handgun. Any way you look at it, whether you load out of pure economic necessity or simply because it is an enjoyable pastime, the practice is creative in a real sense; and, it does provide satisfaction in a degree seldom encountered in other leisure-time activities.

As we said in the beginning, these are only a few of the reasons you'll discover for handloading. As you work your way through the pages and chapters to follow, you'll find dozens more. As you gain experience and familiarity with all the processes, and the results obtained therefrom, you may even find, as many have, that handloading can become not just a necessary part of shooting, but a highly rewarding end unto itself.

CHAPTER 2

Handgun Ammunition History

HISTORICALLY SPEAKING, the reloadable, center fire cartridge—for all firearms, not just handguns—is a relatively young device. In this country, it dates from only about 1870, though a lot of preliminary work went on before that which we must take a look at in order to have a reasonable understanding of when, how and why.

The earliest handgun cartridges came with the development of the percussion revolver, and they were not metallic nor entirely self-contained. They were, in fact, "combustible" cartridges consisting simply of a paper container (though other materials were used), attached to the rear of the bullet and holding the propelling charge of black powder. These cartridges were simply rammed solidly into the cylinder chamber of the revolver from the *front,* after which a percussion cap was placed on the nipple at the rear—the outfit was then ready to be fired. When the flame of the percussion cap flashed through the nipple vent into the chamber, it ignited the powder charge and the flame consumed the cartridge material. This material did not inhibit ignition because it was nitrate treated, thus so flammable and frangible that the cap flame would not only penetrate it to ignite the powder, but ignite the paper "case" as well. It wasn't really a cartridge as we know it today, and it certainly wasn't reloadable, but it made reloading the gun much more rapid and convenient than manipulating a separate flask of powder and a handful of loose balls or bullets.

By the time the combustible cartridge for revolvers came along, workable metallic cartridges were almost on the scene. A Swiss inventor, Jean Samuel Pauley, had earlier devised a reasonably successful self-contained cartridge which was primarily used in shotguns. A Paris gunsmith named Flobert devised a simple and cheap low-power cartridge with a copper rim fire case in 1846. It resembled a conventional percussion cap with its closed end bumped out a bit to form a hollow rim, then loaded with a lead ball or bullet. The hollow rim contained a minute amount of fulminate which would ignite when crushed by a firing pin, thus producing flame to ignite the small charge of black powder contained in the case. Flobert's patents were granted in 1846 and 1849. At about the same time copper-cased pin fire cartridges began to appear. The die was cast. The metallic cartridge was on its way, but it just

Not until the coming of the successful percussion-ignition revolver as *fully* developed by Colt, was there a suitable vehicle for the adaptation of any handgun mechanism to the use of a self-contained, obturating, metallic cartridge. Very early in its life, the percussion revolver was provided with combustible cartridges containing propellant and bullet loaded into the muzzle of the chamber. From that point, with the development of the Flobert and Pauley cartridges, it was a logical progression to Rollin White's patent for a bored-through cylinder which resulted in Smith & Wesson's introduction (in 1857) of the first successful American cartridge handgun.

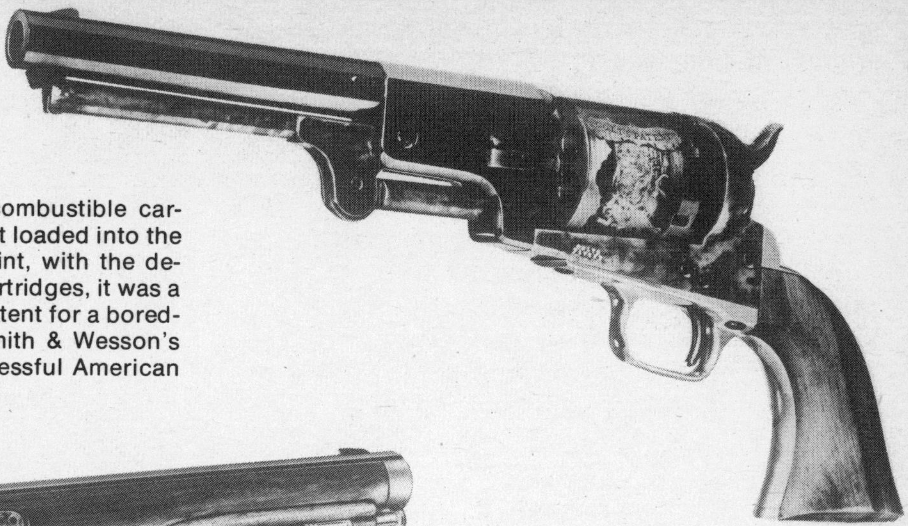

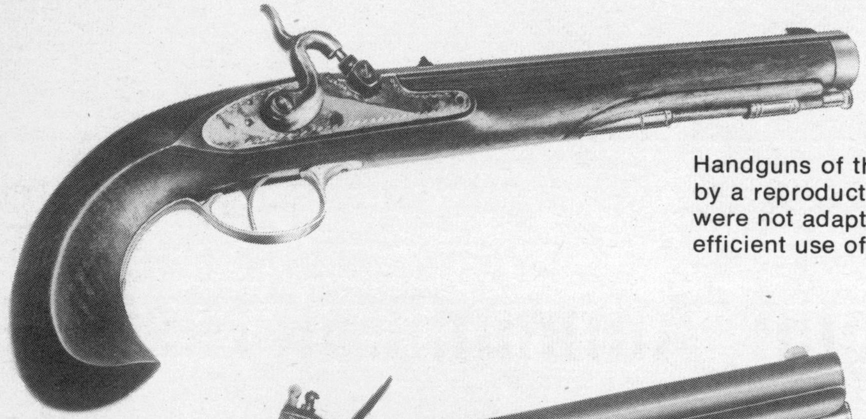

Handguns of the percussion-ignition era, typified here by a reproduction or replica "Kentucky" pistol, simply were not adaptable to modification or development for efficient use of a self-contained cartridge.

The flintlock pistol, the standard for roughly three centuries, was highly developed in its own right, but in a form that simply did not permit adaptation of the use of self-contained cartridges. The Collier revolver was of flintlock persuasion, but was not truly successful; the Hall breech-loading rifle was also flintlock in its original form, but even so, was not adaptable for further development to accommodate self-contained cartridges.

This Flobert, single-shot pistol (designed to use the Flobert .22 rim fire cartridge) appeared on the market in the middle 1840's; it was this particular cartridge which S&W improved to produce what we know today as the ".22 RF Short." From the .22 RF Short others developed larger rim fire cartridges (up to even .58 caliber) which served during the transition from muzzle-loading arms to the successful center fire cartridge and repeating arms utilizing it.

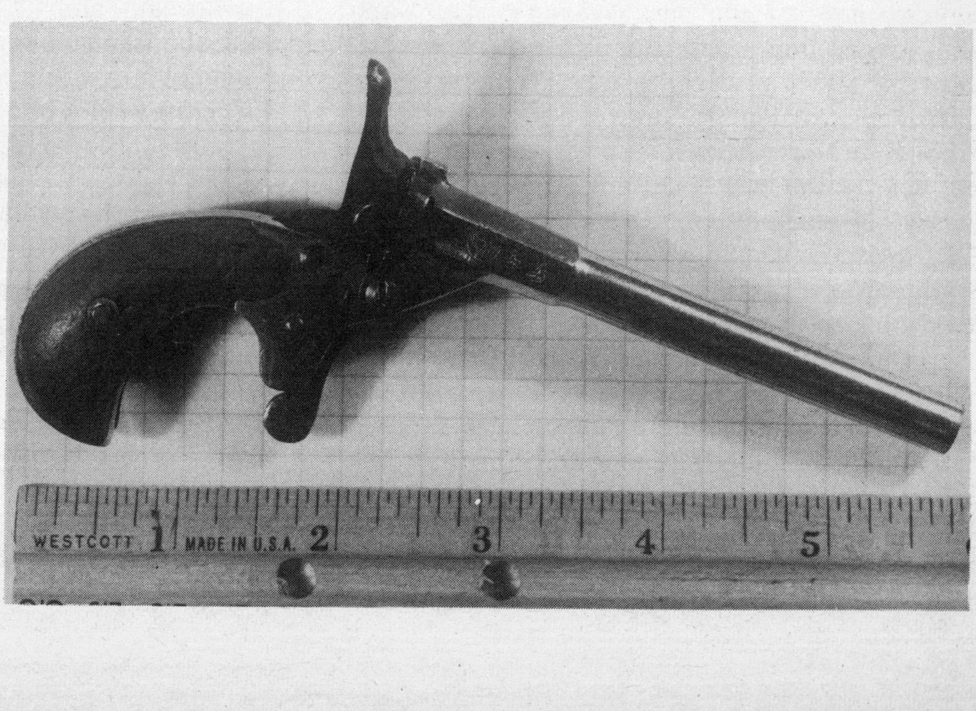

hadn't quite reached its first genuinely practical and worthwhile form.

In 1855, one Rollin White secured a revolver patent containing one particular claim which few observers of the day recognized for its true value. In fact, Colt refused to even consider it. This claim covered a cylinder with the chambers bored completely through from front to rear. Naturally, the percussion revolvers of the period had the chamber bored into the cylinder from the front, but not all the way through to the rear. Perhaps it is significant of the thinking of the times, but none of the numerous revolver patents issued previously gave any consideration to a chamber other than the closed type common to the percussion ignition system. White changed all of that, perhaps unintentionally, with his bored-through chamber which permitted loading and extracting an as-yet-nonexistent, self-contained metallic cartridge from the rear of the cylinder. Smith & Wesson acquired rights to the White patent in late 1856.

Smith and Wesson had been building "Volitional Volcanic Repeating" rifles and pistols during the early-middle 1850's; and, because of the unsatisfactory "Rocket Ball" ammunition used in this design, they were seeking a better cartridge. They investigated improving the Flobert rim fire cartridge, and upon acquisition of the White bored-through cylinder patent, made the logical connection and put the two together in revolver form. The definitive S&W revolver cartridge was, as a practical matter, identical to what we know today as the .22 rim fire Short. Thus, at one fell swoop, Smith & Wesson acquired the exclusive right to manufacture revolvers with chambers open at the rear, *and* a practical self-contained cartridge to use in such cylinders.

With this combination, introduced in 1857, Smith & Wesson was off and running. The revolver design for the cartridge was the Model No.1 First Issue, a very small hinged-frame, single-action type with a seven-shot cylinder. It was quite unusual by today's standards in that the tip-up barrel was hinged to the top strap on the upper rear of the frame and latched in the closed position at the bottom front of the frame. It was diminutive in size, and the cartridge possessed very little power. Nevertheless, Smith & Wesson was almost immediately inundated with orders for both the gun and ammunition; in fact, from that point until after the Civil War, S&W

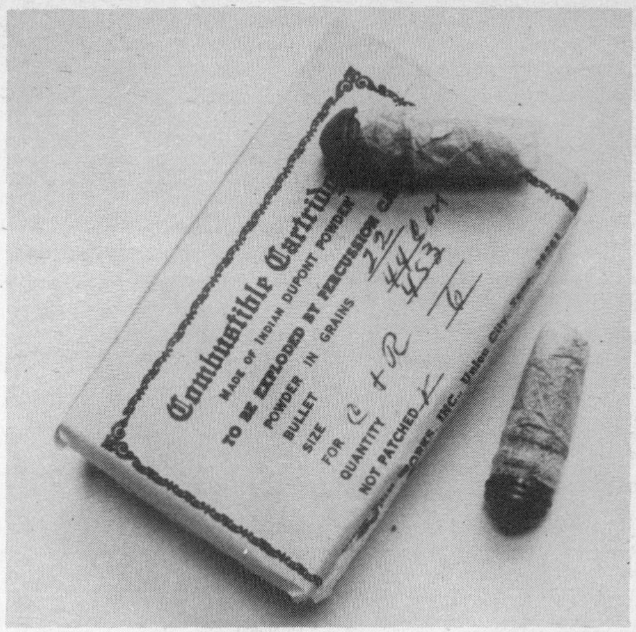

The very first handgun "cartridges" were of the "combustible" type, quite fragile, and did not contain a primer or other ignition device. Though it may be surprising to some, such cartridges are again available (from Dixie Gun Works), being currently manufactured for use in replica cap-and-ball revolvers.

was never able to even approach catching up with demand.

In those early years, Smith & Wesson manufactured not only the revolver, but the cartridges, eventually setting up a separate factory to produce the latter. This plant produced only rim fire cartridges and was sold in order that S&W could concentrate on revolver production.

Though it may not look like much today, this little revolver and pip-squeak cartridge became tremendously popular, and with no other company being permitted to bore chambers completely through the cylinder, Smith & Wesson had a clear shot at the market. Though the cartridge was nowhere near powerful enough to meet any military requirement—that was taken care of almost exclusively with big-bore percussion revolvers—it was a favorite among civilians and was highly desired as a personal side arm (usually concealed) by enlisted men and officers alike. Recognizing this shortcoming, Smith & Wesson introduced its .32 rim fire caliber, still small by today's standards but vastly superior to the .22 Short. Naturally, a larger revolver was designed to accommodate this cartridge and was introduced in 1861, right after the beginning of the Civil War.

Left is a typical early outside-lubricated cartridge, followed by a similar cartridge with the bullet pulled to show its "heel" type of construction with a reduced-diameter section inside the case. Right is a modern inside-lubricated cartridge where bullet diameter is essentially the same as case inside diameter and all lubricant is enclosed by the case.

The most popular then, the most popular now. In the late 1850's and early 1860's, the .22 Rimfire Short developed by S&W was certainly the most popular self-contained, metallic, handgun cartridge; in fact, for a few of those years, it was the *only* such cartridge readily available. On the right then, we have the ubiquitous .38 Special first introduced by S&W shortly after the turn of the century; and since then, doubtless more center fire handguns have been manufactured in this country for it than for any other caliber.

A typical big-bore, rimfire, outside-lubricated cartridge of the 1860's. Not reloadable, of course, and relatively low in power and lethality, this ".41 Rimfire Short" gained immortality as the cartridge for which Remington chambered its famous "double derringer." Though long, long, out of regular production, freshly-made ammunition in this caliber still shows up occasionally.

From the very beginning, though, the shortcomings of the rim fire case were recognized. The rim had to be thin enough and soft enough to be indented by the firing pin thus crushing and thereby igniting the priming compound contained therein. This requirement prevented making neither the walls nor the head of the case strong enough to withstand high pressures of heavy charges of powder behind heavy bullets. Partially for the same reason, the cases were made of soft copper and this could cause difficulty in extracting the fired cases under almost any condition less than ideal. Other people did develop much larger and more powerful rim fire cartridges for use in rifles, principally B. Tyler Henry's .44 rimfire cartridge for use in the famous Henry repeating rifle of 1860. In the years immediately following the Civil War, countless attempts were made to develop a case which had its primer centrally located in the head so that it could be made stronger. By 1868 the center fire Boxer case was used in the manufacture of the British .450 Revolver cartridge. Adopted for the Adams revolver, this cartridge was used in the first of the double-action, solid-frame Webley revolvers. In 1870 Smith & Wesson introduced its center fire .44 cartridge in its Model No. 3 revolver. Development of this cartridge amounted to converting the .44 rim fire to center fire, retaining its short length, bullet-diameter case body, and exposed lubrication of the bullet. It was what we now call "outside lubricated" because the entire major diameter of the bullet and the lubricant it carried was exposed beyond the case mouth. The entire brass case was of thin, weak construction, and the rim was both thin and relatively small in diameter. Old ballistic tables show it drove a 218-grain, soft-lead bullet at about 660 fps, though this velocity claim may well be optimistic.

Also, in 1870 Smith & Wesson developed what is now known as the ".44 S&W Russian" car-

Some of the early revolver cartridges of the black powder era assumed rather unusual forms by today's standards, though they were quite acceptable and successful in their day. At right is an unusual bottleneck employing a metal-jacketed bullet (quite uncommon for the day) while the others simply illustrate shapes of case and bullet, all of which are outside lubricated.

tridge. This center fire number was partially the result of a Russian requirement. The Russian Government adopted the basic Smith & Wesson, single-action .44 revolver for its military establishment; but, in addition to requiring gun improvements, the Russians also specified certain improvements in the .44 cartridge. The result was the .44 S&W Russian cartridge with its thick sturdy rim, centrally-located, external, removable primer, cylindrical case and round-nose lead bullet whose lubricating grooves were fully enclosed by the case mouth. To do this, it employed a case of greater length and diameter.

The .44 Russian set the standard for American revolver cartridges. As a practical matter, revolver cartridge design has not changed since; with the exception of the introduction of smokeless propellants around the turn of the 19th century. The .44 Russian demonstrated greater and better all-around performance than any of its predecessors. Others followed the pattern it set, particularly the .45 Colt of 1873.

All of the basic revolver cartridges we know today were developed *before* 1910, and because they were designed around the only propellant generally available (black powder), they used rather large cases. Now, don't jump up and tear the pages out of this book, shaking your fist and telling me that you know of plenty of revolver cartridges developed in recent years. Remember, I said the *basic* cartridges were in existence by 1910. Cartridges, *new in name* and new in that they offer higher levels of velocity and energy, have been developed since then, but the basic design, function and configuration has not changed. We'll get into all that a bit farther along.

Some earlier cartridges had used the case construction we now call "folded head;" this meaning simply that the basic case was drawn just like the rim fire type, from thin sheet brass or copper, with a hollow rim which was then squeezed flat to eliminate the hollow and give the *impression* of solid construction. The external primer pocket was formed by simply puncturing the outer face

At left we have what was originally called "solid-head" (or "Balloon Head") revolver cartridge-case construction wherein it was required that the relatively thin but solid-brass head be bulged forward to accommodate the primer pocket. This form of construction was used in almost all revolver cartridges until after WWII, excepting some special loads and the .357 Magnum. Obviously, the modern solid-head case (right) possesses greater head strength, though the importance of this is often over-emphasized. Since revolver chambers fully enclose the case right up to the forward face of the rim, the extra brass in the head lends little *effective* strength in view of the fact that it yields at a far lower pressure than the surrounding steel chamber walls.

The coming of the "Magnum" revolver cartridge in 1935 with the "S&W .357 Magnum" did not represent any particular improvement in ammunition. The high performance level of this cartridge was due only to the employment of chamber pressures triple those normally used in revolver cartridges. The loading techniques used to produce those pressures had long been established in medium-bore rifle cartridges such as the .32-20 and .44-40. What *did* make the .357 Magnum (and all subsequent Magnum revolver cartridges) possible was the refinement and development of manufacture and materials which gave the S&W N-frame design (.44 size) adequate strength to withstand those pressures. Shown here is the present-day counterpart (Model 27) of the original "S&W Magnum" introduced in 1935.

of the head. This form of construction carried with it most of the weaknesses of the rimfire case when made from copper and was little better when done in brass. The solid-head case came next, and the .44 Russian featured this type of construction. It was drawn from thick sheet brass, first punched out into discs, and formed into cups. This allowed the case head or *web* to be made much thicker and the rim to be made solid. The thickness of the original brass stock was not sufficient to make the head *solid* in the sense we use the word today. Its head had to be punched forward to form an inward protrusion to contain the primer pocket. This solid-head case was vastly superior to all that had gone before it. Even though it was developed before smokeless propellants, this case proved quite suitable to their use; it remained in production in most revolver calibers until after World War II. Later, after the advent of the autoloading pistol cartridge, what we now know as the solid-head cartridge case was developed, and the type originally known as "solid-head" was given the new name of "balloon head." The modern solid-head case was made from brass thick enough to accommodate the depth of the primer pocket without being bulged inward. Technically this made the case stronger, but for ordinary revolver cartridges, it offered no particular advantage over the earlier solid-head type. It did require fewer operations in the making, and since this type of construction was used for autoloading pistol cartridges, it was more practical to eventually apply it to revolver cases as well, thus streamlining the manufacturing process.

Assuming the same caliber and chamber dimension, the folded-head case had the greatest volume, therefore the greatest propellant capacity, of all types. The thicker metal in the balloon-head case reduced this internal volume (external dimensions being limited by the gun chamber) and thus the case contained less propellant. The coming of the true solid-head case reduced internal capacity even more. Consequently, a cartridge such as the .44-40 which might contain a full 40 grains of black powder in folded-head form may contain only 37 grains in balloon-head design, and no more than 35 grains (perhaps even less) in modern solid-head construction.

The original solid-head case design offered much greater strength, easier extraction, and

Right, above — A good deal of confusion existed about primer types and sizes during the developmental period of the center fire handgun (which means revolver, considering the time frame) cartridge. Here on the left we have a modern case head showing the Boxer-type primer pocket and case construction standardized in this country by the late 1880's; at right we have a European case of the same caliber showing Berdan-type case construction and primer pocket with its twin, small-diameter flash holes flanking a conical anvil made integral with the case head. Right, below — An inside view of Boxer (left) and Berdan case heads shown in the previous photo.

made reloading much more practical inasmuch as the primer received a stronger support, and there was less expansion near the head on firing.

The .44 S&W Russian represented the first widespread use of "inside lubricated" construction or assembly. Earlier revolver cartridges utilized a case of essentially the same outside diameter as the bullet, there being a reduced-diameter portion of the bullet provided for insertion in the case mouth. With this type of construction, lubricant was supplied over the bearing surface of the bullet ahead of the case. Initially, no particular nomenclature was applied to this type, but it was eventually to be called "outside lubricated." The principal disadvantage of this type was that the exposed lubricant (particularly the types available at that point in time) was easily rubbed off in normal handling and transportation. If not rubbed off, it gathered dust, dirt and grit with attendant adverse effects on the gun and accuracy. This problem was avoided in the .44 S&W Russian by first making the case larger —its inside diameter was equal to the major bullet diameter. Then, the bullet was made cylindrical throughout its bearing surface or shank, and the lubricant applied in circumferential grooves, with the lubricated bearing surface of the bullet placed entirely inside the case. Thus there was no exposed lubricant to be lost or contaminated, and this type became known as *"inside lubricated."* This later type offered other advantages in that the entire rear portion of the bullet (from the ogive to the base) was cylindrical and of full barrel groove diameter. This obviously provided a greater bearing surface and better stability than could be obtained with the "heel type" bullet of outside lubricated cartridges. While the reduced-diameter heel might sometimes upset sufficiently to fill the rifling, such action was by no means a certainty, and accuracy suffered as a result. The .44 Russian quickly developed a reputation for being the most accurate revolver cartridge of its time, and there is no doubt that this single feature contributed more to that reputation than did any other.

So, in all honesty we must say that the 1870 .44

Above, left — Boxer (top) primer compared with the Berdan type. Note the presence of its own 3-legged conical anvil pressed into the Boxer cup, while the Berdan cup contains only the pellet of priming compound, preformed with an indentation into which the conical anvil in the primer pocket fits as the primer is seated. Incidentally, the Berdan example shown is of a large rifle size rather than the typical .177-inch diameter pistol size, in order that the detail be more apparent; pistol sizes are identical in construction.

Above, right—When the self-contained, metallic revolver cartridge was developed, only black powder propellant was available. As can be seen at the right, the 40 grains weight of this propellant required to give the .45 Colt cartridge its tremendous (even by today's standards) performance bulks up rather large and fills the case completely. Today the same level of performance is produced by the barely one-quarter the weight of smokeless propellant as shown at the left.

S&W Russian cartridge represented the definitive development of the reloadable revolver cartridge. To repeat its characteristics, it consisted of a drawn brass, solid-head case of cylindrical form, possessing a sturdy rim which contained a removable, centrally-located, external primer in its head. It utilized a solid lead, round-nosed, internally-lubricated bullet. The bulk of the revolver cartridges manufactured today differ only in two respects from the 1870 .44 Russian—they use smokeless propellant instead of black powder; and, the case is of *modern* solid-head construction and therefore case volume is slightly less than it would be otherwise. So, it is for this reason that I say—and at considerable risk of being hanged in effigy by advertising and public relations types representing both ammunition and revolver manufacturers—that the common revolver cartridge has not progressed in design since 1870.

True, we have the "Magnum" calibers in .357, .41 and .44. They differ from their counterparts (conventional revolver cartridges) only in that they are loaded to much greater chamber pressure in order to drive their bullets at higher velocities and to develop greater muzzle energies. The ability to withstand these greater pressures is built into the *revolver, not* the cartridge case. Therefore, as a practical matter, the magnum cartridges do not represent any significant development in the cartridge but in the guns adapted to use them. As a matter of fact, the magnum revolver cartridges were developed by simply applying loading techniques and working-pressure levels developed many years earlier for medium-bore rifle cartridges of comparable size. Thus, the .44 Magnum revolver number represents no particular advance or advantage over the .44-40, high-velocity rifle cartridge which was developed years earlier and frequently employed (against the manufacturer's recommendation) in revolvers. That some revolvers employed with such high-velocity rifle cartridges failed is a reflection upon the gun and not the ammunition.

During the evolution of the metallic revolver cartridge, various types and sizes of percussion primers were used. The Berdan type was developed in this country by Colonel Berdan and was of open-cup design, containing a thin wafer of fulminate which in functioning was crushed against a conical anvil integral to the case head. One or more (as many as three) tiny flash holes were drilled around the base of the conical anvil to lead the primer flash into the propellant charge. This type of primer functioned exceedingly well and became, as a practical matter, the standard small-arms ammunition primer of nearly all the rest of the world. It remains so to this day. Unfortunately, though cases assembled

Left — Here is one of the several very good reasons that successful autoloading pistols and cartridges were not developed until the coming of smokeless powder. This .45 ACP case lays alongside the amount of black powder that would be required to produce standard ballistic performance for this caliber; obviously, this short, fat case simply couldn't hold that much powder, even if one-third of its interior was not taken up by the bullet shank.

Above, left — Clearly indicative of the advantage of smokeless propellant over black is this comparison of a .38 Special (left) and .38 Super Auto cartridge. The .38 Special was introduced at the end of the black powder era, and its case was dimensioned around the amount of that propellant necessary to give a 158-grain lead bullet approximately 850 fps; the .38 ACP was developed at about the same time, but in order to function in an autoloader, it was forced to use smokeless propellant.

with this type of primer *can* be reloaded, the process is more difficult and time-consuming than with the Boxer type primer developed by a Britisher, Colonel Boxer. This type employed an open cup and thin wafer of fulminate but was then fitted with a conical anvil of its own. The anvil was pressed tightly into the cup mouth with its apex lightly compressing the fulminate against the inside of the cup. It required no separate anvil in the case and permitted the use of a single, large-diameter (up to .080-inch) flash hole located centrally in the case head. It was, and is, ideal for reloading inasmuch as a simple concentric punch is all that is required to remove it.

Early in the game, the Boxer primer was adopted as a virtual standard for handgun ammunition in this country. As different cartridges evolved, a *small* and a *large* size of Boxer primer eventually became standardized. Ultimately, a nominal diameter of .175-inch became known as "Small Pistol" and was (is) employed in all calibers of 9mm/.38 and under; a larger size of nominal .210-inch diameter is known as "Large Pistol," and is used in all larger calibers. While in the early days there was no standard-setting body to say that these particular sizes would be adopted, it became quickly apparent to ammunition manufacturers that it would be to their collective advantage to use primers of the same form and dimensions if not necessarily of the same performance characteristics. It may not be well known among casual shooters, but even today there is a great deal of components transfer between companies. Company A may buy primers from Company B and/or bullets from Company C, either to augment its own production capabilities or simply as an economy measure. Today there are even small loading plants which do not actually manufacture *any* of the components they assemble into completed ammunition. In addition, there are independent manufacturers who produce only components for sale to loading plants. Obviously, then, this interchange of components is possible only with standardized sizes of primers.

The earliest center fire primers contained various fulminates, eventually standardizing more or less upon fulminate of mercury. The mercuric primers performed exceedingly well with black powder ammunition with the single exception that combustion of the fulminate released free mercury which amalgamated with the brass of the case (in some instances copper), making it exceedingly brittle and destroying its malleability. This mattered not in the least until one attempted to reload a case fired with a mercuric primer. Then, in spite of the dilutent effect of black powder residue, the case was often so brittle that it would crack or crumble during resizing. The next generation of primers employed chlorates rather than fulminates. Chlorate primers did not damage the cases, but their combustion deposited a film of a salt, akin to common table salt, on the surface of the bore. While not particularly corrosive in and of itself, this residue was (is) highly hygroscopic, and if not removed promptly and thoroughly by aggressive cleaning, it would absorb moisture from the atmosphere

Below—With the ubiquitous .22 LR rim fire cartridge at left for comparison, we show here the three "pocket-type" autoloading pistol cartridges which had become popular and widely distributed by 1910. From right to left, the .380 ACP or 9mm Browning Short introduced in 1908; the .32 ACP or 7.65mm Browning of the turn of the century; and the .25 ACP or 6.35mm Browning introduced in 1906. These three cartridges differ hardly at all in general characteristics and appearance or function, except that the .380 is truly rimless while the .25 and .32 have a vestigal protrusion of the rim beyond the face and are thus classified as "semi-rimmed." The semi-rim form of case was in vogue only up through the first decade of this century, and no new autoloader cartridges of the type have been introduced since; everything else is of true rimless design.

Above — At left we have the early 7.65mm Borchardt cartridge which developed into the highly successful 7.63mm Mauser of 1896; center is the 7.65mm Parabellum (.30 Luger) that eventually developed from the Borchardt cartridge when the pistol of that name was extensively redesigned into the Parabellum or Luger. At right is the most successful autoloading pistol cartridge of all, the 9 mm Parabellum of 1904. The latter did, in fact, develop from the 7.65mm Parabellum in order to provide a more powerful load for military use. As can be seen, the earlier bottle-neck types were already being superceded in 1904 by the straight or slightly-tapered or rimless case of the 9mm.

and promote rapid rusting. Barrels left uncleaned in a fairly humid atmosphere (like, for example, New Orleans) overnight after firing with chlorate primers would accumulate a thorough coating of rust. Naturally, this would etch or pit the bores. Prolonged exposure could completely ruin a barrel in only a few days under conditions of high humidity. As a result, chlorate primers were called simply "corrosive primers," but the name wasn't applied until the problem (unidentified for many years) was recognized and the so-called "noncorrosive" styphenate primer developed. Clorate primer residue is soluable only in water, so the prescribed military method of cleaning a barrel 3 days in succession with hot soapy water produced the best protection.

Commercial handgun ammunition has not been loaded with mercuric or chlorate primers since well before World War II. However, most U.S. military .45 handgun ammunition employed the corrosive type up through the early 1950's. Virtually all foreign military ammunition utilized chlorate primers until the late 1950's or early 1960's. Consequently, virtually all World War II military ammunition, which flooded this country for decades after that conflict, can be injurious to handgun barrels if the problem is not recognized and solved by prompt and thorough cleaning with a water solution. The exceptions to this are U.S. military ammunition in nonstandard calibers (9mm Parabellum, .38 Special, .380 ACP, etc.,) purchased during the same period; those calibers, even when made for military use, contain standard, noncorrosive, primers.

But, back to handgun cartridge development. As I think we've shown clearly by now, the revolver cartridge reached the peak of its mechanical development with the 1870 .44 Smith & Wesson Russian. During the late 1880's and 1890's numerous arms inventors throughout the world were feverishly at work in attempts to produce *practical* autoloading pistols. Early on they discovered that the basic revolver cartridge type did not lend itself well to autoloading mechanisms because of its length, soft lead bullet, rimmed case and black powder propellant. Its length caused feeding problems; the black powder left too much gummy residue to be tolerated by those first-generation designs; the lead bullet introduced other feeding problems; and the rim made magazine feeding quite complex and difficult.

The availability of smokeless propellants in the early 1880's solved one problem by eliminating the residue, and another by simply offering much greater energy for its bulk. In short, in a cartridge like the .45 Colt, the amount of smokeless propellant required to give acceptable ballistic performance occupied roughly half the space required for black powder. This permitted reducing cartridge length (while maintaining desirable ballistic performance) and thus reducing the bulk of the gun and simplifying feeding. In fact, it was the coming of smokeless propellant that made it possible to contain autoloading pistol magazines inside the butt in the configuration we consider standard today. Other feeding problems were eliminated by using a rimless cartridge case after the fashion of the M1888 German service rifle cartridge, (though this wasn't the first use of the principle) so that cartridges would stack neatly together in a magazine with no overlap. Problems in feeding, caused by deformation of a soft lead bullet upon impact with the guide surfaces in the mechanism, were solved by employing full metal jacketed bullets of the type later adopted for military rifles.

Thus, before the turn of the century, the autoloading pistol cartridge developed as quite a different type from that which evolved for the revolver. The earliest successful autoloading pistols employed tapered or bottle-neck cases, not only to burn more propellant within a given length of case behind a given bullet diameter, but also to facilitate reliable feeding from magazine to chamber. A cylindrical cartridge requires very precise guidance to enter a matching chamber smoothly, while a tapered or bottleneck cartridge is literally *funneled* into the chamber and can deviate a good deal from its specified path and still chamber correctly. Examples of this are the Borchardt pistol of 1893, the Mauser of 1896, the Parabellum of 1900, and other design by Mannlicher, Bergmann, Schwarzlose and others. However, improvements in propellants and the desire to use larger-diameter bullets than could be accommodated in a bottle-neck case of acceptable base diameter caused the autoloading pistol cartridge to evolve primarily with a cylindrical or very slightly tapered rimless case.

In any event, the autoloading pistol cartridge evolved into its definitive form during the first decade of the present century. Though minor variations exist, this form consists of a rimless, solid-head case with a length ranging from 2 to 3 times its head diameter. At the mouth, its inside diameter is nominally that of the bullet diameter, and of course it uses smokeless propellant. The bullet is normally full-metal jacketed, the jacket closed at the nose with the core inserted at the base and is therefore of the "open base" type. Bullet shape is generally round-nose, with a cylindrical bearing surface seldom exceeding the measurement of the bullet diameter itself. Its overall length will range from about 1½ to 2½ times the bullet diameter; and, the overall loaded cartridge length generally ranges from 2½ to 3½ times the same measurement.

Thus, by 1910, we have two distinctive classes of handgun cartridges—one for the revolver and one for the autoloader.

There is one other very important difference between revolver and autoloader cartridges, a difference that even today is seldom recognized for its true worth. Typical non-magnum revolver cartridges work at chamber pressures hardly ever exceeding 15,000 CUP. In fact, many factory loads of today fall well below that level. On the other hand, autoloader cartridges hardly ever

In the search for genuinely effective expanding bullets for handgun cartridges, various individuals attempted many approaches. Here we have a variety of jacketed bullets but, of them all, only the one to the right of the third cartridge in line would provide reliable feeding in autoloaders, the others show far too much exposure of soft lead core which generates feeding problems as it strikes the various guiding surfaces in the gun.

These are just a few of the bullet profiles and types which are found today in factory-loaded autoloader cartridges.

work at pressures that low. With the exception of the .45 ACP, virtually all of the larger autoloader calibers work at pressures ranging from 30,000 to 35,000 CUP. The .45 works at a lower level, generally under 20,000 CUP, nearer the level of the smaller cartridges which work at around 15,000 CUP.

Relatively early in the history of the autoloading pistol, attempts were made to produce expanding-type bullets. Both the soft point and hollow point type were offered commercially, but even in the highest-velocity cartridges they offered little increase in lethality over the full metal jacketed types. Such bullets followed conventional expanding *rifle* bullet designs and simply were incapable of reasonable expansion at handgun velocities. During the same period of time, there were also attempts to reduce barrel wear while retaining reliable functioning in autoloaders through the use of a lead bullet with its nose capped by a partial jacket. The object of this design was that the portion of the bullet exposed beyond the case was jacketed and would therefore feed smoothly through an autoloader, while the lubricated lead bearing surface behind it would be "easy" on the bore. Other special-purpose bullets were developed, principally in the law enforcement and target-shooting field. In revolver calibers .32 S&W Long and .38 Special, the unique mid-range wadcutter appeared. This consisted of a cylindrical, soft-lead bullet, usually with a hollow base to shift its center of mass forward. The bullet was seated entirely within the case and driven at a velocity of only 750-770 fps by a very small charge of fast-burning propellant. Such loads were developed to a very high degree of accuracy and are with us yet in .38 Special caliber. After World War II, comparable mid-range, low-velocity target loads were developed for the .45 autoloading pistol. They followed the same lines, using a lighter-than-standard bullet at relatively low velocity. However, the cylindrical wadcutter bullet could not be used because of feeding problems, so a modified semi-wadcutter form was employed. Though initially some such loads employed swaged lead bullets, they were never truly satisfactory, and match ammunition offered by the factories today employs the jacketed type.

In most autoloading handguns the bullet on the left with moderate lead exposure functions reasonably well, while the lead SWC on the right will seldom feed satisfactorily. In the middle we have the truncated-cone Super Vel type which offered (and still does) reliable feeding combined with dependable expansion.

By the early 1960's, the changing social order and its attendant problems and increasing crime rate caused law enforcement agencies to finally recognize the simple fact that conventional revolver and autoloader cartridges simply did not possess sufficient lethality for efficient use upon human criminals. Actually, arms students, inquiring handloaders, and numerous other classes of pistoleros had long recognized this problem and had made rather limited attempts to solve it. One approach had been to use a very hard zinc-alloy bullet of quite light weight, driven at velocities from 1,600 to 2,000 fps. Even a non-deforming solid bullet at those velocities (given a proper tissue-destroying and energy-transferring nose shape) is far more lethal than the conventional bullet loping along at well under 1,000 fps. Another approach was taken, using dead-soft, pure-lead bullets of lighter than standard weight and driven at the maximum velocities that could be safely obtained. Such bullets deformed readily and extensively upon impact and, again, were much more lethal than the conventional type. Unfortunately, they leaded bores very badly, and thus produced poor long-term accuracy. Cup-type copper gas checks were used to improve their performance as were zinc washers and bases. Eventually, they were made with short copper-cup jackets which reduced but did not eliminate leading. None of these types were offered in standard commercial loadings but were assembled by customloaders for sale in small quantities.

Eventually, one Lee E. Jurras developed his "Super Vel" bullet in two different types (revolver and autoloader) in order to attain maximum terminal ballastics and lethality. Essentially, Jurras' bullets were of a lighter-than-standard weight for the caliber (in .38, 110 grains instead of 158) and consisted of a very thin, soft, copper jacket enclosing a dead-soft, pure-lead core. In revolver types they were cannelured for crimping and exposed a portion of the core at the nose to facilitate rapid expansion. For autos they were uncannelured, of truncated-cone shape with a large-diameter meplat, exposing the core only within the meplat. The latter was essential so that no portion of the lead core could contact guide surfaces in the gun during feeding. Generally speaking, these bullets were driven at velocities ranging from 20 percent to 75 percent higher than conventional factory-loaded ammunition. Satisfied that he had achieved his goal of greatly increasing the lethality of existing handgun cartridges, Jurras placed them in production at the Super Vel loading plant near Shelbyville, Indiana. The plant operated for a number of years, supplying tens of millions of rounds of this new super-effective ammunition to law enforcement agencies and officers (not to mention astute individuals) throughout the world. A few years ago the Super Vel venture failed, but not before it had accomplished its goal. Super Vel's efforts demonstrated quite clearly that, with relatively little intense research and testing, modern techniques could vastly improve the lethality of all handgun ammunition. A significant factor in the end of Super Vel was the copying of its bullets and loaded ammunition by all of our domestic manufacturers and a few foreign ones as well. It is perhaps worthy of note that under new ownership the Super Vel operation has been resurrected and is currently producing and shipping large quantities of ammunition from its new plant in the vicinity of Greensburg, Indiana.

In reference to this particular aspect of handgun ammunition development, it can be truthfully said that Lee E. Jurras is the father of the modern "high-performance" handgun cartridge. Had he not persevered (and succeeded), I think it extremely doubtful that the major manufacturers would ever have deviated from the traditional type of ammunition they have produced since the 1870's.

Expanding-type bullets in handgun calibers are by no no means new. At top is a jacketed soft-point loading of the .38 ACP which was probably loaded about the time the the author was born. Unfortunately, the bullet designs and materials of those days expanded poorly or not at all. Below is the modern version of the .38 ACP, the .38 Super Auto, loaded with a 112-grain JHP bullet driven at a velocity well over 1400 fps to produce explosive expansion at close range, this making it extremely lethal on live targets of any sort.

In recent years an unusual bullet has been marketed under the name "Glazer Safety Slug." It was designed specifically for police use, to possess maximum lethality against human targets and to minimize ricochet probability. Both goals are accomplished by filling a thin bullet jacket with very small lead shot in a wax or plastic matrix. When the bullet strikes a target, the jacket ruptures and the shot disperses to form *many* secondary projectiles which create a very massive wound of limited penetration. When the bullet misses and strikes a wall, street surface, vehicle, etc., it breaks up similarly; in this instance, though, the mass of each individual pellet

Truly effective expanding bullets for handgun ammunition depended upon the developments of Lee E. Jurras in the 1960's. The two basic types of revolver bullets are shown here, large-cavity hollow-point on the left and essentially the same bullet without the cavity on the right, other factors being equal, the hollow-point gave the greatest expansion while the soft-point produced greater penetration.

or fragment is so low that it possesses little energy and decelerates rapidly. Thus, the fragments pose no significant hazard to bystanders or property.

The cost of the Glazer Safety Slug precludes wide use—at present it is available only in custom-loaded cartridges, the typical .38 Special load costing *$3* per round. Police officers vitally concerned with weapons effectiveness and safety may buy 6-12 rounds at that price, but few others will.

Early in this chapter I indicated we'd get back to a further discussion of the magnum revolver cartridge. In the middle 1930's Smith & Wesson introduced the first magnum—the ".357 Smith & Wesson Magnum." Actually the term applies not to just a cartridge, but to a cartridge/gun combination. Over a period of years, the .357 Magnum cartridge had been developed as an *improvement* in ballistic performance over the existing .38 Special with its 158-grain bullet at about 850 fps. Phillip B. Sharpe and other prominent arms and ammunition authorities played a part in this work and eventually developed a .38 Special loading driving bullets of 145- to 150-grain weight at velocities of around 1500 fps from extra-long barrels. To do so they used special propellants which put chamber pressures in the vicinity of 40,000 CUP. However, this startling (for the day) performance was obtained with the standard .38 Special cases and perfectly conventional lead bullets of semi-wadcutter shape. What really permitted this level of velocity and energy to be produced was raising the working pressure of the cartridge to around 40,000 CUP. None of the other elements of the cartridge had any significant effect upon this; the .38 case was adequately strong to stand the pressure in a properly chambered gun and the lead bullet could also stand the gaff. At that time, existing .38 Special revolvers (and also some obsolete .38 Long Colt revolvers which would chamber the .38 Special cartridge) were designed for the standard load producing less than 15,000 CUP. Consequently, it was considered very unwise to introduce this new load in a form that could be fired in ordinary .38 Special revolvers. They simply couldn't stand over three times the normal pressure. So, to eliminate such problems, the cartridge case was lengthened approximately 1/10-inch and given a new name ".357 Magnum," the very first modern revolver cartridge whose official designation actually indicates its *true* bullet diameter.

With the new load and cartridge and the final factory development done by Winchester, Smith & Wesson redesigned its N-frame, .44-size Hand Ejector revolver to handle it. The .44-size gun was chosen because it had the weight to dampen the considerable recoil of this load. It also possessed a large enough cylinder so that thick chamber walls could be maintained to help support the case against those 40,000 CUPs. By using the big gun, chamber walls were nominally .125-inch thick as opposed to .090-inch in the

Right — The .38 Special, mid-range, full wadcutter target load alongside the bullet from an identical cartridge. Note the three very shallow, knurled grooves for lubricant, and the unusually long bearing surface. While it appears that this bullet takes up more than one-half of the internal volume of the case, it actually does not because of the substantial cavity in its base.

Above — Probably no handgun cartridge has earned a better reputation for superb accuracy than the mid-range, full wadcutter, .38 Special factory load. Here the bullet is seen from the nose (left) and from the base; in the latter, the deep cavity in the base, which shifts the bullet's center of gravity well forward, can be clearly seen.

smaller .38 size guns. Coupled with this, S&W very carefully chose materials and heat treatment to insure that the high pressures would be safely confined. The gun was also improved cosmetically and in the sighting department; and, every effort commensurate with the times was made to produce a genuinely elite gun. It was priced accordingly, too, selling for $60 at a time when the standard N-frame guns were priced at $37.50. In short, it cost 60 percent more.

Thus, it was not *cartridge* development that made the .357 Magnum possible. Similar techniques had already been used to upgrade the performance of small rifle cartridges (.32-20, .44-40, etc), and were well known. It was *gun* improvement that made the whole thing possible. Thus I stand by my statement that the introduction of the magnum revolver cartridge did not result from ammunition improvement.

Reading contemporary literature and the recollections of pistoleros of the 1930's, one might come to the conclusion that the .357 Magum revolver-cartridge combination was tremendously popular. Not really. From the presentation of the first .357 Magnum to J. Edgar Hoover (even then, Director of the F.B.I.) on April 8, 1935, up until temporary cessation of manufacture at the beginning of World War II, a total of 6,642 revolvers were produced. Even so, this exceeded S&W's expectations, for at the beginning of the period they were only able to finish (and remember, this was a hand-finished, deluxe gun) only about 120 pieces per month. Today, S&W or any other revolver manufacturer wouldn't give the slightest consideration to a model which couldn't be expected to sell a hell of a lot more than 1,000 pieces *per year*.

The .44 Magnum and .41 Magnum cartridges were developed in essentially the same manner as the .357 in the early 1950's and early 1960's respectively, and that same N-frame revolver was adapted to their use. In both instances it was again simply a question of providing a gun capable of withstanding the higher chamber pressures. The cartridges themselves displayed no new or exciting innovations other than their high level of performance, obtained by vastly exceeding normal revolver chamber pressure limits.

This same "magnum approach" was taken in the early 1960's to develop the Auto Mag series of autoloading pistols and cartridges. The gun and cartridge were developed concurrently, the former to withstand chamber pressures in the 50-55,000 CUP range, and the latter to work at that level. Thus the increased performance (over other autoloading pistol cartridges was obtained simply by employing higher chamber pressures and abnormally long barrels. The cartridges themselves displayed no significant differences from existing designs dating back to the 1890's. More recently, actually as this is being written, the Wildey 9mm and .45 Magnums are being developed in exactly the same manner in companionship with the Wildey gas-operated pistol.

To date, all of our so-called "modern" handgun cartridges still represent the basic .44 Russian of 1870 or the 9mm/.38 numbers of about 1900, in some instances coupled with loading techniques developed over 50 years ago for small and medium rifle calibers.

With the .44, as in .38/357 20 years earlier, the magnumizing process involved nothing more than lengthening and strengthening the cartridge case to provide more propellant capacity and to prevent chambering in obsolete guns, and then loading to roughly triple existing revolver-cartridge chamber pressures. Here on the left we have the traditional .44 S&W Special of 1908 compared with the .44 S&W Magnum of the middle 1950's.

Since the first draft of this chapter was written, new bullets have been devised to improve handgun lethality for defensive use. One type is named "Hydra-Shok" and is of wadcutter profile with a wide, deep nose cavity with a central stem rising from its base. The combination of this "cored cavity" shape and soft lead allows consistent expansion of considerable magnitude at impact velocities as low as 700-800 fps. Due to its shape, this bullet will not feed in most autoloaders. Its principal advantage is the production of adequate expansion at velocities that can be reached safely in lightweight revolvers with short barrels while minimizing recoil.

Then we have the so-called "explosive" bullets just introduced. These bullets employ a small bursting charge which reportedly explodes on impact to create a very large wound. The nature and amount of the charge are such that the cartridges are not (yet) classified by law as "Destructive Device Ammunition." Lacking first-hand experience with this type bullet, we simply can't say at this time how practical it might be. Explosive bullets are by no means new, dating back to the muzzle-loading era—but thus far they have a poor reputation for safety, reliability and practicality.

Doubtless, we *will* see significant developments in the future. However, I suspect they will fall within the general framework of the cartridges we already know—no technological breakthroughs are iminent, no caseless cartridges, no "rocket bullets," no electromagnetically-propelled projectiles nor any practical liquid-propellant systems.

Anyway, what we have is pretty good, all things considered.

CHAPTER 3

Auto/Revolver Cartridge Differences

IN THE PREVIOUS chapter we skimmed over the developmental processes of both revolver and autoloader cartridges. The more obvious physical differences are apparent there, but by no means do they tell the entire story, or even the most important part of it. To be sure, the two types of cartridges *look* quite different in most instances. The revolver cartridge is typically assembled in a cylindrical case carrying a rather thick, sturdy rim with the case mouth crimped or turned-in heavily upon a round-nose lead bullet. The typical autoloader cartridge—whether born in 1900 or 1979—utilizes a cylindrical or very slightly tapered *rimless* case and is loaded with a metal-jacketed bullet of round-nose form, and the case mouth *is not* significantly crimped on the bullet.

There are, of course, exceptions to the foregoing descriptions. The earliest successful autoloader cartridges were of bottle-neck case shape for various reasons, some of which we cannot ascertain at this point in time; but, certainly those reasons included chambering and feeding ease and the burning of as much propellant as possible behind a small-diameter bullet without resorting to an excessively long case. It is perhaps significant that no new bottle-neck autoloader cartridges have been introduced commercially since 1900. In addition, of those early bottle-neck numbers, only the 7.65mm Parabellum (.30 Luger) is still in production (at least in this country). Various attempts have been made to use bottle-neck cartridges in revolvers, but they are best noted for their lack of success. Smith & Wesson has manufactured its Model 53 K-frame revolver for the sharply tapered and bottle-necked ".22 Remington Jet" cartridge since 1961. No other revolver/bottle-neck cartridge combination has ever achieved commercial success or shooter acceptance—and in fact, the good sales record of the S&W M53 is (in this writer's considered opinion) *in spite* of its functional problems and is actually the result of the current-decade mania for collecting (as opposed to using) Smith & Wesson "Magnum" revolvers.

As a practical matter, then, we can say that all 20th century, genuinely successful revolver and autoloader cartridges use cylindrical or very slightly tapered cases. Let us take a bit of time here to examine the reasons for this situation. First of all, in both revolvers and autoloaders,

Two of North America's best-loved handgun cartridges; left, the venerable .45 Long Colt, circa 1873, and on the right the .45 ACP adopted in 1911 by the U.S. Army and a perennial favorite ever since. These two cartridges represent the largest revolver and autoloader *calibers* (caliber meaning bullet diameter only) for which production handguns have been manufactured in this country.

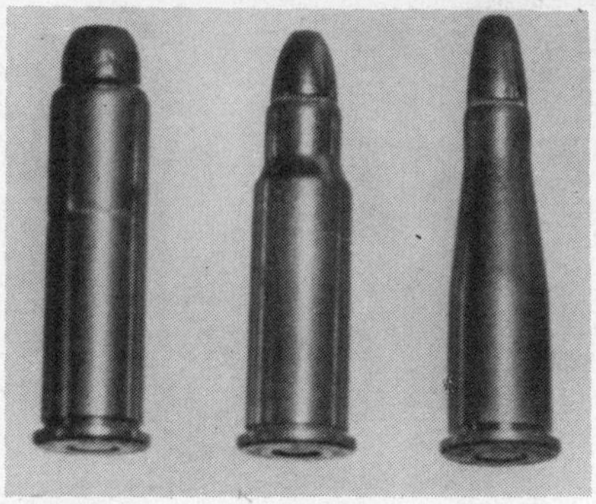

At left is the standard .357 Magnum cartridge; next is the .256 Winchester Magnum, a bottle-neck number originally intended for use in production Colt and Ruger revolvers, but those plans were cancelled because of functional problems described in the text. On the right is the Remington .22 Jet for which the S&W M53 is chambered. Both bottle-neck cases are simply the .357 necked down.

This S&W Model 53 is the only production revolver ever to be manufactured in quantity chambered for a bottle-necked cartridge. Even the 53 experiences a fair amount of functional problems when the tapered case wedges tightly against the recoil shield and interferes with cylinder rotation. Perhaps the only limitation on revolvers is their inability to function with bottle-neck cases loaded to relatively high chamber pressures.

cartridge capacity, whether in a cylinder or box magazine, is restricted first by the physical dimensions of the gun as they relate to its weight, portability, compactness and to its ability to be manipulated quickly and efficiently by one hand. Second, capacity is limited by the diameter of the case and by the overall length of the assembled cartridge. As a practical matter, the upper limit of cartridge diameter is nominally .475- to .480 inch. Given a cartridge significantly larger in diameter than that, the traditionally requisite number of six chambers cannot be accommodated in a revolver cylinder of reasonable size, even with modern materials and technology which permit adequate strength to be achieved with relatively thin chamber walls. (As a matter of interest, the upper limit of cylinder diameter is represented by the 1.7 + inches of the Smith & Wesson N-frame design.) In autoloaders, even a rimless case of greater diameter causes not only the grip or butt housing the magazine to become over-thick, but introduces extra thickness (and therefore weight) to the barrel-breech area. Chamber walls at the barrel breech must be maintained of adequate thickness to withstand reasonable chamber pressures; and, insofar as all known practical designs are concerned, this area of an autoloader must further be enclosed (at least partially) by a slide whose walls are also fairly thick. The result is that when practical costs and manufacturing factors are taken into consideration, a .45 caliber cartridge produces a total gun thickness over the barrel breech area of approximately 1 inch in the latest designs. Even at this level, the working chamber pressures of .45 caliber cartridges are held down to around 20,000 CUP because of strength and safety considerations. The short-lived Auto-Mag pistol which sought to work with pressures in the 50,000 CUP range, utilizing a case of nominally .45 size, was much, much thicker at the barrel breech and thus quite heavy and unwieldy.

So, as a practical matter, both revolvers and autoloaders are limited to cartridges of .45 caliber or less. Larger calibers, up to .60, have been attempted in revolvers in the past, but even with limited cartridge capacity, cylinders and related parts became so large as to make the gun quite impractical insofar as we understand handgun usage today.

At left we have an S&W N-frame cylinder chambered for .38 Special. This leaves chamber walls both on the outside and between chambers which are quite thick and strong. On the right is an S&W N-frame cylinder chambered for the .44 Special which produces much thinner chamber walls in all areas. Obviously, the walls of the .44 chambers possess a good deal less strength than those in .38, all other factors being equal. Both theoretically and practically—although commercial loading of the cartridges doesn't necessarily indicate it—the .38 caliber cylinder can withstand much greater chamber pressures than the .44.

The Colt Government Model .45 caliber autoloading pistol was originally designed to accommodate the .45 ACP cartridge working at pressures *around* 20,000 CUP. Later, it was adapted to the .38 ACP in the hopped-up loading called ".38 Super Auto." On the right we have the .45 caliber barrel for this pistol, viewed from the rear, showing its relatively thin chamber walls. The barrel on the left is in .38 Super Auto caliber and its much thicker (and therefore much stronger) chamber walls are quite obvious. Because of this greater strength, the .38 Super is normally loaded to working pressures in the 33,000 CUP range. This is a classic example of a single gun design possessing vastly different pressure-handling capabilities when chambered for different cartridges.

Overall cartridge length is another matter. Technically, length presents no particular problem in revolvers; longer cartridges can be accommodated quite readily by simply lengthening the cylinder and frame appropriately. When this is done—and some custom makers have produced guns with cylinders nearly 3 inches long—a substantial amount of length and bulk is added. As a practical matter, this is probably a good deal less important than appearance; with an overly-long cylinder (by present-day standards) the revolver takes on a clumsy and even ugly appearance instead of the graceful flowing lines to which we have become accustomed in the last 140 years. A Texas friend of mine once said of a long-cylinder, custom-built .45-70 revolver, "It sure looks like somebody beat it with a ugly stick." I'm inclined to agree. Aficionados of the big-bore six-shooter have become accustomed to and apparently appreciate cylinders with lengths ranging from 75 percent to 95 percent of their diameters. Considering the other limitations we have already discussed, this establishes a practical-cartridge limit of approximately 1.6 inches, which just happens to be the overall length of the .44 Magnum. At the other end of the scale, there is no functional requirement for a minimum cartridge length in revolvers. That is, there is no minimum length/diameter ratio simply because cartridges are chambered manually and at no point in the functioning cycle are they unguided.

Insofar as cartridge length is concerned, autoloaders present a different set of problems. The .380 ACP (9mm Browning Short) has a length/diameter ratio of nominally 2.6; a case diameter of .375- to .380-inch and a cartridge length of .98-inch. This is the lowest ratio of all successful autoloading pistol cartridges, and while others might differ slightly, I would suggest that a minimum L/D ratio of 2.5 is required if reliable feeding is to be obtained with any practical pistol design that may be produced at reasonable cost. In all existing production autoloading pistol designs, the cartridge is relatively—in some cases

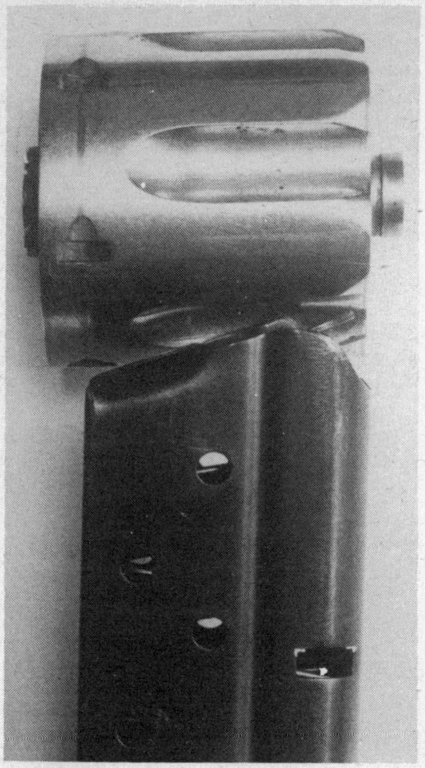

Left—This photo provides a clear demonstration of the difference in length requirements in revolvers and autoloaders placed upon cartridges to be employed in them. The auto magazine below is in .38 Super caliber, and it must be kept in mind that the cartridge length it can accommodate is limited by its *inside* dimensions, not the outside length you see here. The revolver cylinder is from an S&W .44 Special; and, in this instance, it can accommodate a cartridge slightly longer than its own length because of the fact that the case rim protrudes at the rear.

Right—As this is written, the .38 Super cartridge on the right, is the longest for which production autoloaders are chambered. The .44 Magnum cartridge on the left is the longest standard round for which revolvers are chambered, thus illustrating the difference in the length limitations imposed by the two different types of guns. Actually, there is no mechanical or handling reason that the revolver could not be made for an even longer cartridge.

almost completely—unguided during a portion of its travel from magazine to chamber; and in some designs, this travel distance may be double the cartridge length. High-speed photography shows that the path of the cartridge is quite erratic during feeding, and is controlled primarily by repeated bouncing between various guiding surfaces. In order to complete this travel and enter the chamber smoothly, the cartridge requires substantial length, (relative to its diameter) to maintain direction and develop some stability. Thus, even if propellants efficient enough were available to provide the desired ballistic performance in a much shorter case, a cartridge shorter than two and one-half times its diameter would not be practical without a complex and costly special feeding system, Incidentally, such feed systems do exist in prototype form, as do pistols incorporating them, but to date none have ever achieved production status, nor does it seem likely they will in the near future. Further, so long as autoloaders are made in the traditional and accepted form with the magazine contained in the butt, the maximum overall length limit is a pure measure of length, not related in any way to diameter. Our longest production auto cartridges are the .45 and the .38 ACP measuring nominally 1.28 inches. When these cartridges are enclosed in a metal magazine, we add to that length its wall thickness and working clearances and the front and rear thicknesses of the enclosing gun butt, plus space for working parts such as mainspring, hammer strut, etc. When all of those are added up, we have a front-to-back length of the butt/grip portion of the gun measuring roughly $1^{15}/_{16}$ inches. That represents a practical maximum for the *average* shooter's hand if he is to obtain a secure grasp which will enable him to absorb recoil and achieve adequate control over the gun in rapid fire. As a practical matter, then, we say that the upper limit of autoloader cartridge length is about 1.3 inches. Certainly the Auto-Mag pistol uses a much longer cartridge, but even though it was especially designed to do so, the butt/grip is far too large for the typical shooter and almost impossible for those with small hands except when resorting to two-handed shooting. As we've said before, a handgun is not practical for general use unless it can be fired accurately and controlled well with one hand.

If for the sake of comparison we want to assign maximum L/D ratio to autoloader cartridges, it will be nominally 3.5 in 9mm/.38 caliber and 2.7 in .45 caliber. Others will, of course, be comparable. If there were any practical reasons for stretching the .25 ACP to the maximum length, its L/D ratio would be on the order of 4.7 and in the .32 the L/D would be about 3.9.

Above—The Argentine-manufactured copy of the Colt M1911A1 and its commercial counterpart Government Model is typical of the basic Browning, tubular-slide, locked-breech type. In this example, the barrel, recoil spring plunger, barrel bushing, and slide weigh 17 ounces and if the gun is to function reliably the recoil impulse of the cartridge must overcome the inertia of the 17 ounces *and* the force of the recoil spring which at rest (in battery) will normally exert about 10 pounds forward pressure on those recoiling parts. There must still be enough reserve force in the recoil impulse to give those parts sufficient momentum that they will travel roughly 2 inches to the rear, compress the recoil spring and hammer spring, and further overcome all frictional resistance. Any cartridge, which does not have sufficient recoil impulse to accomplish all that simply *cannot* produce reliable feeding. No such minimum-recoil requirement exists for revolvers.

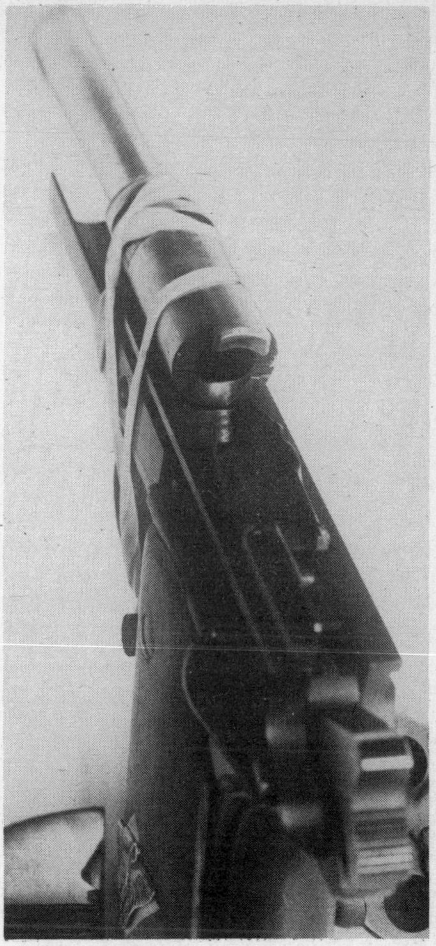

Right—Here we see some of the factors which limit the size and shape of autoloader cartridges; the example is a Colt .38 Super pistol. In moving forward under the influence of the slide, the cartridge first travels in a horizontal line then, at about the same time it clears the magazine feed lips, it is forced upward by the follower, and immediately the bullet nose strikes the feed ramp beneath the chamber mouth; the cartridge is then deflected upward (while still moving forward rapidly) and hopefully into the chamber mouth, where the bullet nose again strikes hard against the chamber wall and is then deflected downward. At this point, the cartridge is aligned approximately with the chamber, and is thrust home by the slide. Deviations in cartridge length and bullet shape can seriously affect the capability of the gun to feed a cartridge with complete reliability.

There are a good many other differences, based principally upon the functional differences between revolvers and autos. First, the revolver is not dependent in any way upon the power, pressure, velocity or recoil impulse of the cartridge. All functioning is manual, with the shooter's hand providing the power. Consequently, any revolver cartridge will still be functional even though loaded so lightly that a bullet of any workable weight is barely ejected from the muzzle; and, at the other end of the scale, any amount of pressure and/or recoil that can be safely accommodated by the gun and the shooter is acceptable. Not so with the autoloader. Any autoloading pistol mechanism is designed to function in and of itself, (aside from initial loading) by use of power generated by the cartridge. Obviously, whether a mechanism is designed specifically to be operated by a certain minimum amount of power or whether that amount of power is established after the mechanism is in being, a cartridge which cannot produce that power will not cause the gun to function. Thus, a functional autoloader cartridge must drive its bullet fast enough (in very few designs, generate sufficient gas pressure) that sufficient recoil impulse is generated to cycle the action. The principal factors of bullet weight, propellant weight and bullet velocity determine the magnitude of the recoil impulse. In the gun, factors of recoiling-parts weight, friction loads, extraction load, feeding forces, spring forces and rigidity of support (by the shooting hand) determine the amount of recoil force necessary to make the gun function properly. It is obvious, then, that if the cartridge does not produce sufficient recoil, the gun simply won't work. Thus there is minimum recoil impulse (plus a safety factor) which must be met, but in layman's terms we simply cannot assign a numerical value to it.

Two variations of the S&W M39 feed ramp seen from the rear. The one at left has two different angles and works quite well with standard full-jacketed round-nose bullets. The other has a single slope and was developed to provide better feeding with current-generation jacketed expanding bullets. The slight difference between these two ramps indicates quite clearly how critical autoloaders are to bullet shape in regard to feeding.

Here, the cartridge is substantially shorter overall than the space provided for it in the magazine body. The magazine was originally dimensioned for the standard military ball 9mm Parabellum loading. To achieve 100 percent reliability of feeding in one particular gun with extra-short cartridges, it was necessary to place a spacer in the rear of this magazine (visible beneath the case head) and thus position cartridges just exactly right for initial forward feeding movement toward the feed ramp. This demonstrates the relatively low tolerance autoloading pistols (at least some designs) have for variations in cartridge length.

At the other end of the scale, the autoloader cartridge is more limited than that of the revolver. Even though a pistol may have adequate strength to withstand the chamber pressures to drive a given bullet at very high velocity, the reeoil impulse produced by firing that bullet may *eventually* cause damage to the gun. In a revolver, no greater force may ever be applied to the moving parts than that which the shooter's fingers or thumb can generate; but, in an autoloader, stresses on moving parts increase very rapidly as the recoil impulse rises. In chasing improved ballistic performance then, we rapidly reach the point where the mechanism is "overdriven" even though it handles chamber pressure and pure firing stresses easily. Excessive recoil means excessive velocity given to the moving parts, greatly increasing friction loads, inertial loads, and impact loads. When the gun is overdriven, parts breakage becomes much more probable,

So, for functional reasons, the autoloader cartridge must be loaded to operate within relatively narrow limits of recoil impulse; no such limitation applies to the revolver.

We've already indicated that since all functioning operations of the revolver are accomplished manually, cartridge length and shape have no affect whatever on feeding. We feed the revolver with our fingers, or in some instances, a speedloading device. The autoloader is an entirely different feline. During the tiniest fraction of a second immediately following the firing of each shot, various parts in the autoloader must move a cartridge upward and two to three times its own length forward from magazine to chamber. The cartridge is partially guided during the early stages of this movement by the magazine feed lips and to a lesser degree by the breech face, but once free of those lips, it is moving forward very rapidly and virtually unguided except as it bounces off of various surfaces intended to insure that it at least begins entry into the chamber. While it might seem theoretically and technically possible to guide the cartridge precisely throughout this entire travel, no practical autoloader design does so. Designs which will do so have been proposed and built into prototype form, but have never achieved production status. *If* the gun could be designed and manufactured (also the cartridges) so that every surface, every angle, every position, every impact, every deflection was exactly the same from one shot to another, feeding would be simple. Unfortunately, the realities of both gun and ammunition manufacture won't allow that. Individual parts differ, individual complete guns differ, magazines and their feed lips and their relation to other gun parts differ, and cartridges differ.

To overcome all these difficulties in feeding, the cartridge must have an adequate length/diameter ratio, and in particular, its front end or leading edge (the bullet) must have the shape which is *least* critical and provides maximum guidance from magazine to chamber. Over the years it has been determined that this shape is the round-nose which has become so familiar to us. In addition to shape, it has been found that the surface of the bullet which impacts against the various guiding surfaces must be relatively hard; at least a good deal harder than lead revolver bullets. Soft lead has a tendency to deform under feeding impact and also to "dig in" where any surface irregularities exist. So, as a general statement, autoloader cartridges are limited to bullets long enough to produce standard cartridge lengths, to very hard lead or metal-jacketed bullets, and to bullets of round-nose shape. Deviations from this shape have been used, in some instances quite successfully, but for them to feed with acceptable reliability, they must approximately duplicate the points and angles of impact produced by the "standard" bullet. Truncated-cone, flattened round-nose, and modified semi-wadcutter forms have all been reasonably successful, though in many instances requiring careful handwork on individual guns. Bullet weight is also critical to some extent in that it must be sufficient to produce adequate recoil impulse, and it must be sufficient to make the bullet long enough so that the assembled cartridge will not be beneath the minimum length required for satisfactory feeding.

Generally speaking, revolver and autoloader cartridges headspace differently in their chambers. All cartridges originally intended for revolvers contain a sturdy rim which halts forward movement of the cartridge in the chamber by contacting the chamber mouth. Thus the position or shape of the case mouth has no relation to proper headspacing and primer support during firing. Of course, if the case is too long for the chamber, it can cause trouble, but that's another matter. Since the rim seating against the chamber mouth provides all the necessary cartridge support, it is practical to give revolver cases a very heavy roll crimp to secure the bullet. The only exceptions to typical revolver headspacing are found in the adaption of revolvers to fire rimless auto pistol or rifle cartridges. The best-known example of this is found in the Colt and S&W 1917 Revolvers (and commercial counterparts thereof) made for the U.S. Army during World War I. In this particular instance, the rimless cartridges were combined in groups of three by special half-moon, spring-steel clips which not

This spread of handloads has been assembled to show eight different types of bullets that can be used in an auto chambered for the .45 ACP. From left to right: Cast 240-grain SWC; a lighter, swaged SWC (185-grains); an almost identical *cast* SWC; a short-jacketed slug; a Sierra 185-grain JHC; (jacketed hollow cavity); a 190-grain JHP (jacketed hollow point); a round-nosed hollow soft point with a fair amount of exposed lead and a 230-grain hollow soft point with a lesser amount of exposed lead. Be it plinking, target or defense, this selection of bullets will do the job providing the gun is properly tuned. However, the selection itself is quite small when compared to the unlimited variety of shapes that can be accommodated in a revolver.

Cartridge headspacing surfaces vary. Left, is the .38 Special with the front face of its rim which arrests cartridge movement into the chamber. You'll note that the rim of the .38 is even with the squared-off mouth of the autoloading cartridge on the right which serves the same purpose.

At left we have the 7.65 mm Parabellum (.30 Luger) with the case mouth roll-crimped into a cannelure around the bullet, as is typical of bottle-neck handgun calibers. At right, for comparison, is the .38 Super essentially without crimp and with no cannelure in the bullet.

Though the .38 Super case (left) is designated as "semi-rimmed," and does possess a rim that extends slightly beyond the base, the difference between it and the comparable *rimless* 9mm Parabellum is hardly visible to the eye. This is just one of the reasons why the semi-rim case does a poor job of head spacing.

only speeded loading, but served to support cartridges for firing and provided a surface against which the extractor could operate. Cartridges could be fired without the clips, but the extractor would not remove the empty cases; when this was done, headspacing was controlled as described a bit farther along for autoloaders. More recent examples are the chambering of the Ruger single-action revolvers for the .30 Carbine cartridge which is a rimless number headspacing by its mouth meeting a corresponding shoulder in the chamber. In this instance, the lack of a rim presents no extraction problem because of the rod-type extractor/ejector system of the Ruger. An even more recent example (and the only other one familiar to me at the moment) is the Ruger double-action Security Six revolver chambered for 9mm Parabellum, another rimless autoloader cartridge. In this instance, headspacing is as for auto pistols by the case mouth against a chamber shoulder, but a unique springwire ring in the extractor engages the extractor grooves in the case heads and permits the otherwise conventional revolver extractor to function.

Revolver fired-case extraction, with the exception of certain single-action types as already mentioned, is handled by a star-like plate that is manually pushed rearward to bear on the undersides of the rims and force the cases from the chambers.

Again, autoloader cartridges are different. The earliest bottle-neck cartridges were supported against firing-pin blow and extractor feeding pressure by the shoulder of the case seating against a corresponding shoulder in the chamber; though this may have been aided in some instances by the case mouth also coming into contact with a slight shoulder at the front of the chamber. These were generally high-performance cartridges such as the 7.63mm Mauser. The small pocket-pistol type cartridges of lower power (.25 and .32 ACP/Browning) *appeared* to be of rimless design, but in reality their extraction rims protruded slightly beyond case

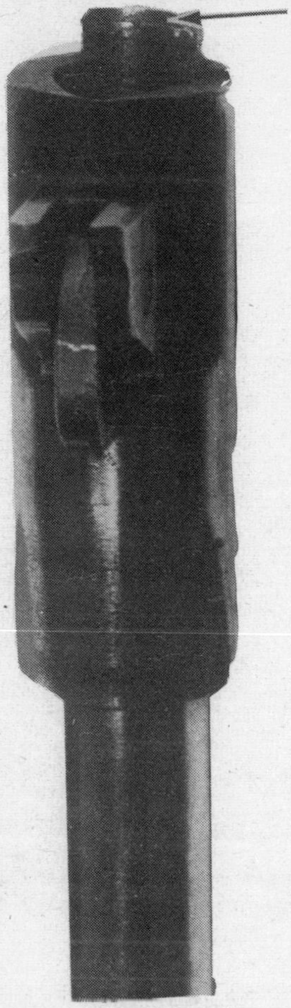

Left—The rear of the .38 Super Colt barrel has the extended barrel tang characteristic of the GM design, but more important, it has a shallow recess cut in its underside at the rear (arrow) to accept the slight semi-rim of the .38 ACP case. Theoretically, the cartridge headspaces upon its rim in this recess, but as a practical matter, the rim often slides right by the recess with the cartridge supported against the firing pin blow by the extractor claw and/or the mouth of the case seating against the front of the chamber. Most guns chambered for the semi-rim cartridges (.25 ACP, .32 ACP, 9mm Browning Long, and .38 ACP) contain a narrow shoulder of this sort for headspacing purposes, but as a practical matter, the guns function well *in spite of* rather than because of it.

Right—This cartridge has entered a S&W M39/59 chamber well beyond the rear face of the barrel tang which establishes gun headspace; this indicates the case is too short. While this too-short case will appear to function normally, proper case length is critical to *consistent* ignition; and, in this instance, better and more uniform ignition would be obtained if the case were long enough to permit its head to be flush with the barrel tang.

head diameter. Thus these cartridges headspaced, albeit somewhat indifferently, by this very small rim projection seating upon a corresponding surface partially encircling the chamber mouth. Unfortunately, insofar as headspacing is concerned, this surface was invariably cut away to a large degree around the lower half of the chamber to permit feeding. The .38 ACP, a high-performance cartridge developed during roughly the same period, also possessed this thin "semi-rim" which was intended to headspace the cartridge correctly upon a very narrow and small surface on the upper tang of the Colt/Browning barrel. Unfortunately, more often than not, manufacturing tolerances in gun and ammunition combine to prevent the semi-rim from giving the cartridge any support whatever. In the Colt/Browning gun, this problem is offset somewhat by the fact that the extractor grips the case rim before chambering is complete, and will therefore hold the cartridge sufficiently to the rear for the primer to be adequately ignited, even though no true headspacing contact was made. The 9mm Browning Long military cartridge also employs the semi-rim headspace method but is long obsolete and seldom encountered today. That cartridge/gun designers learned quickly is evident in the fact that all subsequent autoloading pistol cartridges (most of which were developed during the first decade of this century) were of true rimless form and are headspaced by the case mouth contacting a shoulder at the front of the chamber. When this method is employed—and is the best encountered to date—a heavy roll crimp to secure case/bullet assembly is *not* practical.

Though it isn't apparent until one examines the functioning of the revolver and auto very closely, recoil forces act quite differently upon loaded cartridges remaining in cylinder or magazine. In the revolver cylinder, recoil drives the entire gun rearward as a unit, and this movement is transferred virtually without loss to the cartridge cases through the rims. As this occurs, the bullet—generally quite heavy and thus possessing con-

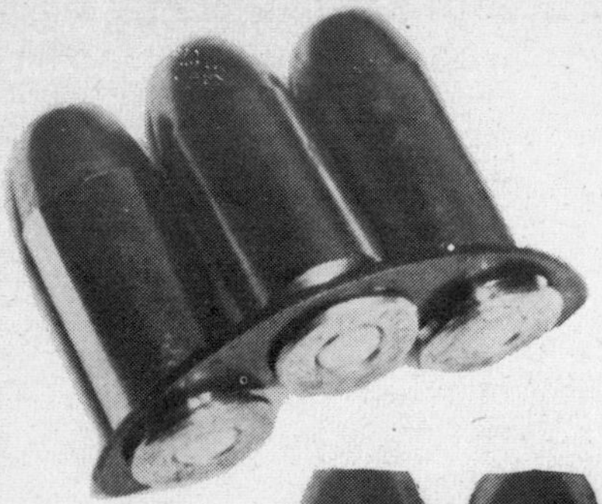

Left and below—Invented by one of the second-generation Wessons in 1917, the three-round, half-moon clip was devised in order to allow rimless .45 ACP cartridges to be fired in and extracted from a revolver. In reality, once the cartridges are snapped in place in the clip, the flat front face of the clip substitutes very nicely for the integral case rim normally found in revolver calibers.

Right—Here we have the two standard crimps which the factories normally apply to handgun calibers: At left is the .38 Special with a heavy roll crimp, and at right is the .38 Super with what appears to be no crimp at all but which is actually a vestigal taper crimp applied more to remove case-mouth flare than to crimp.

siderable inertia—is constrained to move rearward with the cartridge case *only* by the friction grip of the case upon it and by the case mouth itself being crimped. The result is simply that when recoil is severe, case/bullet assembly tightness is less than ideal; and, when the bullet possesses relatively high inertia, the case moves rearward *off* the bullet which tends to remain in place as the gun recoils rearward. This can result in the bullet protruding beyond the front of the cylinder, preventing cylinder rotation, and making further firing impossible until the affected cartridge(s) is removed.

The situation is different in the autoloader in that the problem in bullet assembly is to prevent the bullet from being forced *deeper* into the case as it impacts against the feed ramp and/or other guiding surfaces during its high-speed movement from magazine to chamber. With the pistol slide slamming back and forth at high velocity, the cartridge is driven very rapidly forward during feeding and when its bullet strikes the feed ramp, there is a great deal of force behind it. The case may be simply telescoped forward over the bullet. When this occurs, a feeding malfunction is highly probable. Cartridges remaining in the magazine during firing are greatly influenced by recoil. However, in this instance, as the gun recoils rearward, cartridges are restrained in the magazine only by the front of the magazine striking sharply against the bullet point. Thus, the tendency is for the bullet to be driven deeper into the case, directly opposite the recoil effect upon revolver ammunition. This condition is aggravated by the fact that during recoil and counter-recoil the cartridges bounce back and forth between front and rear of the magazine and are subject to repeated impacts, each of which tends to drive the bullets deeper into their cases.

In both revolver and autoloader, we have the problem of repeated recoil impulses applied to cartridges remaining in the feed system. In a six-shot revolver, assuming all cartridges are fired before reloading, the last cartridge is subjected to five recoil impulses; it must retain the integrity of case/bullet assembly until it is fired. Some people reload before all cartridges are fired, so conceivably that *last* cartridge could be forced to absorb even more recoil impulses. In an autoloader, with far greater cartridge capacity (as many as 18 in some models) one begins to wonder just how that last cartridge could possibly retain its original configuration under 15 or

Below, left—Bottle-neck cases do not always carry a roll crimp into a cannelure on the bullet. This Czechoslovakian-manufactured example of 7.63mm Mauser (left) uses a 3-point *stab* crimp produced by indenting the case neck, about mid-way between mouth and shoulder, with a pointed punch. Two of the "stabs" are visible here at either side. The .38 Super cartridge without crimp is shown for comparison.

Below, right—The two types of damage caused to loaded cartridges by recoil forces in revolvers and autos. At left is a .38 Special where the crimp was insufficient to retain bullet in the case under the forces of recoil, resulting in the case actually moving rearward off the bullet, leaving the bullet protruding sufficiently that it jammed into gun frame and prevented cylinder rotation. At right we have a .45 ACP lead-bullet load in which case/bullet fit was not tight enough to prevent repetitive recoil impulses from shoving bullet *deeper* into the case while it remained in the magazine. (This can also be caused by bullet impacting against the feed ramp.)

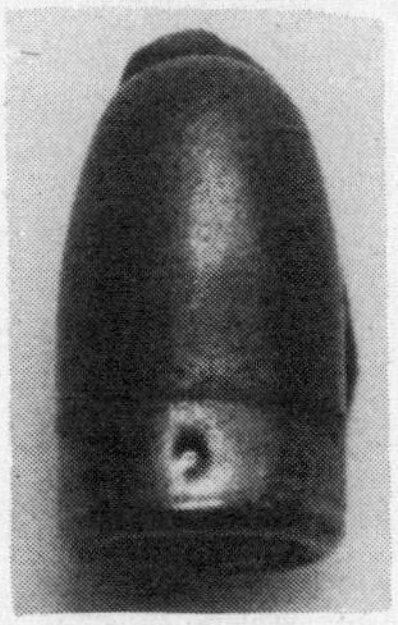

Above, left—Here is a bullet pulled from stab-crimped 7.63mm Mauser ammunition. Note the size and depth of the indentation formed in the bullet jacket when the case neck is indented. There is no evidence whatever to indicate that such indentations have any deleterious effect upon bullet performance.

Above, right—This .45 ACP ball bullet clearly shows the remnants of mouth sealer which remain on its bearing surface. Naturally, this bullet was not fired, but was extracted from the case with an inertia bullet puller.

more recoil blows and the back-and-forth battering attendant to each.

To better resist this, auto cartridges frequently have a cannelure rolled into the case, located so that it supports the base of the bullet against rearward forces. While cannelures are also seen on revolver cases, they do not serve this purpose.

Since a heavy roll crimp cannot be used on most modern auto cartridges, the requisite bullet pull value is more difficult to obtain—and a high pull value is essential for efficient propellant combustion and consistent velocity. A mouth sealing compound is often applied inside the case mouth before the bullet is seated. The sealer is adhesive in nature, but does not harden entirely, so will release the bullet when the proper pressure is developed. Sealer is sometimes used in revolver cartridges, especially those loaded to military specifications. There, however, it is used primarily for waterproofing, rather than to develop bullet pull. Naturally, it *could* be used for bullet pull, but applying a roll crimp is simpler and cheaper.

Internally, conventional revolver and auto cases are of identical solid-head construction today. However, auto cases have been of this type from the beginning, while revolver cases were mainly of balloon-head construction until the 1950's. Prior to the general changeover, only the .357 Magnum and some high-velocity revolver cases were of true solid-head construction.

Those are the essential differences between autoloader and revolver cartridges. Other differences of minor nature may crop up occasionally. For example, uncannelured FMJ bullets are used in some military .38 Special loads.

By understanding the differences you'll encounter less confusion in assembling your own ammunition. The result will be better ammunition and better shooting in the end.

CHAPTER 4

The Basic Tools

OBVIOUSLY, ASSEMBLING something seemingly so complex as a loaded firearms cartridge requires more than one's fingers and the battered screwdrivers and pliers found in most households. The tools, machinery and related equipment which the ammunition factories use to crank out millions of rounds in a relatively short period of time are terribly complex and sophisticated to one not involved in the industry—and they cost hundreds of thousands of dollars. In fact, a single automated machine for merely assembling the propellant and bullets into primed cases—at a rate of several thousand per hour—can easily cost that amount alone. Of course, the ammunition factories face the problem of manufacturing the cartridge case which the handloader does not. Certainly, the case presents the most complex and costly manufacturing process of all the components that go into any cartridge and that includes shotshells. Fortunately, we handloaders don't have to concern ourselves with case manufacture. We are only concerned with producing a few hundred rounds from our fired cases in a few hours spent leisurely at the loading bench.

LOADING PRESSES

Discounting the portable hand tools, which will be covered in considerable detail in Chapter 20, the basic handloading tool is the *loading press*. Of course, it would be difficult to say that the press is more important than the die and the other tools, but, without a press, conventional, standard-thread loading dies are useless. As a practical matter, the press is the tool to which all the other dies and accessories are assembled or directly related.

The basic loading press is a sturdy framework, usually of cast steel though occasionally made of aluminum alloys. Sometimes it is built up of several cast and machined components instead of being a single casting. The purpose of the press is to provide, at its upper end, a threaded seat for the various dies which are needed to perform the several required operations upon the case. At its lower end we find a *ram,* which is in its most common form simply a long cylinder of steel, sliding more-or-less vertically in a close-fitting hole in the press frame. Connected to the lower portion of the frame and to the bottom of the ram is a handle and linkage which permits rotary handle motion (force applied manually, of course) to be converted to linear motion of the ram, moving it toward and away from the die. In between we have the very strong connection portion of the press frame which usually incorporates a substantial mounting flange or plate, drilled with several holes for securing the entire press to a bench top or stand. The area between the top and bottom of the press is left as open and as free from obstructions as the requirements for strength will allow. This "open area" of the press provides sufficient room to handle cases. Nothing is quite so frustrating as to work with a cramped press and bang your knuckles while preparing to seat a bullet, thus dumping the bullet and half the propellant charge on the floor, or perhaps fouling the press mechanism with flying components.

Much has been written and said about the *functional quality* of loading presses. The words fit, finish and workmanship as applied to presses do not refer to the external appearance, but relate primarily to the alignment of the ram with the dies installed in the press. As indicated above, the loading die—which we presume to be made concentrically within close limits—is threaded into the top of the press; then, the ram, through the use of a shell holder, carries the cartridge case into the die and subsequently extracts it. In order for the case to be aligned correctly with the die cavity, it is essential the ram be concentric in all respects with the die; and this means further that the ram must be a close but free-moving fit in its hole in the press and must also be fully concentric with the hole in which the die is threaded. Producing this type of accurate alignment and close fit is not a simple drill-press operation. The better manufacturers use complex fixtures and jigs (in large lathes or drilling machines) to drill and ream these two holes to final dimensions and, also, to thread the die seat as a combined operation in order to insure accurate alignment. But really, it isn't even that simple. Most cast press frames are subject to some warpage. Generally they are allowed to age for a specific period of time—the castings "stabilize" during this aging process—before being machined. However, an occasional frame will still distort a bit after machining, dependent to a large degree upon how carefully it was designed in the beginning to resist warpage.

The quality, and thus the durability of the press, is also determined to some extent by quality and fit of the pins and arms in the linkage. The better presses use closely-fitted, hardened, tool-steel pins and heat-treated linkage—the two fitted accurately together.

The power a loading press can generate when reasonable effort is applied to the handle is directly related to the mechanical advantage of the linkage. Even the simplest linkages found on today's presses offer more than adequate mechanical advantage for all resizing operations (the most difficult of loading operations) and for the swaging of *some* bullets. However, presses fitted with a compound linkage have become exceedingly popular in recent years because of their ability to generate greater power. Most of us prefer the compound linkage primarily because it eases the effort required for resizing, not because its power is actually needed for the job. One can easily differentiate between simple and compound linkages by the manner in which the lower end of the linkage is attached to the press frame. If the lower element of the linkage is pivoted to a fixed point upon the bottom of the frame proper, then it is a simple linkage, i.e., it forms a common toggle or knee joint which is straightened as the ram moves upward, and folded as the ram is retracted. The compound linkage, on the other hand, does not have its lower end attached to the frame proper; instead, it is attached to a pair of long arms pivoted at their upper ends to a point near the middle of the press. The photos on page 47 show the difference far better than we can describe it in words. (The Oriental gent who said a few thousand years ago that a picture is worth a thousand words was pretty close to the mark.) If the compound linkage has a disadvantage, it is that with careless use it is sometimes possible to damage the ram, shell holder, or press frame. It simply generates so much power with relatively little handle effort that you can apply too much force without realizing it.

Presses should always be bolted securely to a thick, solid bench top or loading stand. It can be both frustrating and painful if a haphazardly mounted press pops loose during the power stroke and jumps off the bench to gouge your knees and shins on its way down to smash your toes.

"C"-Press

Loading presses come in essentially four configurations. By far the most common is the "C-

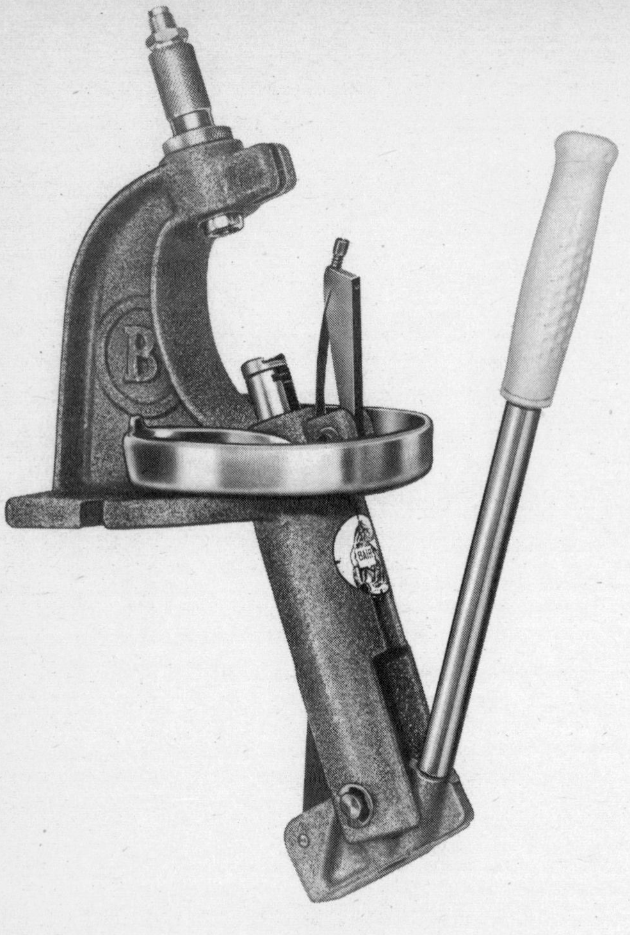

A typical C-type press which differs hardly at all from the original design by Pacific Gun Sight Co., dating back to long before WWII.

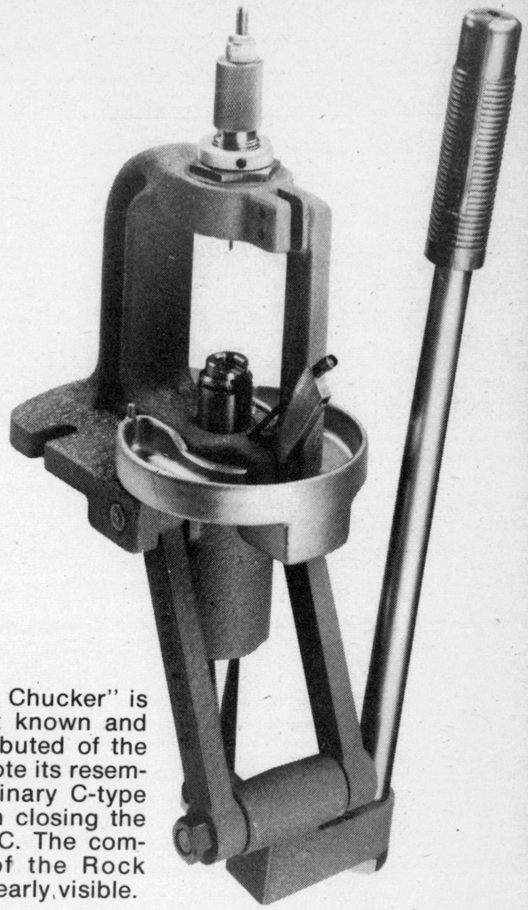

This RCBS "Rock Chucker" is probably the best known and most widely distributed of the O-type presses. Note its resemblance to an ordinary C-type press with an arm closing the open side of the C. The compound linkage of the Rock Chucker is also clearly visible.

type," so named because of its resemblance to the letter "C." The upper arm of the "C" contains the die seat, the lower arm the mouth flange, and then there is a downward extension to house the ram and support the handle and linkage. Modern C-type presses, cast of steel or iron, are quite adequate in strength for all conventional loading operations. They can, however, be distorted by applications of too much force on the handle, even with the simple linkage. This distortion occurs when the ram and shell holder, bearing against the mouth of the die, simply force the upper arm of the C away from the lower. Carried to extremes, this could break the upper portion, but it generally produces only temporary distortion; as soon as the pressure is eased, the upper arm moves back into its original position. The effect of this distortion is simply displacement both angularly and laterally of the die, and since the case is confined by the die and supported by the ram, the case head can be distorted, though the amount is seldom visible, Normally, you'd need a surface plate and dial indicator to detect the change.

The C-type press is likely the oldest press that's commercially available. To the best of my knowledge, it was first introduced by the old Pacific Gun Sight Company long before World War II. Writing in 1935, the late Phillip B. Sharpe stated that he considered the Pacific C-press the most practical all-around tool ever produced and that he had been using one for many years. The C-presses of today are built much more sturdily, including those offered by Pacific. Basic design remains the same, by several makers, but the frame is much more massive and therefore stronger; probably containing two or three times as much metal as the original.

"O"-Press

A logical outgrowth of the C-press is identified as the "O-type." It was originated by Fred Huntington of RCBS many years ago specifically to provide greater strength and less distortion.

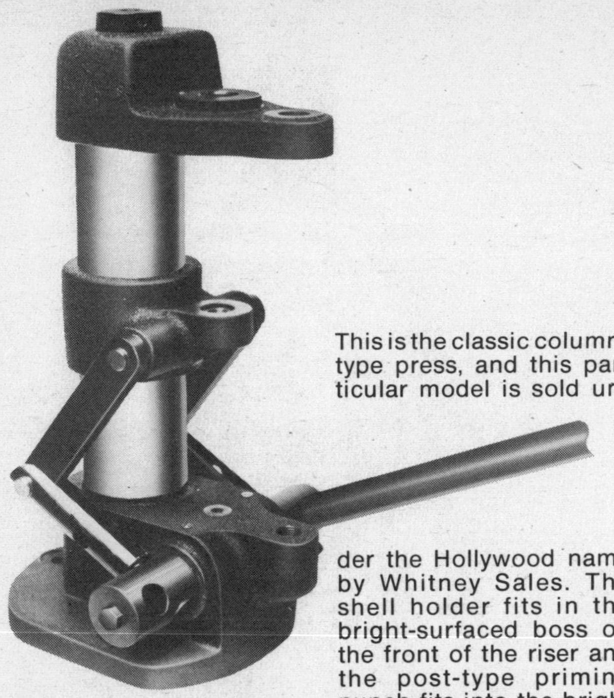

This is the classic column-type press, and this particular model is sold under the Hollywood name by Whitney Sales. The shell holder fits in the bright-surfaced boss on the front of the riser and the post-type priming punch fits into the bright spot on the base directly beneath it. The die head is clamped to the top of the post or column with a nut and carries the die in the innermost hole you see. The hole in the die head to the extreme right allows the installation of a sturdy tie-rod which extends downward into the threaded hole in a corresponding boss on the base. The purpose of the tie rod is to convert the entire assembly in effect into an H-type setup to resist deformation during heavy-duty bullet swaging. The tie rod need not be installed for any ordinary handgun-caliber loading operation.

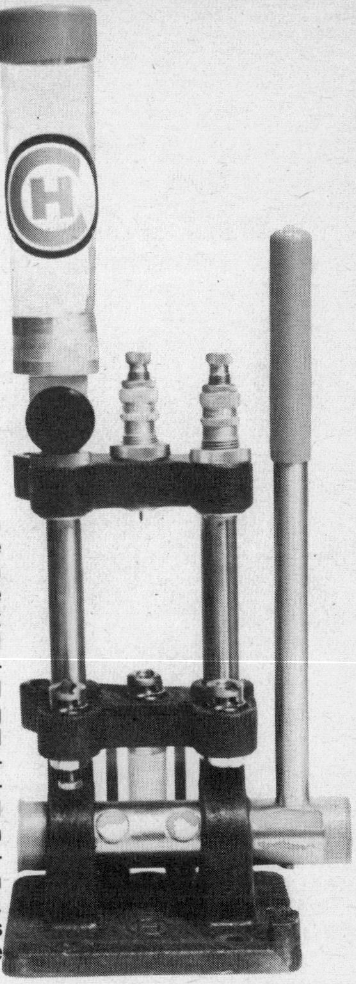

An excellent modern example of the H-type press is this 3-station C-H X333, complete with an entire 3-die set and fixed-charge C-H powder measure installed on the special expansion die. Priming is under the left-hand shell holder, and full-length sizing and decapping is performed at the center station with a container beneath the riser to catch expelled primers. Bullet-seating, then, is done at the extreme right.

The O-press simply has the open side of the original C closed with an additional integral segment angled or offset to the right to provide adequate clearance for one's hands during operation. The C-press absorbs work loads applied by the ram in a completely eccentric manner; the O-press, at least in the better designs, absorbs these loads concentrically. Consequently, the lateral and angular distortion is completely eliminated. Careless use can still produce a very slight linear deformation of the press frame, but this cannot distort the cartridge case; though it can ruin the shell holder and ram. The O-press is by far the most favored type for swaging bullets, forming cases and other operations requiring more than ordinary force; and, it is generally encountered fitted with compound linkage which greatly eases fatigue during operation.

"H"-Press

The H-press is found only (to the best of my knowledge) with simple linkage and takes its name from its configuration. Generally it consists of a cast or machined base which sits upon the bench top. Into the base are assembled a pair of vertical guide rods upon which rides a riser which substitutes for the cylindrical ram and carries the shell holder on its upper surface. This riser is connected by a simple linkage to a large-diameter horizontal shaft passing through the base and fitted with a handle on its right end (though some allow switching the handle to the other end for southpaws). Rotation of this shaft, by means of the handle, therefore moves the riser up-and-down. At the top, the two vertical guide rods are capped by a cast or machined head which maintains alignment of the rods and contains the threaded die seat. In some respects the H-press is simpler to manufacture than the O- or C-type; and, it can be made quite sturdy without resorting to massive castings. In fact, an individual reasonably handy with metal-working tools can make a H-press from steel bar, plate and rod with nothing more than a hacksaw, file,

One unusual variation of the turret type loading press is the one made by Texan. The base of the tool is quite similar if not identical to the very fine progressive shotshell loader produced by the same outfit. It is also a column-type press, surmounted by a 7-station turret. In addition, this Texan turret loader contains a rather unusual manually-operated, semi-auto primer feed.

and a drill-press with threading capability. The H-press has the advantage that with relatively little effort and design change it can be adapted to contain two or more dies in the head with corresponding shell holders on the the riser. This permits it to be set up for a full set of dies and thus eliminates time lost to replacement and readjustment. The most common form of the H-press used today is offered by the C-H Die Company. It has three stations permitting a complete set of rifle or handgun-caliber dies to be kept properly set up and ready for use at all times.

Column Press

We also have the column-type loading press which resembles nothing so much as half of the H-type. It consists of a substantial base upon which is installed a single large-diameter steel column upon which a riser slides vertically, connected to a horizontal shaft in the base by a suitable simple linkage. Then, a head is secured atop the column and contains the seat for the die. Virtually the only column-type press available

This is the unusual Bonanza Co-Ax press which differs from all other types. As can be seen, the handle is at the top connected by linkage to the riser below which carries a unique shell holder and rides vertically on two guide rods. When priming is done on this press, it is done with a conventional shell-holder head and a post-priming punch positioned on top of the press frame and behind the upper portion of the die. Consequently, the mechanism cannot be seen in this photograph. Note the long drop tube and a readily-detachable container for spent primers.

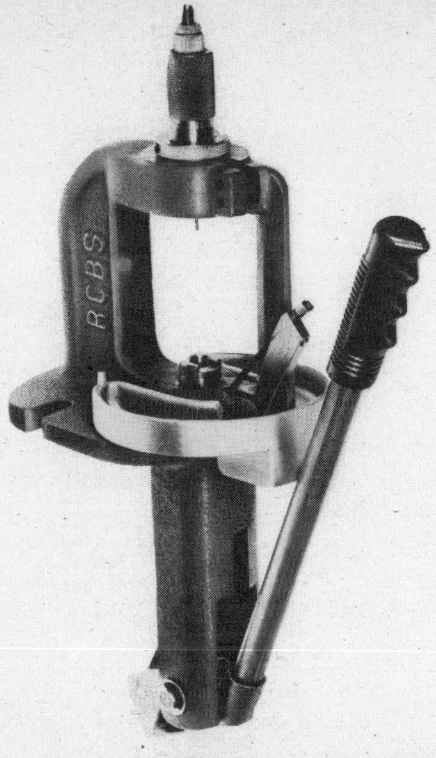

At left we have the economical RCBS "Junior" loading press which utilizes a simple linkage for obtaining ram movement through handle movement. Note the handle fits into a single link pinned solidly to the bottom of the frame which is connected directly to the ram by another single link. At right we have the RCBS compound-linkage system as developed in the now-discontinued A-2 press. Note here that there are two heavy links extending downward from the frame with the handle link pinned between them and connected to the bottom of the ram.

now for many years has been the old Hollywood, now offered by Whitney Sales, Inc. The column-press—like the H-press—sits entirely upon the bench top and does not extend several inches beneath the bench edge as do the C- and O-presses.

Turret Press

C- and Column-presses are frequently offered in a variation intended to speed up loading and/or permit as many as two or three full sets of dies to be kept in readiness at all times. This is the turret press. It consists of the basic press with the single die seat replaced by a thick, heavy, steel, circular plate rotating upon a large vertical shaft. This plate contains four to 12 individual die seats spaced equally around it. A spring-loaded detent serves to align each individual seat with the ram or shell holder as the plate is rotated manually into position. (The detent also serves to prevent inadvertent rotation during operation.) The principal advantage claimed for the turret press is that with all dies and a powder measure properly installed in the turret a case may be placed in the shell holder and not removed until it is completely loaded. It is usually also believed that this offers faster loading than any other type of non-automatic press. I agree with this, however, I don't believe the gain in speed is nearly so great as some advertising has claimed.

Actually, while the more costly O-, H-, Column- and Turrent-presses do have their advantages, any of the good C-types currently available (and that includes lots of discontinued models on the used market) are entirely adequate for loading first-class handgun ammunition. Even with only a simple linkage, it will operate easily and no distortion will occur under the stress loads required to resize handgun cases so long as dies are properly adjusted.

Bonanza Co-Ax Press

From time to time other types of presses show up, and the most notable of these is the unique Bonanza Co-Ax design which bears hardly any resemblance to the basic types we've described. It consists of essentially a H-type frame and corresponding castings which hang about half its

Above — Note that this particular removable shell holder head has a shallow groove cut in the bottom of the case-rim recess; this groove serves to allow the insertion of cases with slightly protruding fired primers. Protruding primers are quite common with light loads in revolvers, and unless this groove is present, difficulty may be encountered in inserting them in the shell holder for resizing and/or decapping.

Below — The various types of rams and detachable shell-holder heads currently in use. At the lower left we have a short ram and long ram, both of the same diameter and employing a wire snap-ring to hold the shell holder in place. Clockwise from above the short ram, the shell holders are: the standard RCBS detachable type with the bottom of its rim-holding recess flat which thereby functions as a gauge or detector for protruding fired primers which indicates — especially in rifle calibers — below-minimum case headspace; ditto shell holder above the long ram, an extension-type shell holder which is needed if some short handgun cases are to be processed in certain case-forming or file-type trimming dies; conventional replacement shell-holder head and (below) a typical shell-holder stud which is really nothing more than a machine-bolt which is assembled to the riser bar of an H-type press and whose head is slotted to accept the conventional removable shell-holder head appearing just above it.

length below the edge of the bench. Its handle is at the top rather than at the bottom, and the shell holder is an unusual universal-type, consisting of two horizontally-sliding blades with V-shape recesses which can open and close over the case rim automatically. A separate priming device is installed in what we might ordinarily consider the head of the press, while the dies fit into a slot in a separate carrier beneath. Even though Bonanza dies are threaded (so they may be used in other presses) they slip into a U-shape channel in the carrier and are positioned vertically by a slot fitting closely over the lock ring. Very unusual but very effective, and the Co-Ax press has much to recommend it.

LOADING DIES

Any reloading press is of very little use, except perhaps as a trot-line weight, without *good* loading dies. While bottle-neck cases are correctly loaded with a two-die set, three dies are essential to the proper preparation of loads in straight or slightly tapered cases. A four-die set was introduced some years ago by RCBS for meticulous preparation of target loads in some calibers, but with special manipulation, the three-die set will accomplish the same goal as outlined farther along in this book.

The resizing die comes first and consists of a body threaded ⅞ x 14 TPI (threads per inch) which is screwed into the die seat at the top of the press. It contains a cavity (concentric with the body) dimensioned to reduce the expanded/fired case below minimum-chamber dimensions for the caliber so that after resizing it will freely enter a *standard* chamber. The interior of this cavity must be very smoothly polished and heat-treated to possess a very hard surface skin; and, the mouth of the die must be radiused or tapered to permit the oversized (fired) case to enter. The resizing die often—but not always—contains a rod passing through the center of the cavity that's fitted with a replaceable decapping punch; the rod is adjustable vertically in the die by means of a threaded bushing and lock nut. During the last portion of the case's movement into the resizing die, the decapping pin enters the flash hole in the head of the case and punches out the fired primer. The ejected primer falls through a hole in the shell holder and passes through a slot or hole in the ram to fall on the floor—unless you have foresight enough to obtain a spent-primer catch-

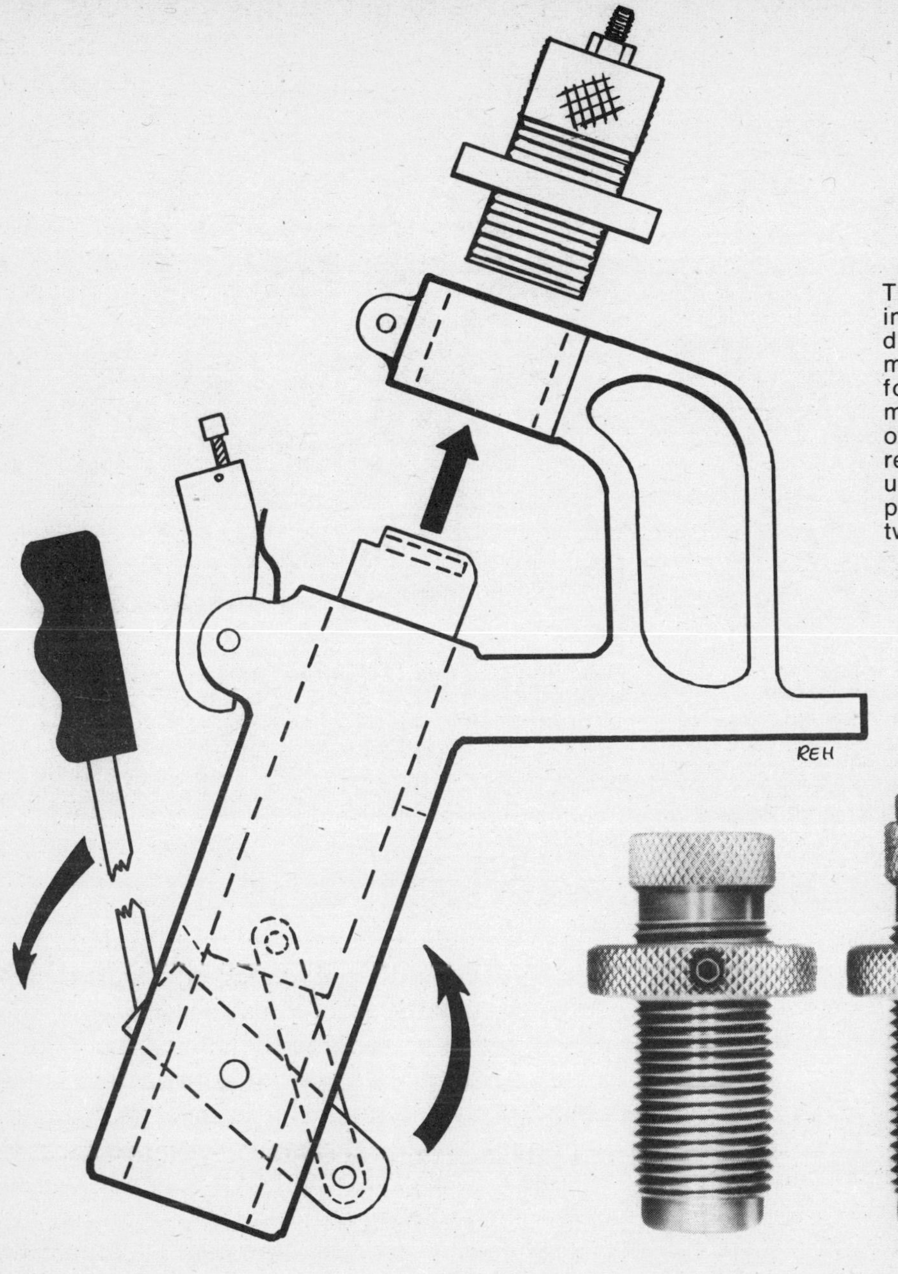

This sketch of the classic C-type loading press shows first how rotary handle movement is converted to linear movement through which cases are forced into the die; and, second, makes it apparent how the upper arm of the C may be bent upward and rearward (to the right in this drawing) under excessive working pressure, possibly destroying alignment between the die and the ram.

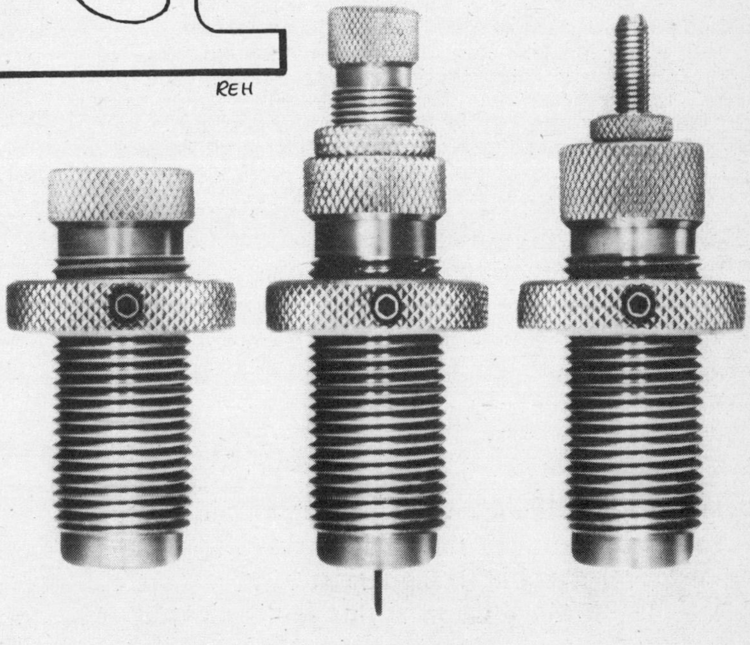

Above — A typical, three-unit, handgun-caliber loading die set. From left to right, the resizing die, the mouth-expansion die containing the expander rod which carries the decapping pin on its lower end, and the bullet seating/criming die which may be adjusted to seat and crimp separately or simultaneously.

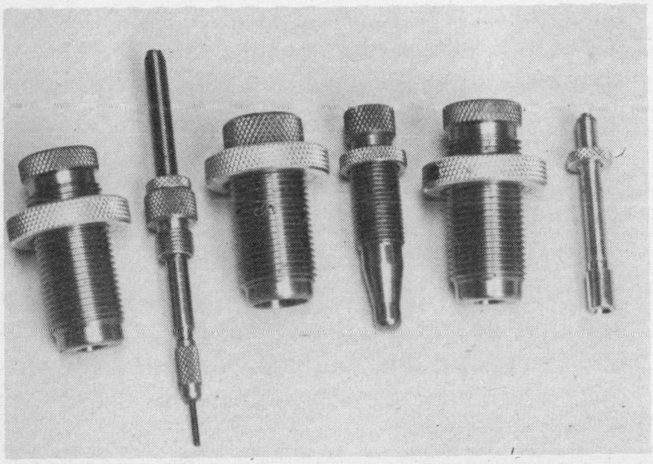

Left — Here we have typical three-unit, handgun caliber, loading die set disassembled. From left to right, the resizing die, the decapping stem sometimes found in the resizing die; the expander die body; the expander rod which often carries the decapping pin at its lower end and thus eliminates the need for a separate decapping unit in the resizing die; the bullet-seating (and crimping) die body; and, the bullet-seating stem or screw which contains a cavity in its lower end which fits closely over the bullet nose.

In this head-on view of a tungsten-carbide resizing die you can see the steel die body constitutes only the outer rim, while the tungsten-ring (cemented inside the die mouth) does all the work on the case. In years gone by, the tungsten-carbine insert was often silver-soldered in place, but today some makers employ simpler and cheaper modern adhesives which do the job fully as well.

Loading dies *should* be stored under cover to keep them from gathering dust and dirt, but if one still insists on storing them in the open, some sort of rack containing holes large enough to accept the die bodies would be a great convenience. This "DiRac" from RCBS is effective and economical. However, if dies are to be stored in this manner, they should be kept covered at all times.

ing device and attach it to the press.

The second die of the typical handgun-caliber set is really not a die at all; it is simply an aluminum shell shaped and threaded externally like a die, but containing a cavity large enough to provide clearance around even the largest handgun-caliber cartridge case. Thus this one size *expanding die* can be used for all calibers. A rod is centered in the cavity and is adjustable vertically like the decapping stem in the sizing die. The lower portion of this rod is tapered or radiused to enter the case mouth, and a cylindrical upper portion follows with a diameter slightly over the desired inside diameter of the case. When this upper portion of the expander rod is pressed into the case mouth and withdrawn, the brass is first expanded and then "springs back" or constricts slightly, stabilizing at the desired diameter. Directly above the cylindrical portion is a flared section which, when the rod is properly adjusted, enters the case mouth only slightly, funneling it outward to slightly more than bullet-base diameter so that at the appropriate time a bullet may be started and seated easily, without shaving the bullet's base.

The third die is usually referred to simply as a "bullet-seating die," even though in most calibers it is actually a combination seat-and-crimp die. Externally similar to the others, it contains a cavity which allows entry of a resized case (and bullet) with only slight clearance; this point is important because the cavity serves to guide the case as the bullet is seated. If too much clearance exists the bullet may be tipped, while too little clearance can result in jamming. The upper end of this cavity is reduced in diameter via a concave shoulder. When the case mouth is forced against this shoulder, it is turned inward against or into the bullet. This action not only removes the original flare, but causes the mouth to grip the bullet tightly. The upper portion of the seating die contains a stem, threaded for vertical adjustment, and the lower end of the die contains a recess which normally *should be* shaped to fit the bullet nose exactly. It is common practice to use a single seating-stem shape for a variety of bullets and thus accept the deformation this produces, but best results are obtained when the cavity fits the bullet shape exactly. When this seat/crimp die is adjusted correctly, a case with started bullet is pressed into the die and, as the seating stem contacts the bullet nose, the latter is forced into the case. During the last 1/32- to 1/16-inch of case travel, seating is completed and the case mouth is turned into the bullet tightly.

The only real difference between the three-die set for straight cases and the two-die set for the bottle-neck variety is that the expander is in the form of a ball instead of a rod and is incorporated in the decapping rod within the resizing die. Therefore the case mouth is expanded over this ball as the case is extracted from the resizing die; thus no intermediate expanding die and rod are necessary or desirable. You might also very well encounter an obsolete design of two-die set which was once the only type available for straight handgun cartridges. In this type, the resizing die cavity is of working diameter only for a very short distance at its mouth; then, above that the cavity is enlarged and an expander/decapping rod extends downward into the cavity. Thus the case mouth is expanded and flared inside the die after passing through the sizing portion. As the case is extracted from the die, the flared and expanded portion is pulled back through the reduced die area and squeezed back down. However, the brass springs back sufficiently so that adequate expansion and flare remain for normal reloading. Prior to the availability of three-die sets we thought this type worked quite well. In many respects, it does. However, case life is greatly reduced due to work-hardening of the brass. My best advice regarding those old two-die sets is to avoid the resizing unit (the seating die is perfectly acceptable) or remove its expander and augment it with a separate expander die to make it a three-die set. With the expander removed, the sizing die works quite well and does not harm the case.

SHELL HOLDER

Some means is required to extract the case from the dies. A case could be forced into the die easily enough with a simple flat on the end of the ram, but then getting it out might present a problem. For this purpose we have the detachable shell-holder head which is readily interchanged by means of a T-slot in the top of the ram and a corresponding tenon on the underside of the holder. Providing parts are kept clean, the holder slips in and out easily and is usually retained during operation by a spring clip, though

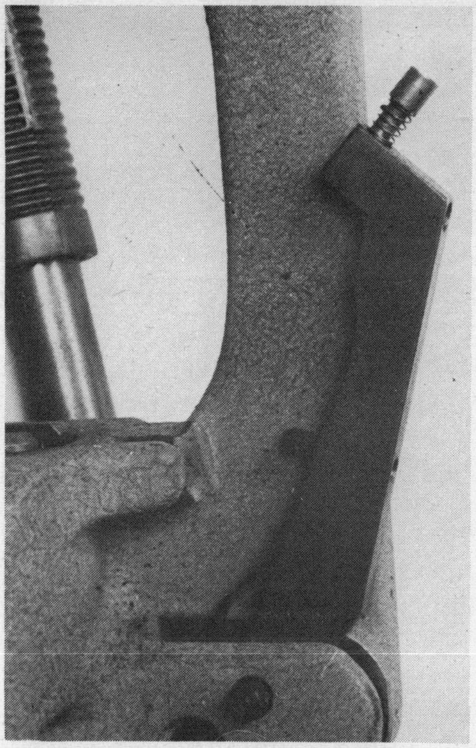

Shown here on the discontinued RCBS A-2 press is a typical swinging-arm priming device found on almost all O-type and C-type presses. This particular priming device contains an adjustment screw which may be set to bear upon inside of the ram to more precisely align the primer with the primer pocket. Shown is the rest position of the arm, and it is held there by a spring.

When in use, the primer arm is pressed forward with the thumb into its slot in the ram, positioning the primer so that as the ram is lowered, the case is pulled down over the primer.

setscrews may sometimes be encountered. The upper end of the shell holder contains a T-slot dimensioned to accept the rim of the case. Thus the detachable shell holder serves to push the case into the die as well as extract it by pressure on the front face of its rim as the ram is retracted. A very high degree of interchangeability exists among shell holders. Due to the commonality of case heads, a single holder may accommodate a dozen or more calibers. All manufacturers supply a table setting forth this interchangeability, and a majority of them now use the numbering system first employed by RCBS, Inc.—the firm that originally developed and introduced the interchangeable shell holder. Prior to that, individual press rams were T-slotted to accept the case rim and had to be interchanged after the fashion first introduced in the Pacific C-press.

PRIMING ARM

Cases cannot be loaded until they have been reprimed, and virtually all loading presses available today are either supplied with, or are made to accept, a priming device (often called a "priming arm"). The device is used as the case is extracted from either the resizing or mouth-expanding die. In all but the H- and Column-type presses this device takes the form of a swinging arm pivoted to the press frame and fitted at its top with a short primer punch surrounded by a spring-loaded bushing which holds a fresh primer accurately aligned over the punch end. A spring holds this *priming arm* out of operation (toward the operator); and, with a primer in place, it is pressed forward into a slot cut in the ram while the case is fully in the die. The case is then carefully extracted from the die just enough to pull it down over the primer, the punch thus seating the primer to the bottom of the pocket. The case is then raised just enough to allow the primer arm to be retracted by its spring, and only then, is the case extracted. H- and Column-type presses use instead a fixed primer-seating punch accompanied by a similar bushing (and positioned directly under the shell holder in the riser) so that when the latter is moved to the lower limit of the stroke, the case is pressed over the primer.

Those on-press methods of priming are entirely satisfactory for ordinary loading with perhaps a single exception: Often there is so much friction and mechanical advantage in the system that one cannot feel the added resistance

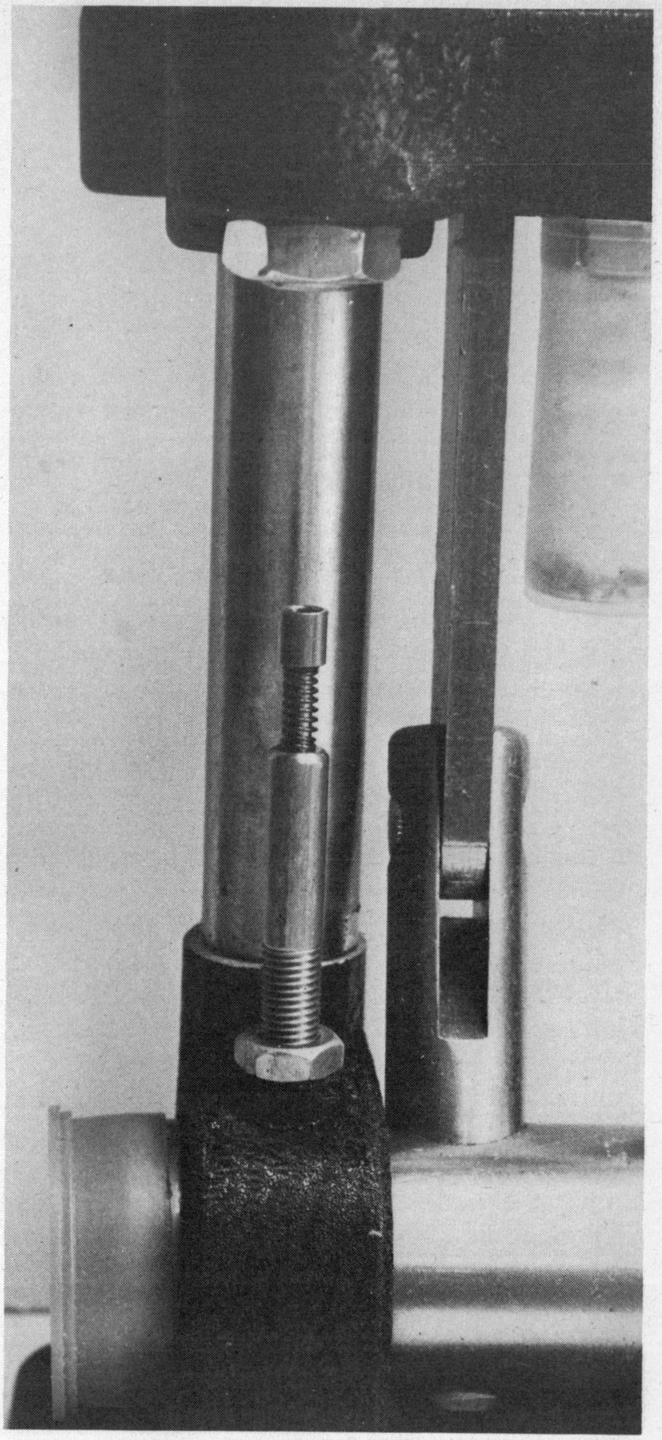

This is a typical post-type primer-seating device as found on the 3-station C-H loading press. A sturdy post is screwed into the press base, and into it is fitted a short primer-seating punch, surrounded by a retractable sleeve or bushing to contain and align the primer. The spring beneath the sleeve serves to keep it in the uppermost position to accept a primer, and yet allow it to be pushed aside by the case head as it is forced down over the primer. Directly above the punch you see the nut on the underside of the press riser which holds the shell holder in place. As the riser moves downward to force the case over the primer, the seating punch passes through a hole in the center of the holder.

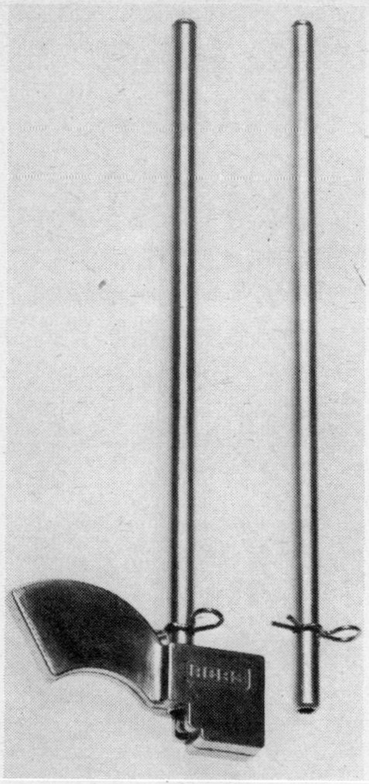

Above — This RCBS bench-type priming tool employs a standard removable shell-holder head through which passes the simple primer-seating punch shown to the right. Punches for both small and large size primers may be readily interchanged—the tool is simple, easy to use.

Above, center — Be it for pistol or rifle cartridges, this unusual priming tool offers very rapid repriming as a separate operation. It differs from other bench priming tools in two respects. First of all, the vertical magazine for the automatic feed places the primers edge-to-edge instead of bottom-to-top. Second, it employs an unusual adjustable shell-holder design and thus is adaptable to a wide variety of cases without the use of a number of replaceable shell-holder heads. The three steel discs on the face of the tool are pivoted eccentrically and rotated to fit a particular case head and rim.

Above, left — If you elect to seat new primers on the loading press, this general type of automatic primer feed is available from several sources and will fit most existing C-type or O-type presses. Nearly all contain a seating slot at the top into which the curved arm on the left of this item's base fits—it's adjusted and clamped by setscrews. The magazine tubes contain upward of 50 primers, and when the device is properly positioned, a primer automatically drops into the cup of the pivoted priming arm as it moves into its rest position.

as the primer is fully seated. Thus it is probable that too much seating pressure will be used and the primer's uniformity of function reduced.

To my way of thinking, it is much more desirable and productive to have more uniform primer function by seating primers separately with a tool possessing less mechanical advantage and friction and therefore greater sensitivity of feel. Several such separate bench-type priming tools are available today—some are of the one-at-a-time type while others feature automatic primer-feeding magazines. Either type does an equally good job of seating primers, but the latter offers much greater speed of operation in exchange for its greater cost. Such bench-type priming tools are available from several makers. The size of such a tool ranges from the diminutive SAS (which can be held in the palm of the hand) up to the semiautomatic RCBS rig several times greater in size. All of these tools except Bonanza use standard removable shell-holder heads, but no separate purchase of same is required since you already own one for the caliber being reloaded—just switch it between press and priming tool. The Bonanza tool, somewhat in common with its companion Bonanza Co-Ax press, utilizes a unique "universal" shell holder consisting of three hardened steel discs pivoted eccentrically. Their retaining screws are loosened, the discs then privoted inward to fit a particular case, and the screws are then tightened. To simplify the operation you can hold the case in position for that adjustment by pressing the primer-seating punch into the primer pocket. Adjustment to another caliber is accomplished in exactly the same way.

While it might seem that a good deal of time could be lost by priming on a separate tool, I do believe the reverse is true: Priming on-the-press requires very careful manipulation to prevent the decapping pin dislodging the freshly-seated primer while the arm is being retracted. This consumes time and causes errors—priming is genuinely faster on a separate tool, even if an automatic feed is not employed.

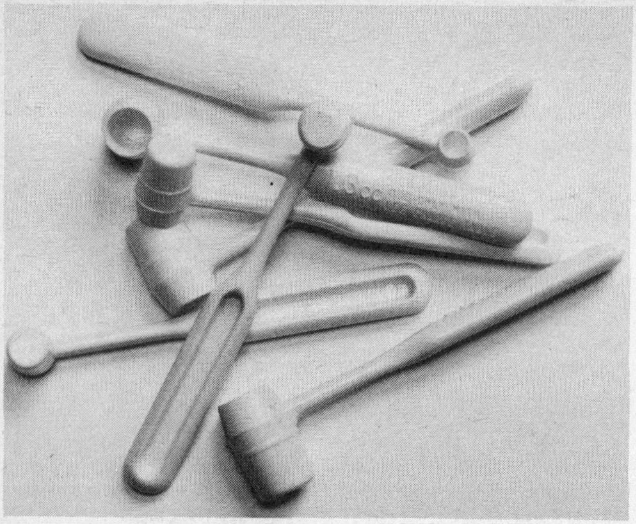

While charge cups are easily made from fired cases and other scrap material, the 7 molded plastic cups shown here are very adaptable to handgun loading and are only a part of a 15-cup set sold for a ridiculously low price by Lee Precision, Inc.

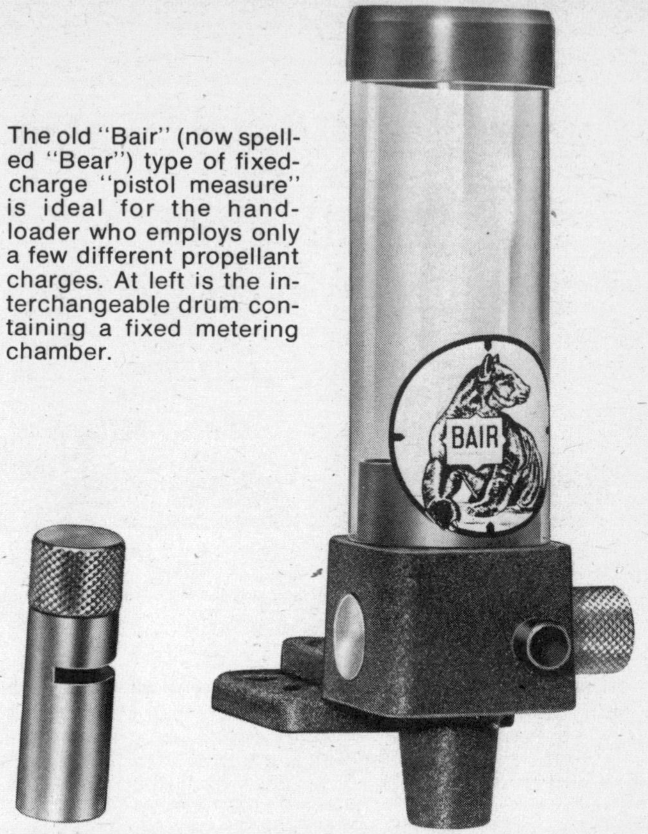

The old "Bair" (now spelled "Bear") type of fixed-charge "pistol measure" is ideal for the handloader who employs only a few different propellant charges. At left is the interchangeable drum containing a fixed metering chamber.

POWDER MEASURES AND SCALES

One other basic operation must yet be provided for and that is the placement of the correct propellant charge in the case. Three basic means are available for this, ranging from a simple charge cup dipped full and struck level to contain the desired amount, up through a simple balance beam type of scale or adjustable-type volumetric powder measures. Charge cups need not be purchased—though the very excellent 15-unit Lee set is one of the best buys in handloading—but can be made by simply trimming a fired case until it holds the correct amount. A powder measure is far and away the most rapid and convenient method of charging cases, but the adjustable type *may not* be used alone. Each setting on it must be verified with an accurate scale, therefore, one must invest in *both* a scale and a measure. On the other hand, a scale may be used alone at the expense of the additional time necessary to weigh out individual charges. In addition, the use of a scale to weigh typical handgun-size charges of 2.5 to perhaps 10 grains of very fluffy, fine-kernel propellant can be very taxing and frustrating. Be that as it may, the single most versatile tool for powder charges is a good-quality balance beam scale, preferably of the micrometer-poise type as currently sold under the RCBS label. The micrometer poise will not shift inadvertently and is therefore much more desirable than the simple sliding poise which is very easily jostled out of position, giving you a wrong—and perhaps dangerous—charge weight.

Next in versatility, and way out front in speed and convenience, is the micrometer-adjustable, volumetric measure. It normally consists of a drum installed in a housing; the drum being rotated upon its horizontal axis by means of a convenient handle. Above the drum is the housing which carries a propellant reservoir and feeds directly into a cavity in the drum when the latter is rotated to the "fill" position. When the drum is rotated to the "discharge" position, the propellant falls from the cavity through a "drop tube" directly into the case. The drum cavity, known as a "metering chamber" is adjustable in volume by means of a close-fitting plug that can be moved in or out. This movement is accurately controlled by a threaded stem provided with easily seen vernier-type or micrometer scale graduations. Most measures are offered with drums containing *either* a rifle-size metering chamber (capable of throwing propellant charges weighing from 75 to 80 grains) *or* a pistol-size chamber capable of throwing charges seldom exceeding 20 grains.

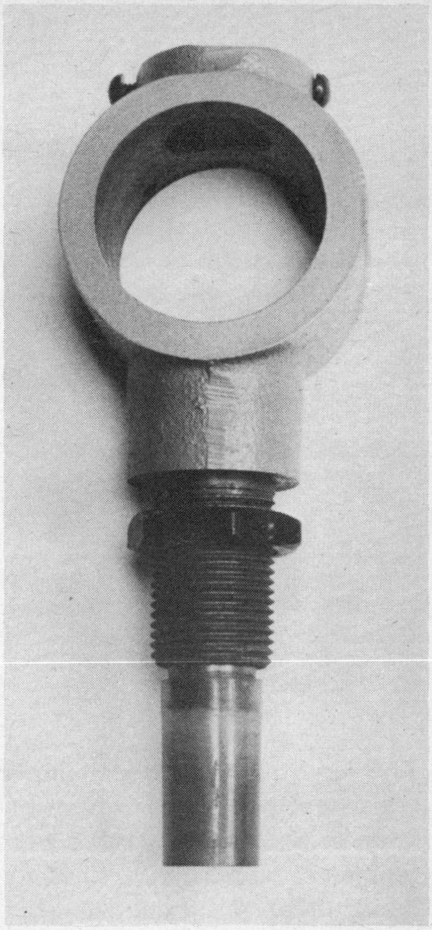

Above — This particular example of a micrometer-adjustable volumetric powder measure, (with the clear plastic reservoir removed) is the now-discontinued Ohaus Du-O-Measure. It differs from most others in that the handgun-size metering chamber and adjustments are on the left end of the drum while the larger chamber for rifle use is on the right end. Both chambers are therefore available at all times without changing any parts; simply removing and repositioning the stop screw at the upper left of the base or housing (above the "S" cast into the housing) allows either chamber to be put into use immediately. (Though this measure is no longer manufactured and sold under the Ohaus name, an improved and simplified version is offered by RCBS, Inc.)

Above — Here we have the stripped housing of the Ohaus Du-O-Measure; note the lateral hole in which the large-diameter drum rotates and the passage connecting it with the reservoir seat above. Threads and lock nut on the bottom of the housing allow installation in a measure stand or a loading press.

Left — In this internal view of the Du-O-Measure, the metering drum with the large rifle-size chamber clearly is visible. The pistol size metering chamber runs at approximately right angles from the lower right, and the large brass adjusting knob with its graduations is visible. The long threaded stem with a large-diameter flat head is the adjustment plunger for the rifle chamber. Note that the drum is hard-chrome plated and that it operates inside the bare metal of the housing, the two dissimilar metals providing smooth and easy operation.

Above — Here can be seen the head of the adjusting plunger (arrow) somewhat off-center of the rifle metering chamber; the pistol chamber works in exactly the same fashion.

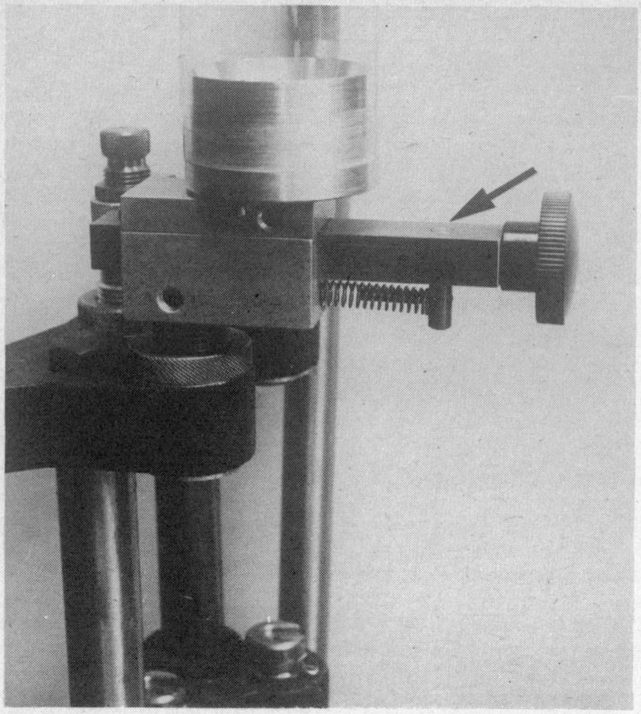

Left — This is another form of fixed-charge measure— this one made by C-H for use on multiple-station loading presses. Its base is made to fit over the C-H combination expander die in its 3-station H-type press. The metering chamber is in the horizontal rectangular bar (arrow). When this bar is pushed forward (left in the picture) it discharges propellant through the hollow expander rod directly into the case; it is then spring-returned to the "fill" position.

Below — C-H offers a fixed-cavity, charge-bar measure that contains a large hole which accepts interchangeable charge bushings (one of which is visible just above the hole).

The large rifle chamber simply will not throw consistently accurate charges of handgun propellants in the weight desired because of its large diameter. Therefore, it is essential that one use a handgun-size metering chamber. Measures may be ordered with either or both, but the RCBS Du-O-Measure contains both sizes in its drum and the changeover may be made in a few seconds by simply repositioning a simple stop and rotating the drum. It is more costly than some other good measures, but one seldom obtains added convenience without paying for it.

Another type of measure is generally available and it is usually referred to simply as "pistol type." It is technically more correct to describe it as a "fixed-charge" measure, for its metering drum (or in some instances a horizontal sliding bar) contains a simple cavity machined therein to accommodate a specific charge of a particular propellant. Some more flexible designs contain, instead, a cavity into which is placed any one of a number of interchangeable bushings, each containing a cavity to throw a particular charge. This is the one type of volumetric measure which may be safely used without being checked each time on a scale. After all, if the manufacturer machines a cavity in either drum, bar, or bushing to contain exactly 3.5 grains of Hercules Bullseye, it will throw that amount of that powder today, tomorrow, next month or 10 years

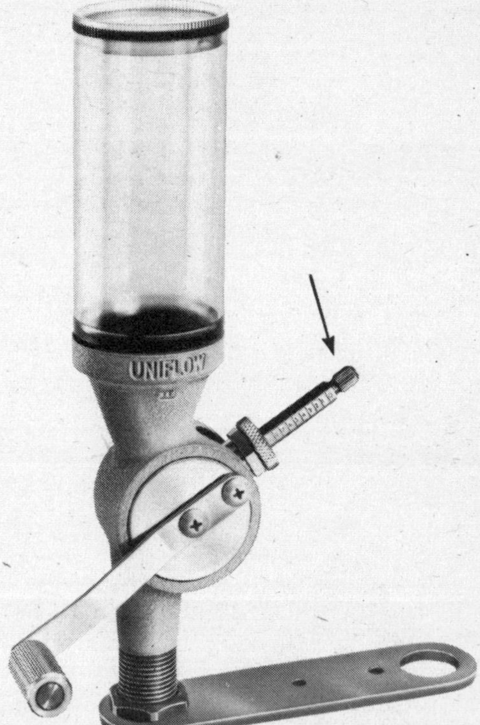

A typical micrometer-type, adjustable, volumetric powder measure as manufactured under the name "Uniflow" by RCBS for many years. This model is offered with small-diameter metering chambers for handgun use and, of course, a larger chamber for use with rifle cartridges. The adjustment stem and scale may be seen protruding from the housing to the right (arrow). Shown also is a simple mounting adapter which may be employed (by means of the two small screw holes seen in it) to secure the measure solidly to a bench/or shelf-top, or, it may be secured to the loading press.

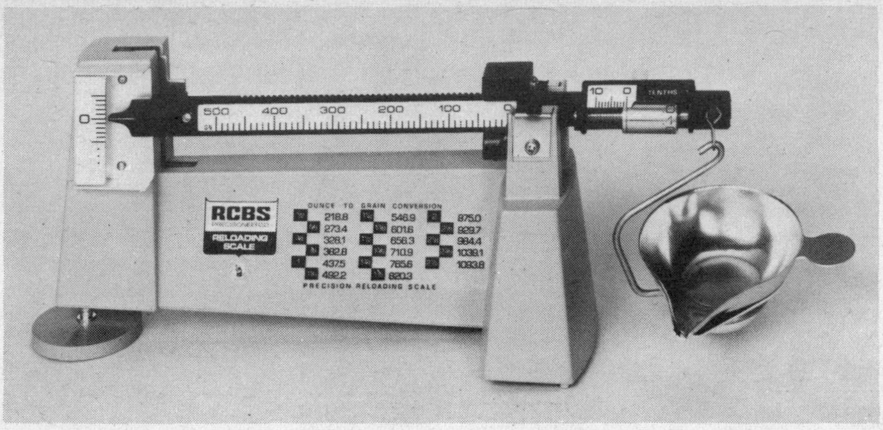

Left — Originally designed and marketed by the Ohaus Scale Co., under its own name, this micrometer-poise powder and bullet scale is now available exclusively from RCBS.

Right — One of the simplest (and yet reliable) powder scales on the market is offered by Redding-Hunter. This model employs two sliding poises, the one on the left of the beam pivot graduated in grains and tenths of grains. The graduations at the extreme left (the pointer end of the beam) also approximate tenths of a grain. This scale is magnetically damped by the bar protruding behind the pointer end of the beam passing through a magnetic field, and is leveled for use by a thumb screw.

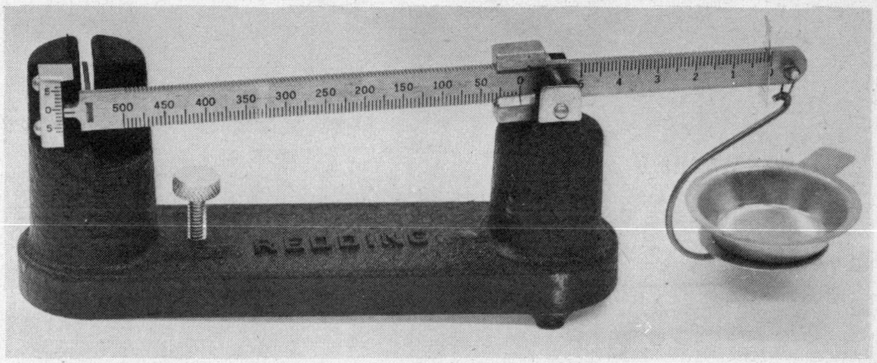

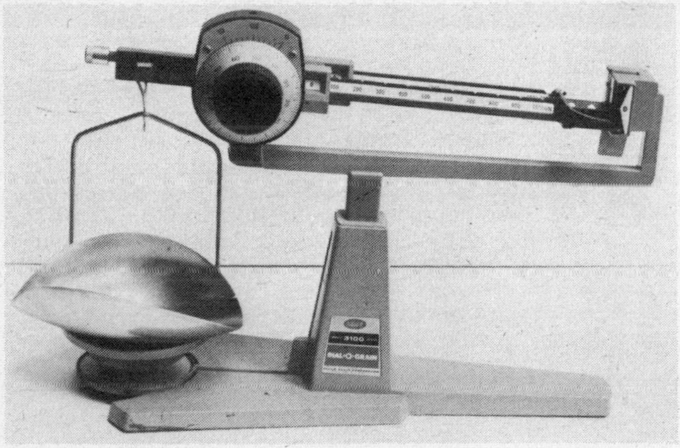

Left — The Cadillac of powder and bullet scales for handloader use is the Ohaus Dial-O-Grain with a drum-type vernier scale. This scale offers adjustment (to 0.10 grains) and a large sliding poise for coarse adjustment in 100's of grains. It is extremely accurate and is quick and easy to use. The Dial-O-Grain unit costs a great deal more than a typical powder/bullet scale, but for the genuine enthusiast, it's an ideal instrument.

from now. There is no possibility of inadvertent over-charge *unless* you make an error in installing a bushing or drum. The individual requiring only a few propellant charges to meet his loading needs is probably most efficiently and economically served by this type of measure, which is available from several makers.

RELOADING ACCESSORIES

The items we've just described consitute the *basic* loading tools and equipment. Of course, there are numerous minor accessories, some of which are almost essential, some of which are simple conveniences, and others which are mostly for bragging about and of little productive use. Among them, more or less in order of desirability and utility, we find the following: *powder funnel,* essential when charges are being weighed and must be poured individually into cases; *loading block,* simply a wood or plastic block containing holes of sufficient depth and diameter to keep cases upright during handling to avoid propellant spillage; *primer-pocket reamer,* necessary for removing the crimp found around the primer pocket of military cases to permit repriming; *deburring tool,* to remove the sharp edges and turned-over burrs on the inside of the mouth of *once-fired* cases to facilitate seating bullets; *stuck-case remover,* to employ in those inevitable instances where, for lack of attention, you manage to jam an unlubricated or dirty case in the resizing die; *case trimmer,* to be used to either reduce over-long cases to standard length, or more likely, to reduce a given batch of cases to

Among the most useful handloading accessories is a *good* loading block. It simplifies handling cases, especially when charging with propellant, more than any other item. This particular example is of molded plastic from RCBS and is thus quite easy to keep clean, a virtue not possessed by those made of wood.

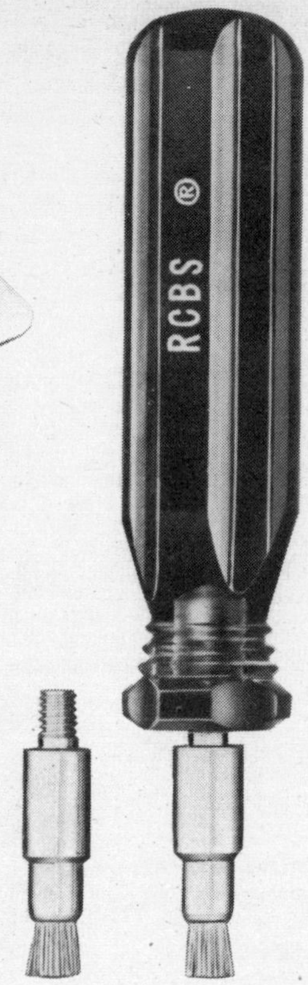

For the meticulous handloader, a primer-pocket cleaning brush, two different sizes, either of which may be attached to the handle.

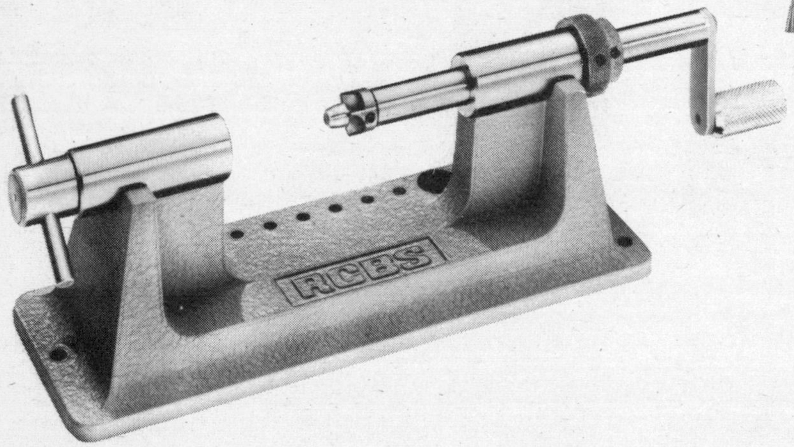

A case trimmer is an accessory usually only of occasional value in loading handgun cartridges. In any event, the case is secured by its head in the collet-type holder at the left, and then the piloted cutter at right enters the case mouth and is rotated by the handle to remove excess brass. A stop collar limits the advance of the cutter and may be adjusted to produce any desired length.

a *uniform* length to permit more consistent crimp in revolver calibers, or more uniform headspacing in autoloaders; and last but not least, *suitable containers* (whether they be plastic bags or costly compartmented boxes) in which to place your loaded ammunition in order to retain the identity of each load and to protect them from dirt, grit and the weather.

This list of accessories is arbitrarily short, and as you read through the rest of this book and, as you attempt more handloading, you'll discover dozens of additional accessories that you either need or think you need.

In addition to the basic equipment and accessories listed herein, there is the extensive equipment available for producing and preparing both cast lead or swaged and jacketed bullets. Frankly, this equipment requires a good deal more treatment than we have space to allocate in this chapter. Instead, we've devoted a complete separate chapter to the subject, and have described the equipment required in great detail.

In any discussion of handloading equipment, there is always the tendency to list and even recommend far more than is actually required to do an entirely satisfactory job. In addition to that, there's an equally strong tendency to imply the recommendation of equipment much more sophisticated and costly than the typical handloader needs. We've tried to avoid these pitfalls in this chapter, and hope that the foregoing pages haven't lead you astray.

You will not need a powder funnel such as this one for charging cases directly from a powder measure but, when using a charge cup or scale, it will be indispensable.

A powder trickler such as this is useful only when relatively large (for handgun calibers) charges of propellant are being weighed individually. Then, though, it greatly speeds up the operation by allowing individual kernels of propellant to be "dribbled" into the scale pan to bring the charge exactly to weight with a minimum of fuss and no over-ride.

When cases are properly flared, de-burring is seldom necessary. However, when cases are trimmed, the sharp, overturned edges do need de-burring, and this tool will do the job nicely. The upper end goes inside the case mouth to cut away the excess, and the lower end does the outside—the pilot serves as a guide.

SETTING UP A RELOADING BENCH

Of course, tools alone are not enough. You need a place to work, and more importantly, you need some sort of bench or stand upon which the press and some other tools can be solidly mounted. The bench *must* be solid, and that means a thick, sturdy top and stout legs. The only suitable loading benches I know of for sale these days cost a great deal of money plus an expensive shipping bill because of their considerable weight and bulk. You can occasionally buy particle-board bench tops at large hardware and department stores for a fairly reasonable price; however, if you try to buy a *really good* hardwood top, it will cost you as much as a new Magnum six-gun. I don't care for that approach. A *thick* (at least 1½-inch) particle-board top will do nicely if you can find, or cut to size, one that suits your requirements. A set of heavy-duty, stamped-steel legs (not the flimsy type usually encountered in discount stores) will complete the bench providing you brace them one to the other and use additional cleatings on the underside of the top to make their attachment as secure as possible. Naturally, if space is no problem, a large bench is preferable. However, the typical handloader, if he's neat and orderly and lays things out well, can get by quite well with a top measuring only 2 feet by 3 to 4 feet (that's only 6 to 8 square feet of floor space being used. If you are forced to use a small bench, enhance its stability by piling a few bags of shot or 100 pounds or so of bullet metal on a plank fitted between the legs. This additional weight will keep the bench from tipping or skittering around for those jobs where you really have to lean on the press handle.

The loading press is the key to bench layout and should be positioned for efficient work while at the same time keeping it from interfering with other operations. On my favorite 6-foot bench the press is installed 2 feet from the right end; this leaves plenty of room to the right for boxes into

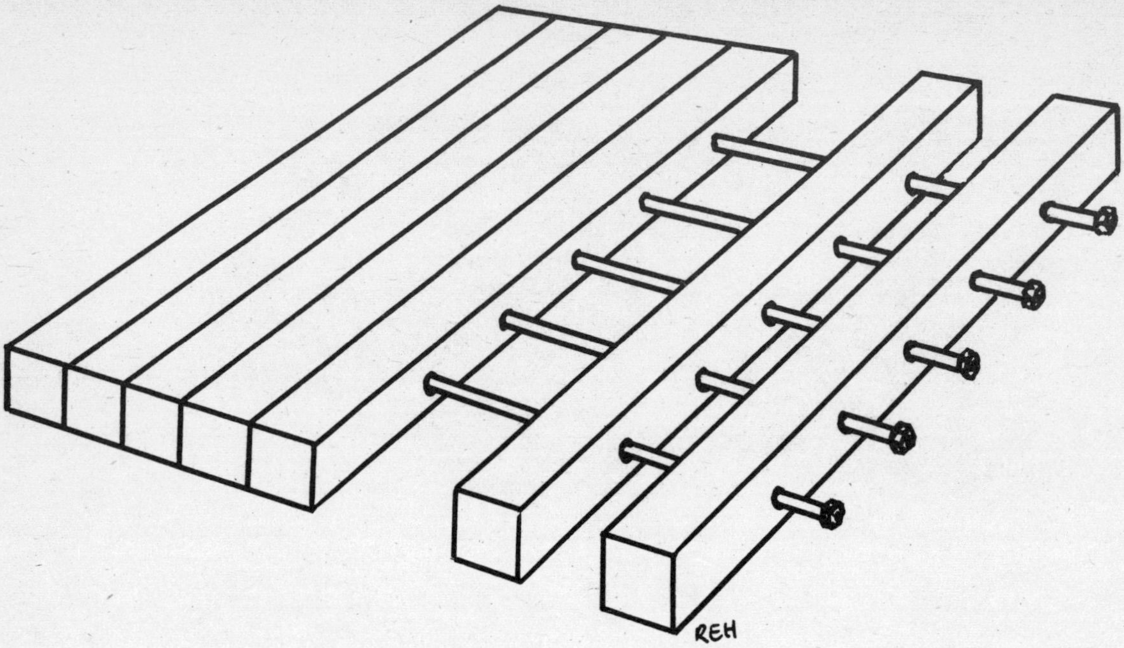

One of the simplest ways to produce a tremendously sturdy and heavy bench top is to drill a series of 2 x 2 or 2 x 4 boards so that they can be slipped onto the lengths of 3/8-inch threaded rod (available at most hardware stores). The boards, cut to whatever length to suit your space are then simply stacked on the rods to achieve the desired bench width, and all clamped securely together by nuts on the rods. The edge boards, at least the front, should be counterbored to accept the nuts so you don't gouge your tender skin on those protrusions.

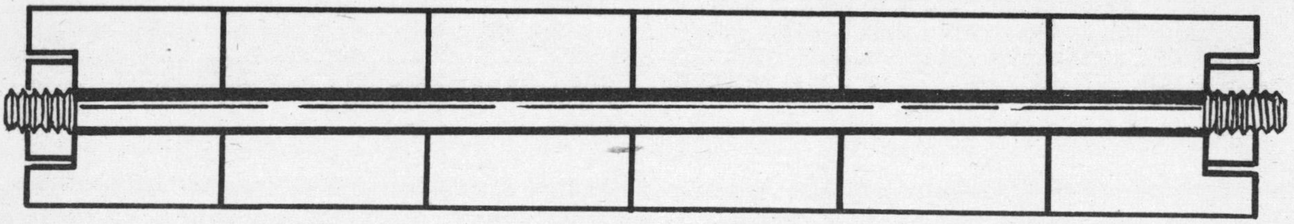

which processed cases or loaded cartridges are dropped as they come off the press. Then, there is plenty of room on the left to set up a powder measure and lubricator-sizer, both of which need to be installed very solidly. And, there's still plenty of room to set up for casting bullets when the urge strikes. If the left end of the bench is clear (mine is), I suggest the lubricator-sizer be centered on it and the powder measure placed about 12 to 18 inches to the right of the front corner.

Install the press solidly. Use bolts and nuts completely through the top of the particle board with large washers or a cleat on the underside to prevent the nuts from sinking into the board. On wood tops less than 1½ inches thick do the same. On thicker, wood tops, use the longest lag screws the thickness will accommodate. If you're fortunate enough to have a maple or other really hardwood top, forget the lag screws and drill completely through it for bolts—turning lag screws into maple will really pop sweat out on your brow.

Most presses are intended to be mounted up against the front edge of the bench with the lower part of the frame overhanging. Position it where you want it, push it back against the front edge of the top, and clamp it temporarily in place. Now, move the handle through its full stroke—you may find a clearance problem, and if so, it is preferable to notch the edge of the bench rather than move the press outward which will reduce its support. When you're sure every thing works freely, mark through the holes in the press mounting flange, remove the press and drill your holes. Bolt it down *tight* and the job is done. Before actually unclamping the press and drilling the holes, you might lay out some components and dry-run through the usual loading operations to make sure the press installation will allow con-

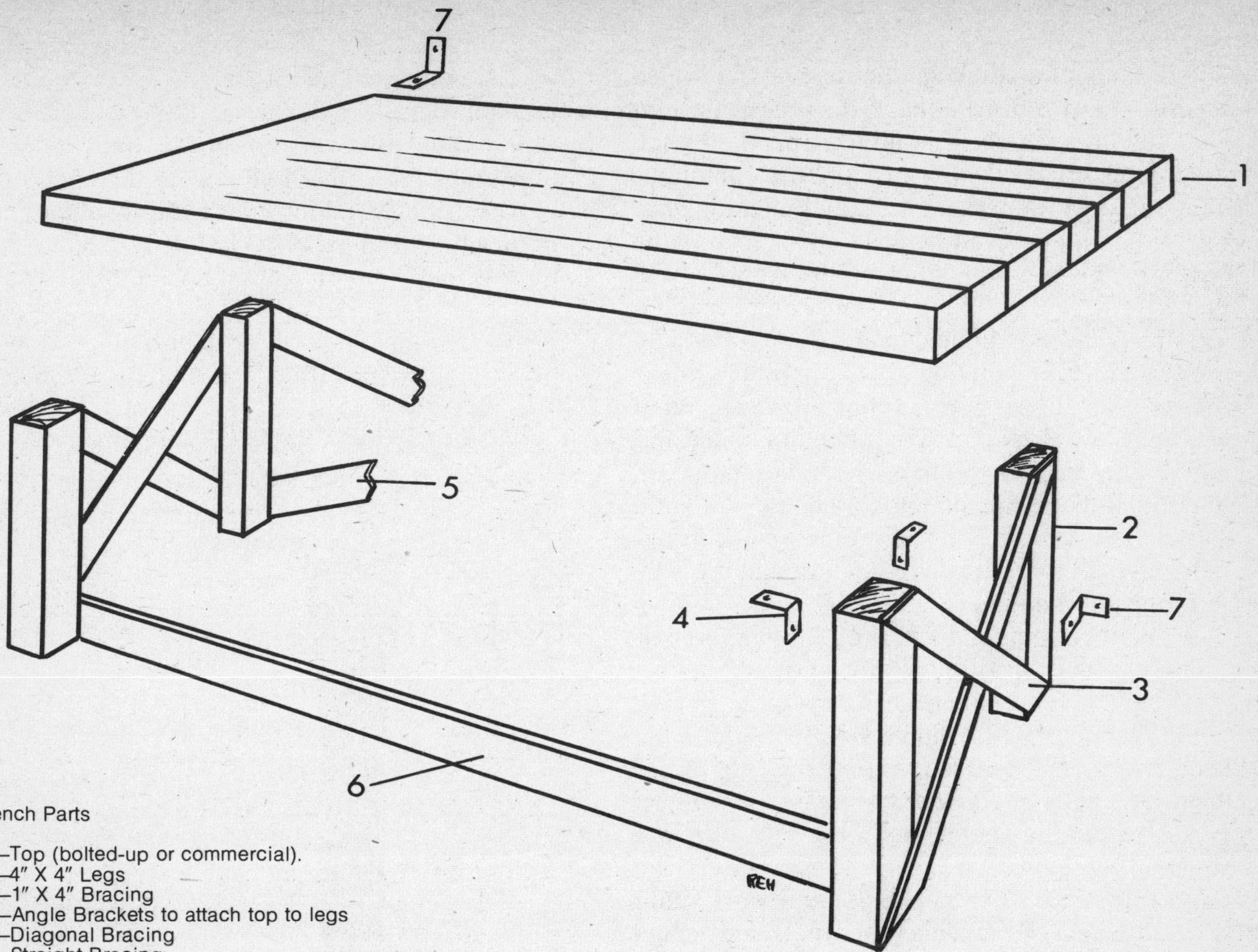

Bench Parts

1—Top (bolted-up or commercial).
2—4″ X 4″ Legs
3—1″ X 4″ Bracing
4—Angle Brackets to attach top to legs
5—Diagonal Bracing
6—Straight Bracing
7—Angle Brackets to attach top and legs to wall

Whether you use our bolted-up top or a commercial bench top, a simple set of legs can be assembled in this fashion from 4 x 4 uprights and 1 x 4 straight and diagonal bracing. If a lighter top is used, it may be a good idea to secure it to the wall or floor to prevent movement when you're putting lots of beef on the press handle.

venient operation (particularly if you're a southpaw). Space permitting, the press should be installed permanently; however, if you feel it must be removable, use large diameter bolts through the bench top in conjunction with *large* wing nuts. If you can't find wing nuts big enough to be easily turned by hand, keep a wrench handy and use common nuts.

The typical powder measure will throw more uniform charges if it is mounted solidly enough that it cannot vibrate significantly during operation. The simplest way to accomplish this is to use one of the several metal measure stands on the market. Secure the stand permanently to the bench top with screws or bolts and then install the measure tightly to the stand. If permanent installation of the stand won't do, then use the method suggested for the press, but always make certain the stand is held tightly to the bench during use. In positioning the measure on the bench, make certain there'll be plenty of clearance around it to allow the use of loading blocks. When nearly all the cases in a block have been charged, you don't want to bang into the press or some other tool and dump propellant all over the place.

If you'll be setting up a lubricator-sizer, it, too, must be tied down very securely. Like the press, it also hangs over the edge of the bench, and you may need to do some bench trimming in order to tie it down tightly and still have adequate clearance for all the moving parts.

As you'll learn farther along in this volume, a powder scale is best used at eye level. Whether you sit or stand for loading, this rules out simply setting the measure on the bench. The most convenient installation I have ever found is simply a hollow box of 1x4 or 1x6 lumber screwed and glued together (and braced in back if necessary to insure rigidity) which is simply set on the bench top in a clear space when measure use comes up. The box is given greater stability by piling a bag or two of shot or a few bullet-metal ingots on the bottom. This box may be anywhere from 12 to 18 inches long (not at all critical, just don't make it too short) and of sufficient height that the beam of the scale is level with your eyes when the scale is set on top of the box. This setup allows the stand and the scale to be removed when not in use and thus leaves the bench top uncluttered. You can, of course, make this or some other type of solid stand a permanent fixture, but remember that it will only add to bench clutter, and get in the way of other operations.

All bench-mounted tools should be covered to protect them from airborne dust when not in use. You can, of course, make fancy clear plastic hoods, but that's a bit much. For decades I have used covers made by cutting off the legs of old jeans or slacks and stapling, tying or sewing (if someone can be found to do the job) one end shut. These covers are simply slipped over the top of the tools when not in use and work quite well at no cost. Regarding other tools and accessories: They (dies especially) should be at all times protected from dust and grit. Most makers supply them in close-fitting plastic boxes these days, and if kept in same, they can be safely stacked on a shelf. Otherwise, supply yourself with some sort of well-closed containers or a snug-fitting drawer under the bench top. For open-bench storage, a board containing a row of $^{15}/_{16}$-inch holes makes a handy die rack and you can make a cloth cover to furnish the desired protection. Drawers under the bench are probably best for storing other tools and accessories, though an overhead cabinet with doors will add a bit of class to your layout. Components require extra attention for storage, and you'll find a separate section devoted to just this subject.

It may be that you're so strapped for space, due to a small apartment or large family, that there simply isn't any place to put even a small bench and layout of the type we've described. Fear not, there are other solutions—they may not offer the ideal in space and convenience, but one way or another they will allow you to set up to load properly. In my office I have what appears to be a small cabinet 15 inches deep, 30 inches wide, and about 3 feet high. It looks like just another piece of furniture. However, at any time I can walk over, lift off the top, slip the top into brackets which hold it onto the back of the cabinet, open the doors and swing up a thick shelf upon which is installed a C-H Auto-Champ progressive loading tool—all set up and ready to go in .38 Special caliber. After locking the shelf in position, I need only to drop into place case and primer magazines (filled ones are inside below),

If you are really short of space, a simple pipe pedestal with folding legs such as this may be purchased; alternatively, you can make same with a length of pipe, two pipe flanges, and pieces of plywood for base and top. The entire assembly will then break down in moments into small pieces which are easily concealed and stored under the bed or in the closet.

set the powder reservoir on top of the automatic measure mechanism, pull a box of bullets off the bottom shelf inside and immediately begin loading at the rate of several hundred rounds per hour—one complete cartridge for every stroke of the handle. Then, in less than 1 minute I can have everything back inside the cabinet, the removable top again in place and with no visible evidence whatever that I've been loading. With minor changes in design, this cabinet could be made to accommodate other types of presses as well, though it works best with those that do not overhang the work surface.

Pedestal mount for a loading press has been given more working surface by adding a 12 x 18-inch piece of 3/4-inch plywood clamped between press and pedestal top.

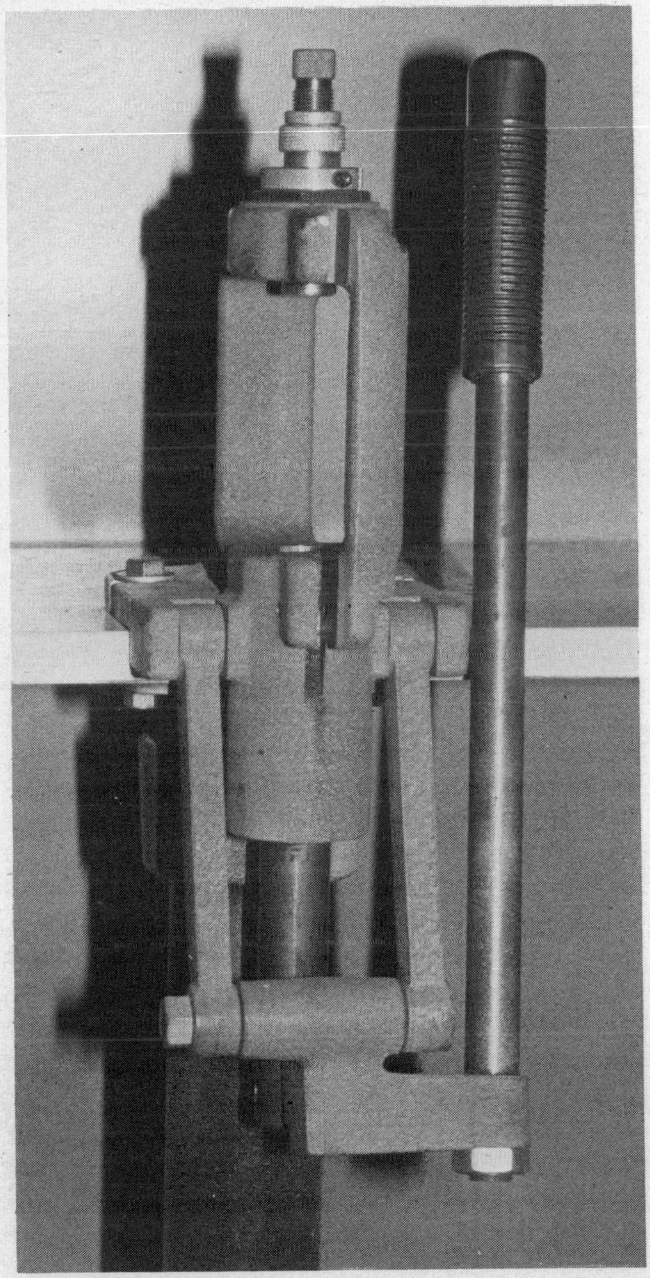

Then, in the storage room there is a simple pipe pedestal upon which I frequently install new presses and tools to try them out. It consists of nothing but a 30-inch length of 1½-inch pipe threaded on both sides to accept large pipe flanges. A plywood base is screwed to the bottom flange and a small work surface is similarly attached to the top. The top is drilled to accept several different makes and models of tools and is just large enough to hold a press and a powder measure. A much larger separate plywood top has been made and shaped so that it is easily and quickly clamped in place without disturbing the tools, and provides sufficient work surface for components, scale, loading blocks, etc. In only a couple of minutes the auxiliary top can be unclamped, the press mount unscrewed from the pipe, and the pipe unscrewed from the base. These four parts can then easily be slipped under a bed or couch or behind the winter clothing hanging in a closet.

There are many other solutions to cramped quarters. I've seen presses mounted between the vertical studs in a garage wall, clamped to a lolly column in a basement, and fitted to a plywood panel which could be clamped by a novel method to the kitchen counter top. One friend took an old-fashioned, solid-oak typing desk with the concealed typewriter well and modified it to hold a loading press. In less than a minute he could have the tools concealed with only an attractive, functional desk visible. Another friend installed a sturdy bench top from wall-to-wall inside an ordinary closet, along with suitable lighting and shelves for storage, so he could open the door, sit down to load, and yet hide the whole thing in seconds by closing the door when he chose to.

If space is in short supply, just use your ingenuity. I have yet to visit a home of any type (and that includes mobile homes and motor homes) where I couldn't find space for unobtrusive and usually concealed installation of an adequate handloading setup if one were willing to spend a little time and effort.

In short, the basic tools and equipment are readily available; and, somewhere in your home you can find space to set them up. Nothing more is needed to get you started on the way to better and cheaper shooting by handloading.

CHAPTER 5

Selecting the Right Load

YOU'VE GOT TOOLS, you've got cases, primers and other components, you've got everything—but you've taken a look at a couple handloading manuals and have found that there are literally thousands of combinations of propellants and bullets listed or recommended therein. Some books show as many as 12 different powders for a single bullet weight, and sometimes three or four charge weights for each propellant in that table—and that's just for one bullet weight. It not only seems impossible to sort out the right amount of the right propellant for it, but there are as many as a half-dozen bullet weights and types listed, each with almost as many different propellant combinations. If nothing else confuses the neophyte handloader, surely the mass of loading data that has been published will succeed in doing so. What we've just described is *only one* manual, and the average individual will be able to lay his hands easily on three or four more, not to mention the paperback and hardcover books published on both handguns and handloading and the thousands of loads they contain.

Where does a handloader start? It would be a lot easier if not so many load combinations were published. The beginner must naturally assume that all of those loads he sees are safe and good, but he has brains enough to realize that thousands of combinations are not necessary, and most of them must be superfluous. Why under the green apple tree do authors and publishers (guilty here!!) include so many loads?

Lots of reasons, really. First is the business of selling those loading manuals. When there was only one loading manual available there wasn't any problem. (It's been a long, long time since that situation existed, back when the Ideal Handbook was the only offering.) We'd just buy that one book and use it. But, since the early 1950's, numerous companies have entered into the handbook publishing business, and today at least a half-dozen are offered by manufacturers of handloading tools and components. Each of these handbooks represents a very large investment in research, compilation of data, and writing—this investment is justified in two ways. First, though not necessarily first in importance, the publisher expects to make a reasonable profit on the sale of the book, even though some of his other activities might include giving away substantial numbers of copies. To be perfectly frank,

Here we have two .44 Special cartridges that appear to be identical. Cases are of the same manufacture and the modified SWC cast lead bullets appear from this angle to be identical. However, one of the bullets is solid and employed with a light charge of Hercules Unique to approximate factory-load ballistics. The cartridge on the right, employs a hollow-point version of the same bullet with a large charge of H110 propellant to drive it at about 1200 fps for hunting or defense use. This demonstrates not only the versatility of handloading, but the wide performance range that can be given to a single basic bullet design. It also points out the necessity for *very* carefully labeling each different lot or load, insuring that its identity is not lost. In this instance, the left load was assembled for practice in the lightweight small-frame Charter Arms Bulldog revolver, while the one on the right was assembled for use in a S&W Hand Ejector .44 Special revolver manufactured in 1926. The light load can certainly be employed safely in the S&W but, while the hot load won't necessarily take the Charter 5-shooter apart, it will certainly produce vicious and painful recoil.

that's why all publishers produce books, and protestations to the contrary notwithstanding, every author I know writes books primarily to obtain the wherewithall to pay his rent and buy groceries. Second, every loading manual I know of is used as a valuable vehicle to promote sales of the publisher's handloading products. Most manual publishers are primarily manufacturers of bullets, and if you examine their offering closely, you'll see that very, very seldom do they list a load for any bullets other than ones they sell. The Speer Manual is a notable exception in that its most recent edition does include some competitive bullets. And, several manuals do include cast-bullet loading data as a matter of reader convenience. Propellant manufacturers also publish manuals, and again, you can hardly expect one company to recommend any propellants other than its own.

Add to this the simple fact that the various manual publishers are competing with one another for book sales, and you see why they cram into each edition as many loads as possible. What better sales pitch to make for a loading manual than to say that it contains "3,752 different loads" or "more loads than any other book."

The end result of this is that the publishers of loading manuals inundate the reader with load combinations for which he has no need. In a few more paragraphs we get into how you can select from that mass of data the relatively few loads that are worthwhile for your particular needs.

The most common complaint we receive about loading manuals goes something like this, "I want to pick the best load for my .357 Magnum Revolver with 6½-inch barrel. I've looked in the Lyman Handbook, and it tells me I can get 1540 feet per second with a 158-grain lead bullet by using 15.0 grains of 2400 powder. Then, I looked in the Speer Manual, and it says I can get only

Every one of these cartridges has been loaded to suit a particular shooting need (real or imagined). All it took was careful perusal of a variety of loading-data sources. In some, economy was of primary consideration; in others velocity or accuracy.

It's a far cry today from the situation of 40 years ago, and these five loading manuals don't represent anywhere near the entire present crop. Each contains hundreds of handgun-caliber loads; and, even if you tried, you'd probably never be able to test them all.

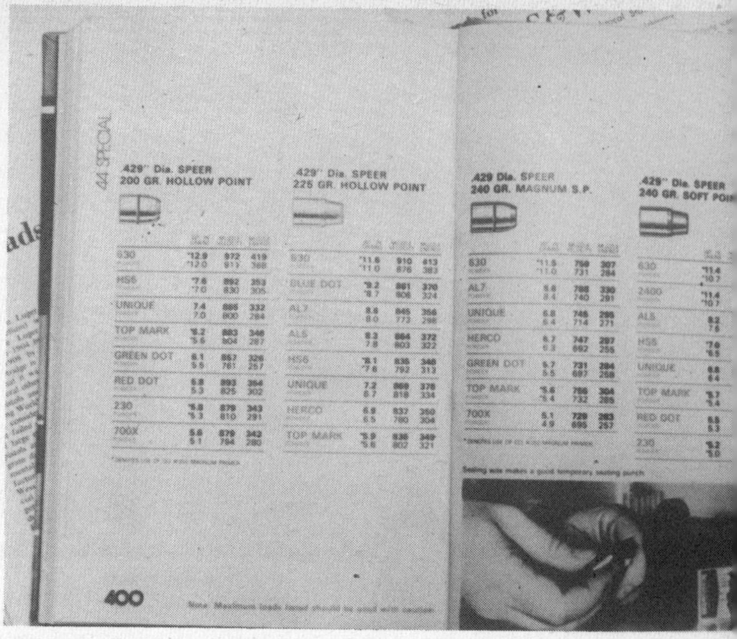

This is a typical page from a handloading data manual—in this instance the Speer Manual No. 9. These two pages do not represent all of the .44 Special data contained in this particular book, but embrace three bullet weights and two different designs of the jacketed type alone. These two pages alone list a number of different propellants and over 60 loads. This certainly points out the versatility of loading you can have at your fingertips.

1338 fps using 14.0 grains of 2400 which it lists as a *maximum* load. Which is right; what's the matter with these people?" The fellow has a right to ask that question; the two manuals do show exactly what he says. First of all, if he had read the fine print, he would have discovered that the Lyman load was tested in an 8⅜-inch *solid* test barrel, while the Speer load was tested in a *Smith & Wesson Revolver* of only 6-inch barrel length. First of all, the test barrel did not have a gap, and therefore no gas escapes, so other factors being equal, it will normally show a somewhat higher velocity than a revolver with its barrel/cylinder gap. In addition, there is a difference in barrel length, and that too will normally produce quite a difference in velocity. What about the maximum-charge indications; well, in the Lyman Manual the load was actually tested for pressure, and the table clearly shows that 15 grains of 2400 with that bullet produced 35,000 CUP. Here, the pressure is actually *known* and is not truly maximum because working pressures for the .357 Magnum cartridge have long been established in the 40,000 CUP range. On the other hand, the Speer load was not tested for pressure. Instead of using costly and complex pressure-testing equipment, the operator made a subjective judgement, based upon such factors as ease of extraction, recoil, fired case and primer condition, etc., that the 14.0 grain charge was maximum in the *particular gun being used for the test*. The result is often quite different from what would be obtained from a pressure barrel.

This points up the fact that *most* published loading data *is not* based upon laboratory-type testing for pressure and velocity. Instead, loads are developed by the author or some other party for individual guns, and while velocities are actually measured with an accurate chronograph, all other results reported are based purely upon the subjective judgement (normally well-founded upon long experience) and thus cannot be precise. Another large variable enters the picture—that of the difference in guns used in generating the data. In my own testing of revolvers I have encountered variations of as much as 200 fps between two different guns of the same caliber and barrel length with the same load. For example, a particular .357 Magnum load might produce 1250 fps in one 6-inch revolver and 1450 in

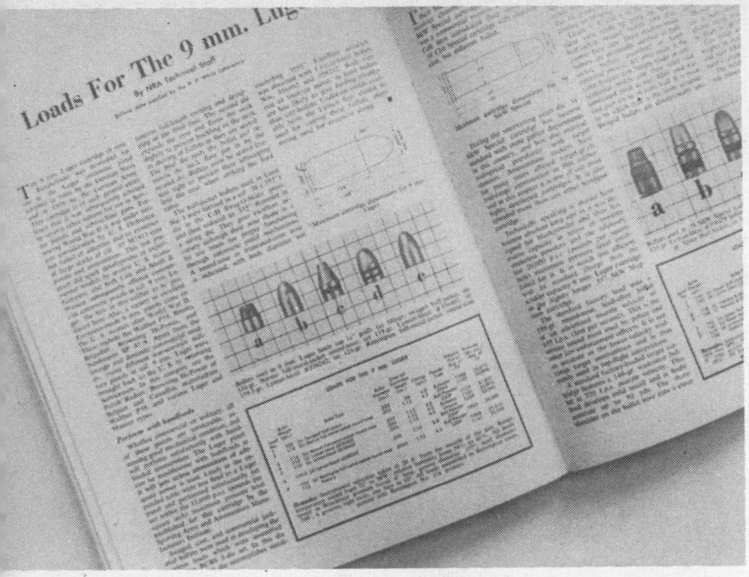

Probably one of the best handgun handloading references available — in spite of the fact that it offers fewer loads and fewer calibers than any other — is the NRA Handloaders Guide. Generally speaking, this book provides objective views and precise testing results — their load testing is done within established industry testing standards.

another. Variations in timing, alignment, chamber and bore dimension, etc. cause a great deal of velocity variations between revolvers. Another factor often responsible for differences in loading data is the effect of the bullet. Even though a bullet used in Manual A is of the same type and weight as one used in Manual B, design and construction may differ sufficiently between them that pressures and velocities will vary considerably even when all other factors (including the gun) remain unchanged.

What it all really means is that in the example question quoted above *neither* manual is wrong and neither is necessarily right. Each is correct to the extent that the load data as shown was correct for a particular set of testing circumstances with a particular operator and gun. Neither result is necessarily what you will obtain from the same load in *your* gun. In short, the data contained in loading manuals is only a *guide:* The results shown are not absolute values. The principal goal of any manual is to present you with at least a reasonable variety of powder types and charge weights for a number of different weights and types of bullets in each caliber. The data published toward this end is safe and functionally reliable insofar as the testing done by the author can make it. Therefore, *never* expect to obtain exactly the results shown in the book when you use a published load in your particular gun and under your particular shooting conditions.

In reality, selecting a load to meet your needs is not necessarily difficult. Since we cover powder selection and bullet choice in considerable detail farther along, selecting a load at this point means choosing the ballistic performance level (velocity) which will do the job best with the least cost and with the least stress on the gun. Let's look at those and a few other factors.

The reloading life of a cartridge case varies indirectly with the pressure level to which it is loaded. Inasmuch as case life is a factor in per-shot cost, pressures should be kept down where possible. Obviously, if 50 new cases, costing perhaps 10¢ each, have a life of only 10 loadings with a high-pressure load and a life double that with a more moderate load, the latter will cut your per-shot case cost in half. The most efficient propellant for a given load is the one which requires the least quantity to drive the bullet to the desired velocity, so long as it does not produce excessive pressure in doing so. For example, it takes 18 grains of IMR4227 to drive a 125-grain .357 Magnum bullet at 1350 fps. Essentially the same velocity can be produced by exactly half that much Hercules Unique. Ergo, Unique is by far the most efficient of the two in this application. Using it instead of 4227 will cut propellant costs roughly in half and at no measurable sacrifice in performance.

While the life of most good handguns will exceed that of their first owner, we all still like to think that we are extending the gun's usable life by not overstressing it. Toward this end, again in the .357 Magnum Revolver, there is no reason whatsoever to use a full-blown H-110 load, driving a 125-grain bullet at 1540 fps for plinking or game hunting when a much more moderate Herco load at 1300 fps will do the job fully as well. The latter load will not stress the gun nearly so heavily (particularly true in small-frame guns) and thus will extend its useful life.

Comfort-of-shooting and gun control also play important parts in selecting a load. Many shooters find the full-charge magnum loads, with velocities in the 1400-1500 fps range, distinctly uncomfortable and thus cannot shoot well with them. Under those circumstances, it is foolish to choose a full-charge load, at least early in the game. You'll be far better off and will shoot more accurately

These two revolvers are of the same caliber, .44 Special — the replica Remington M-1875 (at top) by conversion, since it is not manufactured in that caliber. In view of their vastly-differing barrel lengths, it is impossible to expect similar performance from the same load. Factory ballistic tables and most handloading data sources present .44 Special velocity and energy data from a 6-inch barrel; consequently, the velocity will be different in the 7½-inch barrel of the Remington, but we can't be sure how much or whether it will be lesser or greater. At least with the 3-inch barrel on the shorter .44 Bulldog (below), we know that the velocity of a given load will be substantially less than shown in any standard table we might see. To make the most efficient use of these two different barrel lengths in .44 Special, ammunition for the two guns should be loaded with different propellants. Unique is most likely the best choice for the Bulldog, while W296 or H110 will perform far more efficiently in the long Remington barrel.

(and the only bullets that count are those that hit the target) if you back down to a load in the 1000-1200 fps range. Once you've gained enough experience to control it well, then you can work up in stages to successively heavier loads until you can handle the real, ring-tailed tooters well.

If from the last few paragraphs you've gotten the impression that I am discouraging really heavy loads, well, I am. Ninety-nine percent of most pistoleros' shooting can be accomplished quite adequately with loads driving medium-weight lead bullets at velocities not exceeding 1000-1100 fps. Probably 75 percent or more can be handled just as well with the same type of bullets at velocities well under 1000 fps. Shooting these lighter loads is less likely to offend neighbors, some of which take particular exception to the piercing bark of magnums loaded to the hilt. The lighter loads also make for safer shooting in that they have less penetration, less maximum range, and generally present less of a ricochet hazard; however, in some respects the lower-velocity loads *are more* prone to ricochet. They bounce their bullets more readily, but they just don't bounce them quite as far or as forcibly.

Insofar as autoloaders are concerned, in Chapter 3 we detailed the particular functional ammunition requirements of the type. For correct and reliable gun functioning, you must select a load whose recoil impulse falls between the minimum and maximum of values for a particular gun. Consequently, you can't have a pip-squeak load in 9mm Parabellum and expect the gun to function as an autoloader was intended; and if you choose the hottest and fastest 9mm load in the books (perhaps even jazzing that up a bit), it will over-drive the mechanism, shorten the overall life of the gun and perhaps even break some parts in the future.

One last factor to carefully consider in selecting a load—accuracy. Assuming that both you and the gun are capable of reasonably precise accuracy, then loads should be chosen from the types that have a demonstrable history of good accuracy. It has long since been proven that maximum-velocity loads (with any bullet in any caliber) do not shoot as accurately as those in the middle of the range for which the gun/cartridge was originally designed. Toward this end one must also choose a bullet type of design which has demonstrated a capability for good accuracy. For example, though lightweight, jacketed, expanding, high-performance bullets often shoot unusually well at high velocities, they have a reputation for doing considerably less well in the 750-900 fps range with most target loads.

When we boil it all down, there are essentially three classes of loads with which experienced handloaders expect to meet all shooting needs. At the top of the scale we have the full-charge magnum and magnum-type, high performance loads using relatively light jacketed expanding bullets at the top velocities. Beneath that we have those loads which duplicate factory ballistic performance in all non-magnum calibers, more often than not with lead bullets. The third class consists of those producing less than standard velocities with bullets of standard weight or less. The third class is usually labeled "light loads" and is normally assembled with lead bullets. This class includes precisely-accurate target loads as well as plinking ammunition of indifferent accuracy.

High-performance and magnum loads are seldom actually necessary. They are essential for big-game and varmint hunting, for personal defense and also for long-range shooting of any sort or metallic silhouette matches in particular. They are also required by conditions in bowling-pin matches and are desirable in some calibers for IPSC matches. They are *required* for those uses, but far more are assembled for yet another reason—a combination of shooter ego and self-satisfaction. Many a pistolero gets far more satisfaction and pleasure out of shooting the heaviest loads, because they present more of a challenge (and *are* more difficult to master) than other loads. "Light loads are sissy stuff" is a common attitude, and there is also the simple fact that maximum skill with heavy loads is developed if they are shot exclusively. Many shooters therefore feel, "If I shoot only heavy loads, I'll do better, even with light loads when they become necessary." This seems to be true, for once the ability to handle the "loudenboomers" is developed, lighter loads are a breeze. The opposite is *not* true. For if one shoots mainly light loads, he'll never develop first-class ability with the magnums.

If cost, noise, and other factors were of no importance, then we could simply say shoot the heaviest loads at all times except where rules and conditions prohibit it. But even that wouldn't work all the time, because many shooters (far more that will admit it) simply can't handle the heaviest conventional loads well, much less the full-charge magnums. Individual tolerance for recoil and blast are simply too low in many people, and this can't be entirely overcome by training.

So where the heaviest loads are actually *necessary*, they are the obvious choice. Where the shooter simply *wants* to use them, they are a viable choice if he accepts the disadvantages.

Such loads are the most costly of all and the jacketed bullet represents the greatest part of that cost. Except when targets are live animals (four legs or two), or in metallic silhouette matches,

Normally, we would expect the JHP bullet on the left to expand much more (and much more quickly) than the JSP design on the right. However, when the two bullets are sectioned, it becomes obvious that the JHP has a jacket more than twice as thick as the JSP, and thus the latter can be expected to give greater expansion. Because of this, always base your bullet selection upon reports of known performance, or, lacking that, tests of your own.

This revolver, an older N-frame S&W chambered .44 Special (1950 Model, .44 Military) has its own characteristics which must be considered in load selection. For example, it will take top .44 loads that are a bit too hot for the SAA Colt (and its predecessors) or S&W's First and Second Model Hand Ejectors which are built on the same frame. The barrel length of this N-Frame .44 Special should also be considered because it is merely 4 inches — a load suited to a much longer tube won't perform efficiently in it. Also don't overlook those fixed sights — they can't be adjusted back and forth to accurately suit different loads. You'll have to settle on a load that prints close to point of aim.

Left — At left we have a perfectly good .41 Magnum case which has been loaded and fired at least seven or eight times with a full-charge load; at right is another .41 Magnum case which split badly on the second loading and firing, probably as a combined result of some manufacturing defect and an undersized resizing die which resulted in a good deal of excessive working of the brass.

Right — The magnum metamorphosis; in the beginning, about 1908, the shorter .44 S&W Russian cartridge was lengthened to form the .44 Special (left) and then in the early-middle 1950's, the .44 Special case was again lengthened about 1/10-inch to give us the .44 Magnum (right). Of course, the much greater ballistic performance of the magnum required loading the longer case to much higher chamber pressures. Magnum loads require special care because of those pressures.

cost and gun wear can be reduced by substituting cast lead bullets of similar shape and weight without reducing load effectiveness. Consequently, if you must use the heaviest load, choose two—one with a jacketed bullet for the most serious use and a duplicate with a lead bullet for practice and less demanding uses.

Full charge loads in non-magnum calibers present few problems. Various manuals list many loads in each caliber which duplicate standard factory ballistics. In revolvers, a lead bullet of standard weight is invariably the best choice, though for improved performance on both paper and live targets it should be of semi-wadcutter shape. In autoloaders, a *hard* lead bullet of standard weight and shape is to be preferred for the sake of both cost and reliability.

Since loads in this range are the most forgiving and generally generate moderate chamber pressures, select those which produce the velocity desired with the least amount of propellant. This will help keep costs down without reducing effectiveness.

Light loads, whether for plinking or serious target use, should be chosen carefully. There is a strong tendency to choose loads that are *too light* in the interest of economy and recoil reduction. In autoloaders this results in poor functioning, and in revolvers it may reduce accuracy. Lack of recoil energy causes the first problem, inadequate bullet stability the second. Bullets of standard weight and length are not adequately stabilized at very low velocities, inasmuch as the rifling does not rotate them rapidly enough. Thus, if you want lowest possible velocities, a lighter and shorter bullet will be needed, ideally of pure wadcutter shape.

Consequently, don't choose the lowest-velocity loads unless you are willing to use them merely as a starting point and then increase the propellant charge until performance is satisfactory. It must also be kept in mind that the low-velocity target loads listed in manuals are generally intended for well-tuned target guns, not the over-the-counter service types. While a target .45 auto may work quite well with a load consisting of 3.2 grains of Bullseye and a 185-grain bullet, an ordinary service gun may not function with it.

So, when selecting light loads, don't go below 700-750 fps in revolvers unless a lighter-than-standard bullet is to be used; in .45 ACP don't go below 750-800 fps; in 9mm/.38 autos, stay above about 850-900 fps. Such loads are most efficient with Bullseye and W-W 231 propellant rather than those with slower burning rates, so choose accordingly. In fact, any propellant burning slower than Unique isn't at all suitable for loads of this type. This isn't to say that even lighter loads aren't practical; they are, but they require special treatment and techniques which are given detailed coverage in Chapter 16.

Selecting a load need not be difficult. Mainly it consists of avoiding extremes of velocity, bullet weight and propellant. Keep in mind all we've said, and you should be in clover.

CHAPTER 6

Basic Loading Procedures

RELOADING OR HANDLOADING a fired cartridge case (or a new one for that matter) involves performing a series of relatively simple mechanical operations upon the case. The brass cartridge case is the container for all the other components (primer, propellant, bullet) and ties the entire assembly into a solid, oil and waterproof package, possessing the strength to resist handling, storage, and the sharp, heavy stresses imposed upon it by feeding mechanisms and recoil forces. In order to accomplish all this for even the first firing, the case must necessarily be made strong enough and durable enough that it may be reloaded and reused many times. Because of all this, the brass case represents roughly 50 percent of the production cost of the cartridge.

The loading processes—which must include preparing the case for actual loading—are not difficult to perform or to understand. We'll list the various steps here first, then go on in the following pages to describe the *basic* steps in detail, along with those secondary operations which occasionally become necessary or at least desirable.

Since cases are sometimes subject to damage during use, particularly when being extracted and ejected from autoloaders, and are also subject to being ejected onto the ground, they must first be carefully *inspected* for defects, then *cleaned* by washing or tumbling. In many instances they should then be *sorted* by make or headstamp. The second step consists of *resizing* the case, thus restoring it externally to approximately its original dimensions. This is followed by *decapping* which simply means punching the fired primer out of its seat in the case head. Decapping is often performed simultaneously with resizing, though it is sometimes done in conjunction with neck-expansion, dependent mainly upon the make and design of the dies being used. Decapping at either point in the process is quite acceptable.

Following this, the mouth or neck of the case must be *expanded* to uniform diameter so that it will grip bullets uniformly, and at the same time it must be *flared* very slightly so that the fresh bullet will enter easily. *Repriming* follows, consisting of pressing a fresh primer into the primer pocket in the head of the case. Repriming is usually performed simultaneously with neck expansion; however, it may also be done in conjunction with decapping, depending upon the type of tool

Four typical headstamps of the dozens that might be encountered on .45 ACP cases. While generally there is little difference in internal dimensions between makes, it is still a worthwhile effort to segregate cases by headstamp, especially if they are to be used for precisely-accurate target loads.

Keep an eye out for this when sorting cases (and also after washing or tumbling if calibers are mixed). Here a .25 ACP case is nestled inside a 9mm Parabellum case and, if this combination gets into a resizing die by accident, there will likely be lots of cussing.

being used. In addition, if it is being done as a separate operation, repriming may be done after either decapping or expansion.

At this point, the case is ready for the actual loading or assembly operations. The next step in line is measuring or weighing out the correct charge of the proper propellant and *charging* the case with it. With the propellant charge in place, the bullet must be *seated* to the correct depth in the case mouth, and in most instances the mouth must then be *crimped* to some degree against, or into, the bullet. Seating and crimping may be conducted simultaneously or as separate operations, depending upon the type of dies being used or the manner in which they are employed. Crimping is the final assembly operation, and after that there remains only careful visual and mechanical *inspection* of the loaded cartridge and, in some instances, functional *testing* which may or may not include actual firing of a small random sample of any given batch.

That's putting it just about as simply as possible. Now, let's go through those same procedures in detail.

CASE PREPARATION

The typical pistolero handloader is an inveterate packrat and scrounges fired cases (brass) wherever possible. Neophytes hustle brass almost as avidly because buying factory loads or new cases for starting is just too expensive. As a result, the case supply will usually consist of assorted makes and ages and will include damaged, dirty and corroded specimens as well. All too frequently, calibers will be mixed, and sometimes jammed together with .32's wedged inside the .38's and .38's jammed inside .45's, etc. As a result, they require sorting, cleaning and inspection for damage before any work can be started.

Case Sorting

I prefer to sort cases before cleaning simply because it reduces the quantity in a given batch to be cleaned (usually) and also allows digging out foreign material or stuck cases so that the insides will be cleaned as well. Inspection and sorting can be done simultaneously and should be done under really good light. Naturally, if it's a mixed batch, you'll sort fired cases first by caliber; keeping suitable containers handy for each size so they won't get mixed up again. Secondarily, they should be sorted by type and, if you are of a

Case at right is ordinarily darkened and slightly dirty from firing while at extreme left is a similar case after a modest period of tumbling. Center case really isn't worth salvaging, due to its thick hard corrosion.

Right — This .41 Magnum case split after its initial firing, *probably* because of a manufacturing defect; nevertheless, splits of this sort do develop from repeated reloading and resizing, and their probability is greatly increased if a more-than-necessary amount or resizing and expansion is employed. Regardless of why it cracked, such a crack requires that the case be disposed of.

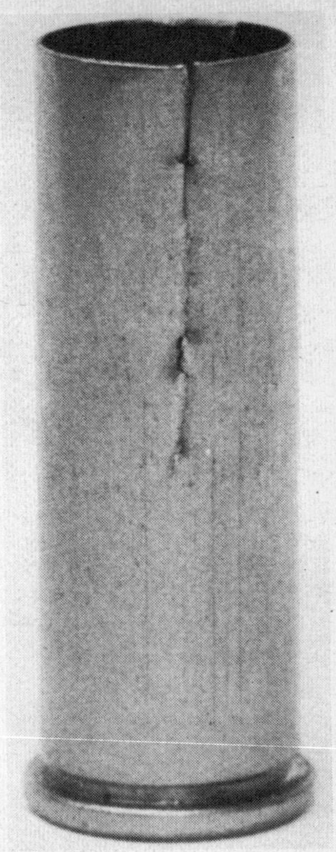

meticulous bent, by make and/or headstamp. Sorting by type consists of segregating military cases with their crimped-in primer from the commercial variety because of the different treatment military cases must often have before being reprimed. It also consists (in revolver calibers) of looking inside to determine whether the case is of balloon-head or solid-head construction. A heavy load, safe in a balloon-head case, may well produce excessive pressures in the solid-head type; and, in any event, the same load will not necessarily shoot to the same point of impact in both types. All recent revolver-caliber case production is of solid-head type, however, older balloon-head cases often show up in range pickups or donated material.

Case Inspection

Inspection requires first that you make sure there is no smaller case or other foreign material inside. Such material will not normally be removed by cleaning and can result in a broken decapping pin, stem, or damage to the case during resizing/decapping. I've found it best to keep a pair of needle-nose pliers and a couple of stiff wire probes handy to clear such cases as they are encountered. Aside from that, examine the cases closely for bent, creased or crushed mouths, usually caused by someone having stepped on the case or, in autos, by the case striking the gun during ejection. If there are no creases, the case can usually be salvaged by methods we'll describe later and should just be set aside. If it is actually crushed, creased or even cracked, it cannot be restored to perform safely as a practical matter, so it should be discarded. Look for excessive bulging ahead of the rim or extractor groove; this is found primarily in autoloader calibers, but can occur in a poorly rechambered revolver or when a cartridge of a particular dimension has been fired in a chamber not designed for the dimensions of that particular cartridge. A moderate bulge that flows smoothly from the normal surface can usually be ironed out by the resizing die without any serious effect upon the case strength; however, if the edges of the bulge are sharply defined, and it extends more than just a few thousandths of an inch above the surrounding surface, the case has probably been seriously weakened inside and should not be reused, even though resizing might remove the bulge entirely. Look for longitudinal splits or cracks, most often found at the case mouth, but occasionally found anywhere between mouth and rim. Such cases should naturally be thrown away unless the crack is just beginning at the mouth, in which case

Above — This is a case-mouth deformation typical of having been stepped upon lightly at the range. This case should be set aside for salvage later because in its present condition it will probably not enter the resizing die mouth; and, any attempt at resizing will crumple part of the case mouth.

Above — This type of mouth damage is typical of that which occurs in the Colt .45 auto pistol during normal ejection. This case will enter the resizing die well enough, but should be set aside for salvage simply as resizing may cause the dented area to fold inward and aggravate the damage. If it is set aside and eventually run over an expander plug to round out the mouth before resizing, it will probably turn out well.

Above — Ejection damage to .45 auto cases is not always at the case mouth. At left is the case shown in the previous photo, along with one which has struck the ejection port well back on the body resulting in a deep, sharp dent. This case can be salvaged by running an expander plug well down inside past the dent before resizing, but the sharp crease at the bottom of the dent has probably over-stressed the case wall and it may crack after only a few more loadings. Consequently, I'd not recommend its use for heavy loads.

Right — This photo shows two things to watch for in sorting cases. First, it is of the older balloon-head construction which, though perfectly serviceable, should not be mixed with modern solid-head cases. Second, there is a substantial amount of internal black-powder residue and, if it has been there any great length of time, the brass will most likely be pitted and weakened beneath it. Unfortunately, it is quite difficult to completely remove black powder fouling once it has hardened completely, so if the cases aren't washed thoroughly after firing, they're best set aside for moderate or light loads.

salvage instructions will be found further along. In magnum calibers you should also keep an eye out for an annular crack or thin bright ring, indicating an incipient crack just ahead of the rim. Cases possessing same are definitely headed for the trash bin. If any cases are discovered without a fired primer in the pocket, it is likely that either the case is too soft and expanded excessively during firing, or it was loaded to excessive pressure. Either is adequate reason to discard it; however, if the pocket can still hold a fresh primer *tightly,* it may be used for moderate and light loads. Primer leaks will also be encountered occasionally, and they make their presence known by sooty smudges of propellant residue around the primer pocket. If, after decapping, the primer pocket will hold a new primer tightly, the case is okay to use, otherwise it's junk. If primers are pierced at the firing-pin indentation or excessively flattened, cases should be okay to use *only* if (after decapping) you find that they'll hold a fresh primer tightly. Incidentally, if you fall heir to some old balloon head cases, it's worth checking the primer pockets carefully after decapping to insure that the bottom of the pocket is still present; they can wear out from either the past use of mercuric primers or from repeated reloading. The bottom of the primer pocket sometimes breaks away, removing support for a fresh primer and opening up a monstrous passage into the propellant. Look at the case-head face for evidence of any distortion such as deep impressions left by the breech face and extractor, or a headstamp marking that appears to be partially "ironed out." All of the above are evidence of excessive pressures. They can also be evidence of excessive headspace in autoloaders. In autoloader calibers, sometimes the extractor can bend the rim rearward, particularly when the chamber is badly fouled; and, in some guns, the ejector may bend the rim forward. A bent rim poses no particular problem if it isn't cracked as long as it will enter a shell holder properly. If it won't, you might as well ditch it because it's hardly worth the effort required to straighten it. Corrosion is seldom encountered on nickeled cases, but the plain brass variety will often be found with crusty green blotches and discolorations. If there is no surface buildup and the discoloration is smooth to the touch, it is simply a stain rather than destructive corrosion. If it's rough and lumpy, it can be removed with proper cleaning procedures, but the case beneath will be roughened and probably reduced in thickness to the detriment of its strength. As a matter of policy, I would not recommend you use heavily corroded cases, even after cleaning. Watch particularly for old cases with a crusty, thick, black coating inside; this is black powder residue, and it has most likely been there long enough to etch the brass deeply and thus reduce its strength. Such cases can be cleaned, but if the surface beneath the fouling is not perfectly smooth, they're candidates for the scrap heap unless they happen to be so old as to be of value to cartridge collectors.

Along this same line, if you encounter any cases of unusual color or surface texture, it's a good idea to set them aside until you can determine just what they are and whether they are usable. If any steel military cases show up, they aren't worth the trouble to reload unless they happen to be U.S.-made .45 ACP cases, bearing headstamp dates in the 1950's. Steel cases can usually be recognized by their corrosion-resistant coating which may be a greenish lacquer or grayish finish in foreign manufacture and a mottled coppery look on the U.S. variety. Some are also copper plated, and there may be others as well, but the color and texture will indicate that they are at least out of the ordinary and should be set aside. Also be sure to pick out any Berdan-primed cases; *they can* be reloaded as we'll describe elsewhere in the book, but they'll break a decapping pin for certain if mixed in with the ordinary Boxer type. Cases bearing U.S. commercial or military headstamps *will not* be Berdan primed, however, virtually all foreign cases of both military and commercial manufacture *will* unless of very recent manufacture. Examine the inside of foreign cases closely, and if they are of the Berdan type, you'll see two (sometimes one to three) very small flash holes drilled off-center. At first glance some Berdan-primed cases may *appear* to have a single Boxer-type flash hole because of an indentation formed in the center of the head; don't be misled by this and look closely for those tiny off-center holes. In donated brass and range pickups you'll usually encounter some cases that have been reloaded before; this is normally evident in minor scratches and striations on the case body and perhaps a burnished ring just ahead of the rim, marking the depth to which the case was forced into the resizing die. You may also encounter cases which are heavily scratched and gouged from being resized in dirty, burred, or otherwise damaged dies. If the case shows the latter evidence of previous use, I'd strongly suggest discarding it; those which still look clean and neat will normally be okay for light or moderate loads, but should not be used for high-performance or magnum loads.

Those are the main things to look for. Nevertheless, always keep your eye peeled for anything out of the ordinary. If a case doesn't look like it should, set it aside until you can question someone else or dig the answer out of a reference book.

Case Cleaning

Cases should always be cleaned. Even if they have been fired only once in your pet revolver and ejected directly into a clean container, they still will have picked up some dirt, grit and probably at least a modest amount of propellant fouling. If they look clean and only slightly smudged, wiping with a clean, dry cloth will be sufficient—*providing* they are protected from dust by a closed container until reloaded. The dirtier cases will require at least washing to get rid of dirt, propellant fouling, grit from the firing line, etc.

If only cleanliness—as to opposed to a bright, shiny, new look—is your goal and/or if the cases are nickel plated, washing will do the job. Put mama's biggest kettle on the fire, fill about two-thirds with water and a goodly portion of detergent. Dump in the cases and bring to a boil, agitating them vigorously with a length of dowel or an old broomstick to loosen dirt. After a few minutes of boiling and agitation, transfer the kettle (carefully, we don't want any scalded feet) to the sink and rinse with the hottest tap water your residence has to offer. Agitate while rinsing and keep it up until all suds are gone and the rinse water runs clear. Pour off all the water and dump the cases immediately on a piece of old carpeting or a long-nap towel. Stir them around a bit to insure that water inside runs out, then let them air-dry. An electric fan blowing over them will speed up drying. If you're really in a terrible hurry to get the cases dry (and they must be completely dry before repriming) spread them on an old cookie sheet in mama's oven and set the temperature control at *no more* than 225 degrees. Actually, 200 degrees is better, though they won't dry quite as quickly. Leave them in the oven no longer than necessary to dry all the moisture out of the heads and flash holes, and whatever you do, *don't* exceed those temperatures. To do so may very well result in partial annealing of the case heads, destroying their ability to withstand normal chamber pressures, and setting the scene for a serious accident. Frankly, I prefer to avoid applying any artificial heat and I wash them far enough in advance to let them dry naturally.

Nickeled cases seldom require any more attention other than washing, unless they have already been so heavily used that the nickel coating is scratched, chipped, and worn. Even so, *tumbling* is a more convenient and quicker method of cleaning cases, and it is the only really practical means of obtaining that bright and shiny "new" look on unplated brass. It's also the only way to remove stains and minor corrosion. Tumbling is done by placing cases in a barrel-like container along with a quantity of ground nut-hulls or fruit-pits, then rotating the drum so that continuous agitation of the cases and the polishing medium gradually scours away all dirt, grease, grit, stains and minor corrosion. The nut-hulls and fruit-pits ground into small, sharp-edged, irregular fragments are only very mildly abrasive, just enough to scour surface deposits away from the case metal. The use of this type of polishing medium is important because, if harsher abrasives are used (for example, steel shot which has sometimes been recommended), some case metal will be worn away reducing both dimensions and strength of cases so treated. Tumbling cleans not only the outside of the case, but when the proper-size granulation of medium is used, it scours the inside as well. Admittedly, cleaning action isn't quite as good inside as outside, but it's good enough to remove all propellant residue by the time the cases are clean and bright on the outside. The inside of the case will never be as bright and shiny as the outside, but by extending tumbling time, all stains can be removed, right down to bare brass. It will also remove from once-fired cases the remnants of mouth sealer which was applied by the factory. Washing won't affect this sealer residue at all, and the only effective method (other than tumbling) is to use a wet swab with a proper solvent (acetone) inside each case mouth—and that's a pretty boring job.

These days case tumblers are available from a

Under no circumstances should either of these cases be used, even though they might be resized in such a manner that the excessive feed-ramp bulges *appear* to be removed. This much expansion over-stresses the brass at the junction of the web and wall and may even have produced an internal crack. Pushing the metal back in place by resizing won't restore the impaired strength, and the case may let go on the next firing, even with a light load.

Even the factories make mistakes that aren't always caught by the inspectors. The case of this high-performance .45 ACP load has been damaged during bullet seating. Note the triangular piece broken out — it is probably inside the case, having been folded inward and torn off by the bullet. This cartridge chambers freely and would doubtless be safe to shoot, but I would be very surprised if it printed within the normal group — the case should probably be discarded after firing.

The head of the case on left shows evidence of high pressures in the "coining" marks produced by irregularities in the slide breech face. These marks are distinct and separate from the impressions made by ejector or extractor — they do not in and of themselves indicate excessive pressure. I would, however, reserve it for moderate or light loads. Case on right has no "coining" marks.

This case is not necessarily defective, however, the large number of striations on its outer surface indicate that it may have been resized and reloaded a number of times and lost some of its strength; and, may have lost even more of its ability to maintain a heavy and consistent crimp. For these reasons, it probably should not be used for heavy or high-performance loads.

This lineup of 9mm Parabellum cartridges displays a variety of headstamps as well as the two different types of primer crimps normally found on military ammunition. At left is 1941-vintage Canadian cartridge loaded with a Boxer-type primer and assembled with a relatively heavy ring crimp; second is also of Canadian manufacture in 1944, but employs a triangular, 3-point, stab crimp; third is a World War II 9mm cartridge manufactured by Winchester under military contract, of Boxer type, apparently employing only the very slightest trace of a ring crimp (so slight that it needs no treatment before repriming the case); last is a recent-manufactured German Geco *commercial* round employing 2-hole Berdan primer and no primer crimp whatever.

Typical Berdan-primed case, looking at the inside of the head or web with the walls cut away for clarity. The "hole" in the center really isn't what it appears to be — it's simply the underside of the anvil (formed in the case) for use with Berdan primers.

number of sources, in particular Gil Hebard guns (Knoxville, Illinois) and J.&G. Sales (Prescott, Arizona). Those sold for cleaning cases are identical to some sold by lapidary supply houses for rock polishing. Consequently, if your local gun shop can't supply a tumbler, the nearest hobby shop or lapidary shop most likely can. The ground nut hulls or fruit pits aren't so widely available, but can be mail-ordered from shops supplying tumblers to handloaders. (In sizable communities they can also be obtained from houses supplying industrial abrasives and cleaning materials.) Because the medium is relatively soft, it loses its sharp edges fairly quickly, and as it wears, a longer tumbling time is required. This can be avoided to a large degree by using the special rouge-impregnated medium available from J.&G. Sales, and perhaps other suppliers as well. The polishing rouge it contains imparts a brighter finish to cases and makes the medium work longer and faster than the untreated variety. The rouge may also be renewed, thus giving the medium an even greater life.

Cases need not be thoroughly dry when tumbled, but excess moisture will just slow down the process. Over-filling the tumbler drum will actually slow down the polishing and cleaning process, so I recommend roughly one-third medium (by tumbler volume), and one-third cases. Unless the brass is unusually dirty or corroded, only an hour or two of tumbling is required with fresh medium; however, I find it more convenient to simply fill and switch on the tumbler at the end of the day and let it run until morning. By doing this every time I generate or acquire fired cases, and doing it every night, cleaning is always ahead of loading and by having the cases thoroughly clean when they are stored, they aren't likely to deteriorate.

The utility of a tumbler does not end with initial cleaning of the cases. They require lubrication for resizing (more on this later) and subsequently that lubricant must be removed. I tumble cases initially, then resize and decap, and then tumble again to remove the lubricant and to at least partially clean the primer pockets. If tumbling time is extended the pocket will be thoroughly cleaned of that gritty primer residue. While dry medium will remove lubricant well enough, it will do so faster and better if very slightly moistened *not wetted* with lighter fluid or similar solvent. Because the medium eventually picks up a fair amount of lubricant, it is best to keep a separate batch (or even a separate tumbler so loaded) for this purpose.

There are other methods of cleaning cases, including manual or power brushing and polishing, and assorted chemical dips. Cleaning individually, even with a power-driven wire brush or buffing wheel is not only very slow (and hazardous to fingers) but presents a very real hazard in that it is very easy to inadvertently remove too much brass and weaken the case. A few liquid chemical cleaning solutions are marketed by handloading suppliers, and while they are very effective—particularly on heavily-corroded cases—their use is not especially convenient. In addition, most of them are toxic to some degree. Most are basically a weak acid solution including a small portion of chromic acid as brightener, and thus great care must be exercised in their use. Some solutions, particularly those for which formulas have been published in older references, work very rapidly, and if cases are left too long in the solution, considerable metal will be removed, and they'll become weakened. Some of the commercial solutions are of the "self-canceling" type which attack only surface deposits and then quit operating when they reach the bare case metal. Almost all chemical cleaners, including the latter type, will lightly etch the surface of the cases. The probable exception to this is found in those jewelry-cleaning solutions one occasionally sees advertised. I've never had occasion to try them on cartridge cases, but if they perform as advertised on copper and brass utensils and costume jewelry, then they should

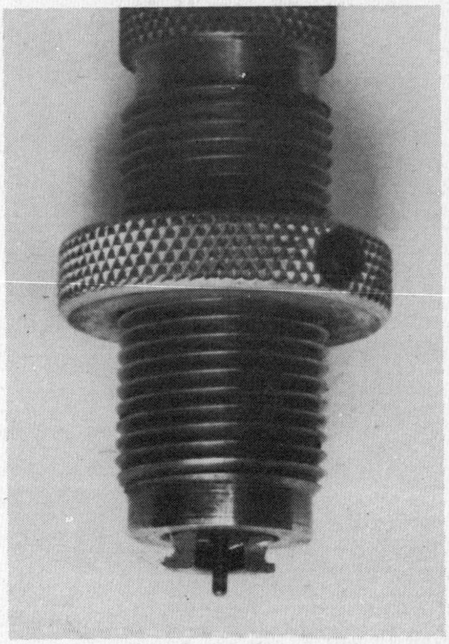

Here we see the shell holder making solid contact with the mouth of the die, as it must in order for the case to be fully resized. (This is also the proper relationship of shell holder and die when the maximum amount of crimp is being applied.)

The .45 ACP case shown here has been sectioned and inserted in a combination resizing/decapping die. Note that when the shell holder makes contact with the die mouth (shell holder left out here for clarity), the decapping pin must protrude sufficiently beyond the case head to insure positive ejection of the fired primer. It is important that the decapping pin be centered in the die mouth (below) to avoid striking the case head and damaging both it and the pin.

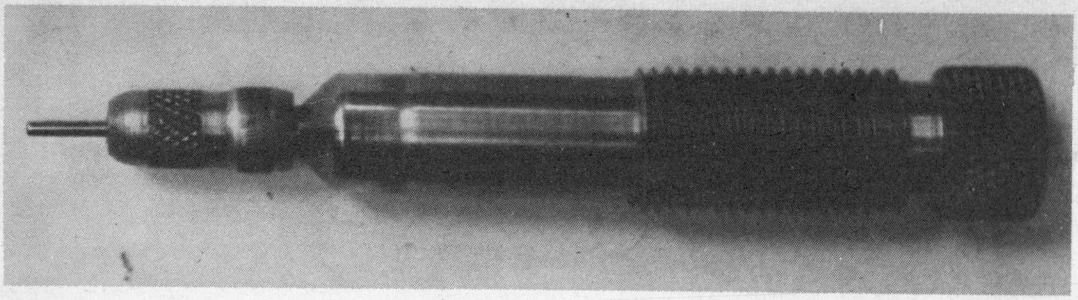

Above—This expander rod combining the decapping pin happens to be for a .45 Colt caliber. Differing from some makes, this unit cannot be used on the shorter .45 cases because the nipple securing the decapping pin is too long; when run into a .45 ACP or similar case, the nipple strikes the case head before the expander rod proper enters the mouth.

Right — Here's how the combination expanding and decapping rod works. Shown in .45 Colt caliber (employing a sectioned balloon-head case), you can see how the flaring shoulder on the rod engages the case mouth, and at the point where adequate flaring is produced. The decapping-pin nipple *must not* contact the inside of the case head, and the decapping pin *must* protrude far enough beyond the case head to insure that the fired primer is ejected.

certainly do the same on brass cases.

There is one simple home-brewed solution which *will not* harm the cases. Get a gallon of the cheapest white vinegar you can find, and to it add a gallon of water and 1/2-cup of common table salt. The result is simply a somewhat salty and weak (2 percent or less) solution of acetic acid which will remove dirt and minor corrosion, but which *will not* brighten the cases. In fact, it will darken brass slightly and will sometimes give the cases a slightly pink color. It's really not all that great, but it's cheap and easily made. Its effectiveness improves if heated slightly with agitation of the cases necessary to help break loose the various contaminations.

RESIZING

This operation begins with installing the proper shell holder in the top of the press ram and screwing the resizing die into the top of the press until it makes firm contact with the shell holder when the ram is at the top of the stroke. The die lock ring is then turned down snug against the press and secured to the die body by its set- or clamp-screw. Of course, prior to this, the die cavity should be wiped out with a clean, dry, lint-free swab or cloth to remove any dirt or grit that might have accumulated there since the last use. If your particular die is equipped with a decapping stem that allows the fired primer to be punched out simultaneously with resizing, then it must be screwed in or out of the die so that is protrudes 3/16- to ¼-inch below the die mouth. The jam nut at the top of the stem can then be turned down tightly to secure the stem solidly—be sure the stem is properly aligned inside the die cavity. If the pin is off-center, it will not pass freely through the flash hole and will be bent or broken from being jammed against the inside of the case head.

Lubrication

For resizing in standard all-steel dies, the clean and dry cases must be lightly lubricated. If they are not, excess resizing effort will be required, cases may be crumpled or scored, galling may occur, and eventually a case will stick so tightly in the die that you can't pull it out. If you exert a lot of force, at this point, the shell holder will pull the rim completely off the case. When a "stuck case" occurs, don't force it, simply turn out the decapping stem and its bushing, then drive the stuck case out with a brass or aluminum rod of the largest diameter that will enter the top of the die. Ruining a case for lack of lubrication won't cost much, but you can easily ruin the die as well, and that *will* cost you a handful of bucks.

You must use the proper kind of lubricant. The usual oils and greases that might be found around the house or at the local service station won't do. They lack the high film strength required by the very heavy contact pressures involved in resizing. Nearly all handloading manufacturers offer a special resizing lubricant formulated specifically for this purpose. They're economical and a

Left — Almost all loading-tool manufacturers now supply their own brand of resizing lubricant and I doubt there is much difference between brands — all of them seem to perform quite well. You may, however, find it more economical to use anhydrous lanolin or oil additive STP; and, in addition, those materials are readily available at neighborhood service stations or drug stores.

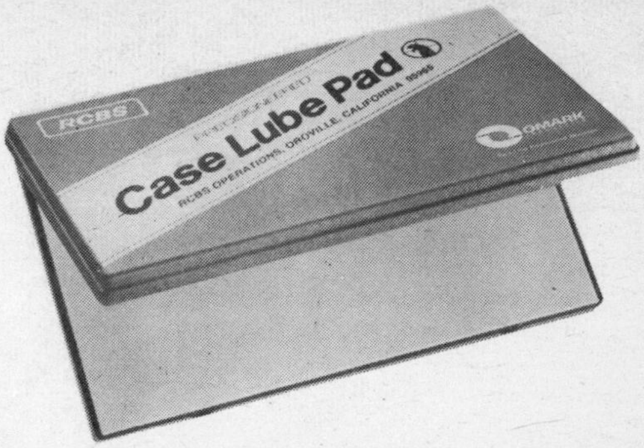

Above — One of the handiest accessories you can buy — and it's not that expensive — is a case lube pad. You simply impregnate the pad with lube, give it a chance to absorb, and then roll the cases across the pad's surface.

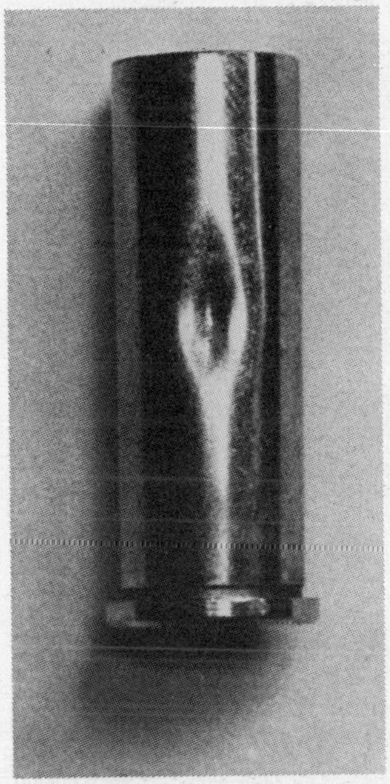

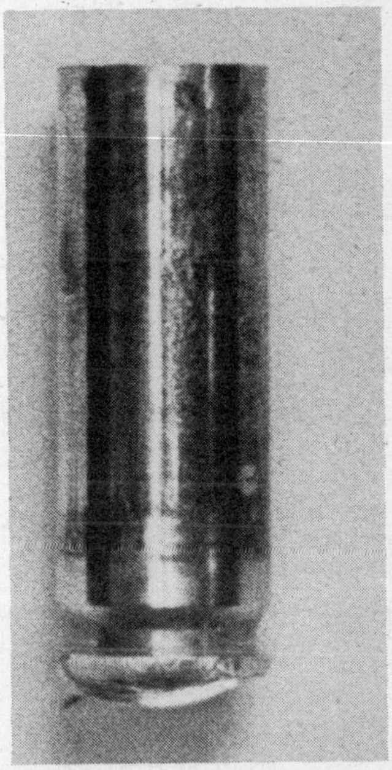

Above, left — Oil dents aren't common in more-or-less cylindrical cases such as this .38 Special, but they *do* sometimes occur. In severe instances, the case may be split or perforated at the bottom of the dent. When this happens, discard the case immediately.

Above, right — This is a .44 Auto-mag case which was inadvertently run into a dry resizing die *without* lubricant; needless to say, it stuck tightly, and efforts to extract it by more beef on the press handle peeled away about one half of the rim. This case is now totally useless, it had to be driven out of the die with a ⅜-inch diameter brass rod after the die was disassembled.

couple dollars worth is enough to lubricate thousands of cases, so there's no reason for playing poor-mouth. However, there are excellent substitutes in case you can't find the specified item locally when you need it. The well known engine oil additive, STP, works beautifully and costs one hell of a lot less than resizing lube. Anhydrous lanolin, available at the local drug store, also seems to work perfectly and was, in fact, highly recommended in almost all references before the tool manufacturers began supplying special lubricants. Green soap solution from the drug store also works well, as will water-softened toilet *soap* (not the fancy soap substitutes or detergents). There are other substitutes, but these should be more than enough to see you through until you can find the real thing.

The simplest method of lubricating cases is to moisten the thumb and finger of your "off hand" *slightly* with the lube, then roll the case between them as you prepare to place it in the shell holder. Only a trace of lubricant is required, so this isn't nearly as messy as it might seem. *Don't* over-lubricate; if you do, puddles of it will be trapped between the case and die, and the case will be dented and perhaps even split. Dents caused in this fashion are referred to as "oil dents." Finger-lubrication is okay when processing only a few cases, but you'll likely want a faster method. The best method I have found is to take an old, long-napped towel, moisten it lightly in the center with lubricant, dump on it a couple hundred cases, fold the edges over and grasp both ends in your hands, then shake the bundle back and forth a couple of minutes. During that agitation the cases will pick up all the lubricant

Above — This is the double-shuffle method (described in the text) of removing the resized case (or complete round, for that matter) and substituting a fresh case with one hand movement.

Right — As the case enters the resizing die, it must be positioned fully to the rear of the shell holder; if it is not, the case will not be aligned with the die, and may be damaged. For this reason, debris of whatever sort must be frequently cleaned out of the shell holder groove which accepts the case rim.

they need from the nap of the towel. Of course, the towel should be clean, free from any grit and, should also not be shedding nap or lint.

Various gadgets are sold for lubricating cases. They mainly consist of cloth stamp pads across which cases are to be rolled, felt tongs and the like. All will do the job, but they are generally less convenient than fingers and much more time-consuming than a towel. If you use the *tungsten-carbide* type resizing die rather than the ordinary steel die, lubrication is not really necessary—though it does ease resizing effort. For this reason, and others mentioned elsewhere in this volume, I prefer the T-C type die and willingly accept its greater cost.

This is a problem that arises when an undersized resizing die is encountered. Even though this case showed normal after-firing expansion at the head, running it through a particular resizing die (an expensive carbide model at that) produced a distinct shoulder on the case as shown. Essentially the same result can occur when an overly-expanded case is pressed into a standard-size die. When this occurs a ridge of a brass is piled up and *may* prevent chambering. If you do encounter a new die which causes this problem with normally-expanded cases, by all means return it for replacement.

If you should carelessly grab a fired case and attempt to run it hurriedly into a resizing die, this is what you'll get. In this instance, the case mouth was slightly out of round, the case may not have been seated fully in the shell holder, and as a result one edge of the mouth caught on the edge of the die mouth, being ruined in the process.

Resizing Straight and Tapered Cases

Take the lubricated case between the thumb and finger of your off hand, and with the press ram fully down, place the case in the shell holder. Make certain the rim seats fully to the rear of the T-slot in the holder and that there is no dirt or spilled propellant compacted there. If the case is not seated *fully,* it will not be correctly aligned with the resizing die and will be damaged if sizing is attempted. If everything is in order, raising the ram (by lowering the press handle) will now cause the case to enter the die; however, it's a good idea to form the habit of keeping your thumb and finger on the case to guide it into the mouth of the die. Then, once the case enters, slip your fingers smoothly out of the way so they won't be crushed between ram and die. Use a smooth, single, forceful stoke of the handle to press the case home in the die; then, without particular hesitation, reverse the handle movement to extract the case. While this is being done, you can speed up the process by picking up another case between thumb and forefinger, having it ready when the ram reaches the bottom of its stroke. When the ram bottoms out, use the middle finger of your off hand to reach around behind the resized case and pull it out of the shell holder, clamping it against the outside of your forefinger. Then twist your hand and slip the fresh case into the shell holder. This cuts down on the time and movement required to remove and replace the case. Done conventionally, you'll first have to pluck out the resized case, drop it into a suitable container, then pick up another and place it in the shell holder. I prefer the quick way. With small cases like the .380 and 9mm Parabellum, some people simply palm a number of them, and let them roll down individually between the thumb and forefinger prior to seating in the shell holder.

There is an alternate method of resizing which is extremely rapid and convenient for those straight rimless cases such as the .380 and the .45 ACP and 9mm Bergmann-Bayard. *It is not* applicable to rimmed, semi-rimmed, or tapered cases. We call it the "push-through method," and it consists essentially of forcing the case head-first completely through a resizing die. Ordinary steel sizing dies which are roughly the same diameter through their full length may be used in this fashion, but the best is the carbide-insert die which is oversized throughout except for a narrow working ring at the mouth. To accomplish this, the die is placed in the press in the usual manner, stripped of any internal parts. Then, using a simple steel pusher turned at its lower end to fit into a shell holder or shell holder slot in the ram, and on its upper end to fit inside the base and to bear against the head, the case is simply pressed through the die. Be not concerned that the case remains in the die; once it's past the working area, the next case will simply lift it upward, and once the die becomes filled, a sized case will tumble out of the top with each stroke of the handle. A cardboard or light sheet-metal chute fitted around the top of the die will direct the sized cases into a suitable container, thus reducing case handling to the absolute minimum. Over the years I have resized many thousands of .45 ACP cases in this fashion with complete satisfaction; and on one occasion, I resized 4,000 .380 ACP cases in one batch without the slightest problem.

Resizing Bottle-Necked Cases

That's all there is to resizing typical straight or slightly-tapered cases. Bottle-neck numbers will sometimes require a different method of die adjustment to avoid creating a condition of excess headspace which will greatly reduce case life by causing head separations. There is considerable latitude in the position of the chamber wall shoulder (which supports the cartridge shoulder) in production guns, particularly among the older models. If your gun happens to have the chamber shoulder relatively far forward, the case will fire-form to fit, thus moving its shoulder forward, upon firing; and, this new shoulder position may be well forward of the position of the die shoulder with the die adjusted as previously described. Thus, it is possible during resizing to push the case shoulder back a substantial amount. The result is that in just two or three reloadings this excessive working of the shoulder can damage the case, and worse yet, the stretching that occurs upon firing because of this can cause the case to pull apart just forward of the head. To avoid this particular problem, screw the bottle-neck resizing die into the press *only* far enough that the resized case will *barely* chamber freely in the gun for which it is intended. The exact position of the die in relation to the shell holder (ram at the stop stroke, remember) can only be determined by trial and error. The best method is to screw it in to a point about 1/16-inch from the top of the shell holder, resize the case, try it in the gun, and then adjust the die successively deeper in the frame until the case chambers freely. Once obtained, this die adjustment is correct only for that particular gun; cases resized in it may not chamber in some other guns, and/or may exhibit excessive headspace in others. If you load for only one gun and this particular bottle-neck caliber, this presents no problem—once the lock nut is secured to the die it will always screw into the press correctly. On the other hand, if you have more than one gun in the same caliber, you'll have to either find a compromise adjustment suitable for all or readjust each time. If you're flush with bucks, you could, of course, keep a die for each gun. Actually, a compromise adjustment suitable for all guns in that caliber is best in that it eliminates the problem of getting ammunition loaded for one gun into another.

After resizing we have no further need of lubricant on the case and to leave it there makes for messy handling and the possible contamination of the fresh primer. This is the time to remove it, either by tumbling as already described or, by the "towel method," substituting solvent for lubricant on a clean towel. However, if you want the primer pockets cleaned as well, you'll be forced to use the tumbling method, and then only after decapping. (Decapping may not occur until neck expansion, depending upon the make and type dies being used.)

NECK/MOUTH EXPANSION

Because of variations in neck wall thickness of the case and diameter variations in the neck portion of the resizing die, it simply isn't practical for resizing to produce exactly the right *internal* case diameter needed to grip the bullet properly. More on that when we get into bullet seating. Dies and cases *could* be made accurately enough for that purpose, but to do so would make them so costly that neither you nor I could afford to shoot. Because of this, the resizing die is made to bring the inside of the case mouth *down below* the desired diameter; then, we obtain the correct (and uniform) inside diameter by forcing the case over a smoothly polished rod (called an expander rod) of the proper diameter.

We do this in what is called an *expander die,* though its major component is not a die at all but simply a fixture holding the expander rod. Most manufacturers make this expander die of sufficient size so that it will handle all calibers (with the proper rod in place) and also make it of low-cost, easily-machined metal. This allows maximum interchangeability and keeps the tool costs down. I'm for that. This die is threaded like the others so that is can be adjusted quickly in the press; and, the expander rod is threaded in the die so that it too can be adjusted vertically. The lower end of the rod is rounded or tapered so that it will enter the undersized case mouth easily, then fol-

Left — Lower left is a complete Boxer-type primer, ready for seating in the case; next is the same primer with the anvil removed, showing the position of a priming-compound pellet inside and the depressed center of that pellet where it was compressed when the anvil was originally assembled to it. Last is the typical 3-legged, conical-type anvil normally employed in Boxer-type primers.

Right — Whether you are priming one at a time on the press or with an auto-feeding setup on either the press or a separate tool, these handy "primer flipper" trays make handling the primers much easier.

lows a cylindrical section of the correct diameter to produce the desired inside diameter after the brass springs back slightly. Above this portion is a conical or flared section which serves to give the case mouth a slight outward turning which is necessary to facilitate entry of the new bullet later. Naturally, the entire working portion of the rod is smoothly polished and hardened to reduce friction and wear.

The die is screwed into the press deep enough that the rod will enter the case mouth slightly. From that point onward, either the die or the rod—sometimes both—are adjusted vertically until, with the ram at the top of its stroke, the rod enters the case mouth just deep enough to produce a barely visible flare of the case mouth. Some bullets require more flare than others in order to enter the case easily, and since excessive flaring will reduce case life, it is best to check the flare with a bullet during adjustment. The correct adjustment is that which will produce *just enough flare* to allow the bullet to start into the case. When that point has been reached, lock the adjustment, and you're ready to roll.

Many modern expander dies incorporate the decapping pin in the expander rod. Doing so simplifies production of the resizing die (which would otherwise carry a separate decapping stem) without unnecessarily complicating production of the expander rod. The decapping pin is removable, held in place by a small knurled nut or collar. Adjustment of combined expanding/decapping rods remains the same; however, you must make certain that the decapping pin protrudes far enough through the flash hole (when the case is fully in the die) to produce positive ejection of the fired primer. If it doesn't, then it must be replaced by a long pin. Don't attempt to compensate for a too-short decapping pin by adjusting the expander rod downward—that will simply over-flare the case mouth and cause early failure by splitting.

PRIMING

If you are repriming cases on the press, it can be done in conjunction with resizing if your particular resizing die contains the decapping stem. If decapping is combined with the expanding die, then you'll obviously have to wait until that operation. While you may want to reprime on the press—this is not only quite adequate, but common—I prefer to decap with resizing, and then reprime in conjunction with neck expansion. If repriming is done at the same time as decapping (and most presses were designed so that it can) we then run the risk of the decapping pin dislodging the freshly-seated primer as the ram is raised to release the priming arm. If an

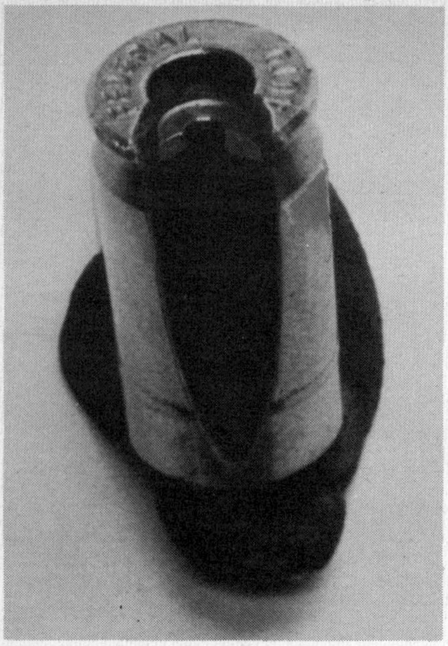

Left — This is a typical large pistol primer pocket shown in a cutaway .45 ACP case. After *many* reloadings, primer pockets may become worn, not allowing a fresh primer to seat in the pocket firmly. If the primer simply won't stay in the pocket, or it seats itself loosely, you should discard the case.

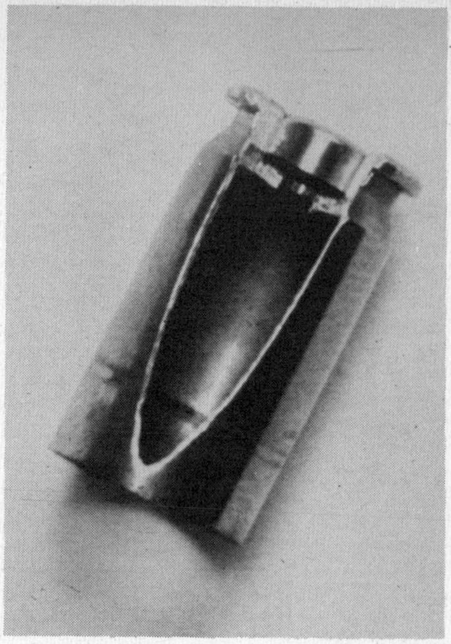

Right — This cutaway .45 case shows a primer incorrectly seated. The primer is not seated deeply enough, even though it *appears* to be flush with the case head (actually it isn't), leaving a significant gap between the "feet" of the anvil and the bottom of the pocket. The primer is thus free to move forward slightly when struck by the firing pin, cushioning the firing-pin blow and thus inviting erratic ignition.

H-press is being used, at least in some designs, we also have the problem of the fired primer and its debris falling down onto the priming punch and fouling up the process.

Most presses employ the spring-loaded, pivoted-arm type priming device. In use, the ram is run all the way up to carry the case into the die; then a fresh primer is placed anvil-up in the spring-loaded bushing or collar surrounding the primer-seating punch. The arm is then pressed forward with the thumb, into the slot in the ram, coming to rest solidly against the arresting surface provided. At this point, the ram is carefully lowered, pulling the case head down over the primer. Close attention is required at this point to insure that the primer is seated fully in its pocket, coming to rest with the feet of the anvil making contact with the bottom of the pocket. The head of the primer cup *must* be flush or slightly beneath the surface of the case head, in order for the cartridge to feed *properly and safely* in the gun. To obtain maximum uniformity of ignition, the primer should ideally be resting against the bottom of the primer pocket. This can be accomplished by using the on-press priming device, however, the high mechanical advantage of the linkage and the friction loads within the mechanism make it difficult to *feel* the primer into the proper position. It requires very careful manipulation of the press handle and thorough concentration on the job. Too much beef on the handle causes the primer to be compressed,

While it may not seem to be protruding much to the eye or the camera, a protruding primer can be easily felt by sliding the finger across the case head as it is withdrawn from the priming device.

fracturing the priming-compound pellet and causing erratic ignition. Too little pressure leaves the primer standing off a bit from the bottom of the pocket and thus it moves forward under firing-pin impact (absorbing some of that impact), which reduces ignition uniformity.

Once the primer is seated, release the priming arm and raise the press ram just enough so that the shell holder clears the priming punch and the spring retracts the arm. Here is where we get into problems with the decapping pin if it is present. Raising the ram to release the arm requires careful control of the handle or the case will be elevated too much, and the decapping pin will push the primer part way (or all the way) out of the pocket. If the primer is moved only slightly, it

Right, above — Note the groove impressed upon the face of this primer cup. If very deep, this type of deformation can produce erratic primer functioning. It is most often produced on a post-type priming device where a piece of foreign material or other debris has fallen inside the sleeve which surrounds the primer. Almost any material can dent the primer cup, but most often it is a sliver of brass produced when a case is resized or decapped. If the mark on the primer is very shallow, it is likely no harm has been done; but if it is deeper than a few thousands of an inch, I prefer to replace that primer to avoid possible ignition problems.

Right, below — Here we see a typical ring crimp as found on U.S. military-contract .45 ACP ammunition. In this instance, a cartridge manufactured by Lake City Arsenal in 1969 as indicated by "LC 69" headstamp. After the expended primer has been removed, you can eliminate the crimp with a primer-pocket reamer, or (be careful) the point of a sharp knife. If the crimp isn't removed, you may not be able to reprime the case.

Above — Priming, as a separate operation, is both fast and easy. The tool handle is lifted to allow the seating punch to drop; a primer is placed anvil-up in the sleeve inside the shell holder; the case is slipped into the holder (all the way, remember); and the handle is depressed with just enough force that the primer can be felt contacting the bottom of its pocket firmly.

won't be noticed but ignition will be affected; if it is pushed beyond the case head, you'll probably notice it.

After the arm retracts, the ram may then be lowered all the way and the primed case removed from the shell holder. You should make a habit of either feeling the primer with the finger tip as the case is removed or examining it visually to make certain that it is seated flush or below the case head. With a little practice, *inspection by feel* is by far the simplest and fastest form of primer-seating inspection.

For convenience in the following operation, cases should be placed head-down in loading blocks after priming.

POWDER/PROPELLANT CHARGING

While "bullet quality" plays, perhaps, the most important role in "cartridge accuracy," proper powder selection also contributes to a round's potential accuracy—or lack of it. It must be said, however, that the actual charging of the prepared cases with propellant is the one operation that is *most critical* to the safety of the completed cartridge and the man who shoots it. Assuming that the proper propellant and correct charge weight has been chosen, it is absolutely essential that no mistakes be made and that a reasonable degree of uniformity of charge weight be maintained. It might seem ridiculous that one would use the wrong propellent after having made a suitable selection from loading data. Nevertheless, *it does* happen with distressing frequency. For whatever reason, we receive all-too-frequent reports of blown up guns caused by handloaders selecting one suitable propellant from loading data and then for some inexplicable reason using a different one that produces higher pressures. As an *example:* If one selects a .357 Magnum load employing 16.0 grains of Hercules 2400 propellant and then inadvertently uses that amount of Hercules Unique or Bullseye, a blown-up gun is quite likely to result. Likewise, if a charge of 9.5 grains of Unique is chosen and a like amount of Bullseye is loaded, the gun may come unglued. These are only two examples of the *many* dangerous possibilities that exist. Therefore, it is *absolutely* essential that the propellants in your loading room or area be kept in their original containers to avoid loss of iden-

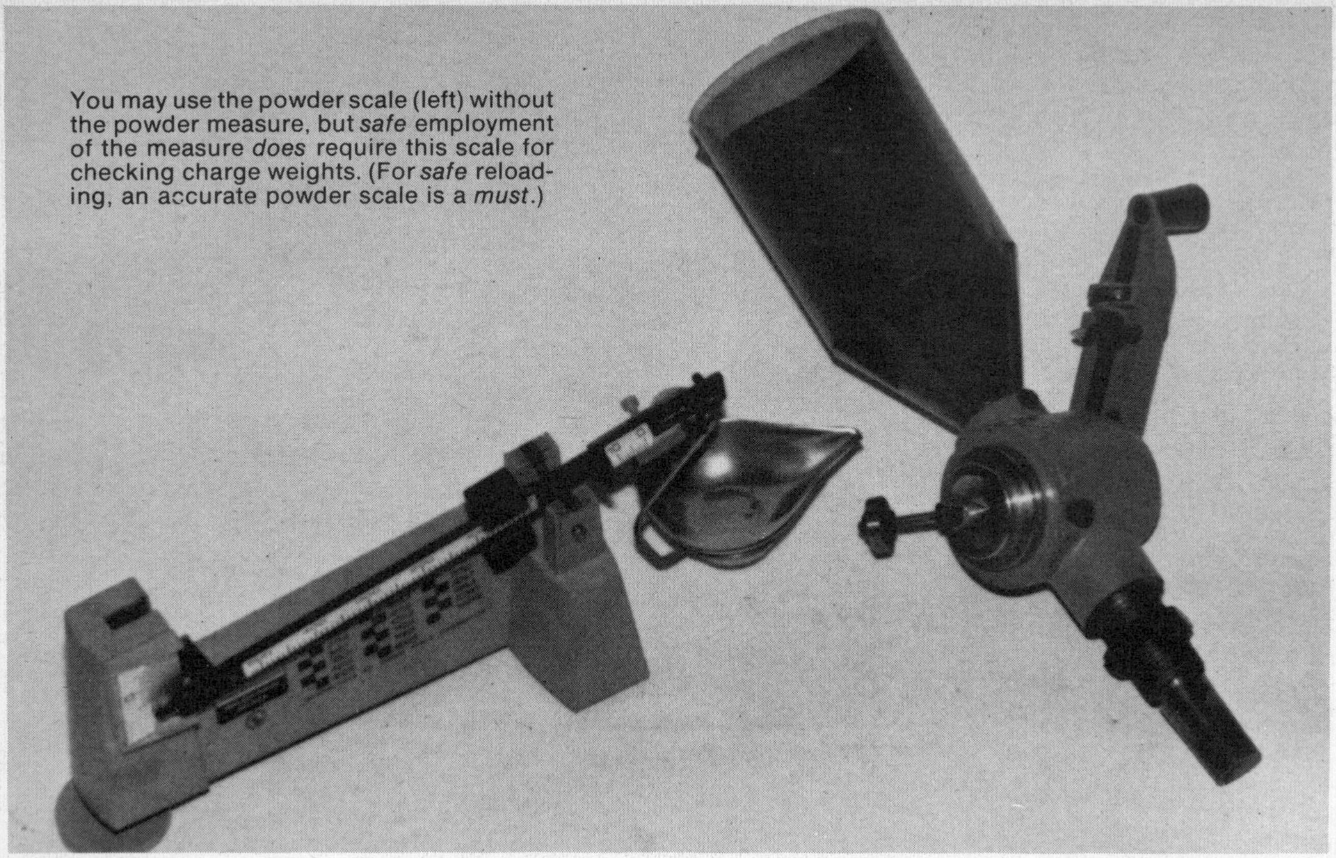

You may use the powder scale (left) without the powder measure, but *safe* employment of the measure *does* require this scale for checking charge weights. (For *safe* reloading, an accurate powder scale is a *must*.)

tity; and, that you double check and re-double check the identity of the propellant as it is taken from storage and poured into the reservoir of your powder measure.

I believe that a particular source of "wrong propellant identification" stems from the fact that many handloaders have a habit of leaving their powder measures filled between loading sessions and forgetting what the contents are when loading is resumed. Not being certain what it is, one attempts to identify it by appearance, and it is almost *impossible* to do this with precision. One should *never* attempt to identify propellant by appearance alone; it is far better (and safer) to flush a few ounces of it down the toilet than to take a chance.

We'll assume that you'll be using an adjustable volumetric measure simply because that is the most efficient and versatile type for general purpose loading. First, this measure must be solidly mounted if it is to throw uniform charges; if it isn't, the vibration and other disturbances of normal operation will cause charges to vary excessively. You'll also need a powder scale with which to set and verify the measure. If this is your first attempt in using an adjustable measure, *do not* assume that the graduations on the vernier or micrometer scale indicate actual weight of charge in grains or any other measure—they do not. For a typical charge of, say, 5.0 grains of Unique, first make certain that you have placed or engaged the pistol-size metering chamber of the measure. You then zero the adjustment scale and look inside the reservoir to make certain that the moveable plunger is at the top of its chamber. Fill the reservoir two-thirds full of propellant (and for the balance of your loading, don't let the powder level fall below about one-third in order to maintain a fairly good head of powder). At this point, get out your powder scale and make certain the beam moves freely. Blow off any dust, and wipe the pan clean. Set the poises on the scale for five times the weight of a single charge; in the example we've cited, that would be 25 grains.

Now (going back to the powder measure), back off the measure adjustment a relatively small amount and lock it in place with the jam nut or whatever securing method is present. Throw five or 10 charges into the scale pan to settle the propellant in the reservoir. On that last stroke, leave the measure handle in the *discharge* posi-

Above — For fine adjustment, the micrometer-type poise employed on the RCBS and now-discontinued Ohaus powder and bullet scales is much superior to the light-weight sliding poise employed by other measures; not only for ease and speed of adjustment, but because it is highly unlikely to move out of position as will one of those feather-light sliding poises. In addition, the lock screw at the top can be employed against the rotating thimble to make doubly certain that the adjustment cannot change.

Right — When charging with powder, holding the case in this manner eliminates the possibility of its slipping through your fingers and falling to the floor, and also allows any spilled propellant to be caught in the hand rather than dumped on the floor.

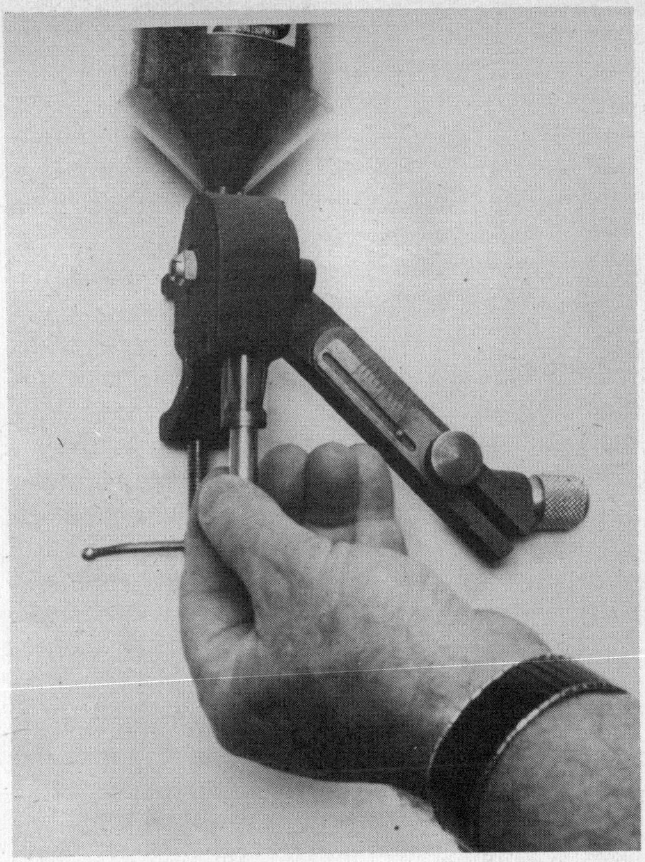

tion and put the pan full of powder back in the reservoir. Now we can start checking the measure, making sure it's throwing that 5.0 grains of Unique. Move the measure handle smoothly but firmly to the *fill* position, making certain that it bumps the stop solidly. Pause for a few seconds to permit the propellant to flow into the metering chamber, then with a single smooth movement, throw the handle to the discharge position, again making certain that it bumps solidly against the stop. Repeat this four more times, for a total of five charges, depositing them all in the scale pan. Transfer the pan to the scale, and you'll be able to tell instantly whether the five charges are over or under the weight they should be. It isn't necessary that the pointer come to rest or to determine the actual weight of those charges; you just need to know whether they are over or under what they should be. Return the powder to the reservoir and make a small adjustment to increase or decrease the charge weight as indicated by the scale. Throw five more consecutive charges into the pan, taking particular care that the measure handle is operated as uniformly as possible with each charge. Again weigh the five charges, and note whether they are over or under. If it seems reasonably close, it might be a good idea to obtain the precise weight so you'll have a better idea of the size of the adjustment you'll have to make on the measure. Repeat the process, throwing and weighing five consecutive charges, until they come out right on the button—in this case, 25 grains. As you approach the final adjustment, you can begin using those graduations, noting with each charge how much one space alters the charge weight. By remembering this—or, better yet, writing it in your notebook—you'll be able to adjust the measure rapidly with that particular propellant in the future. When those five charges finally come out to exactly the weight that five charges should be, you know that the *average* weight is 5.0 grains. Even so, it's a good idea to repeat this a couple times as a double check. Because of the relatively small charges normally used in handgun cartridges, weighing five at a time (or even 10 if you are more meticulous) is more accurate in that it is easier to weigh the larger amount precisely than a single charge. It is also more accurate in that you're automatically determining that the measure is throwing charges that *average* the correct weight. There will assuredly be variation from charge-to-charge, and you can avoid getting stuck on a high or low adjustment by using this averaging method. Recheck everything, including the scale, to make certain you haven't jarred it off its setting, then lock the measure adjustment.

Now (and not until now), you're ready to charge prepared cases with propellant. To facilitate handling, the cases should be in a loading block, though they can be handled individually. (The latter presents some hazard of propellant spillage unless you start a bullet in the case immediately after placing the propellant in it.) With the loading block held *level* in your left hand, press the first case of the farthest row up into the coned lower end of the measure drop tube. In some measures this tube will be made of clear plastic so that you can actually see the propellant as it runs into the case. The block, and consequently the case, needs to be held level so that the case mouth makes full contact in the drop tube and thus eliminates the possibility of propellant loss. From the discharge position, rotate the handle smoothly and solidly to fill; pause a few seconds, and come back down to discharge, dumping the propellant through the drop tube and into the case. I must emphasize here again that it is *uniformity* of handle movement that will add most to the uniformity of charge weight. Some older measures, particularly the Lyman, have a ''knocker'' pivoted loosely to the front of the base so that it may be flipped with your finger. This knocker is intended to settle powder more uniformly in the metering chamber at the fill position, and to insure against bridging (the ''jaming-up'' of powder) during discharge. Actually, the typical small-kernel pistol powders very seldom bridge, though you should still watch to make certain that the charge does fall into the case. I don't recommend the use of a knocker simply because of the difficulty in employing it uniformly. If it is used, I recommend using it only during discharge to jar loose any propellant granules clinging to the inside of the drop tube or base.

Repeat this for each case in the block, paying particular attention that you *do not* put more than one charge in any case and that you *do not* skip a case and leave it uncharged. These errors happen more easily than you might think, especially if you allow any outside activity to intrude and distract you during this operation. An instant's distraction can easily cause a double charge, and if not spotted subsequently, it can blow up a gun. Let's add one more safety factor: After every, say, tenth charge, weigh that charge to insure the measure is still thowing those charges accurately. If the charge is off by more than, say, 3/10 of a grain, recheck your measure (as before) and adjust accordingly.

Before seating bullets—preferably immediately after charging—propellant charges should be closely inspected under good light to

This is what we call the ''Mobile Measure,'' and when this method of charging handgun-caliber cases is employed, you can save a great deal of time and reduce the probability of propellant spillage. A loading block filled with primed and processed cases is set out, and a compact fixed-charge pistol measure is moved from case-to-case, actuating the metering drum over each. Though some might think that this method gives less uniform charges, I have never seen any indication that it does.

ascertain that all of them bulk up approximately the same in all of the cases. This is easily done with a filled loading block, angling it slightly toward a strong light, and looking inside the mouth of each case. With just a little experience, you'll be able to easily spot any charges that bulk significantly more or less than the norm. Any such cases should be emptied and recharged, and if the differences in level still exists, the cases should be checked closely to see if their internal capacity differs from the rest, or whether they might contain some foreign material. Spillage aside, if one case shows less than the normal amount of propellant, the one charged after it probably contains an excess. Failure of a full charge to drop into a case often means that because of static electricity or some obstruction or contamination, part of it remained in the drop tube—with the result that it probably combined with the next charge to put an excessive amount in the following case. If you're loading relatively large charges, and you should accidentally "double-charge" the case, those double charges may well overflow the case and make themselves quite obvious. At the other end of the scale, if you're assembling light target loads, the prescribed charge may occupy no more than one-quarter of the case volume. This makes it difficult to spot double charges in long slender cases like the .357 Magnum; on the other hand, it's easy to spot these problems in something like the stubby .45 ACP. No matter whether it's hard or easy *you must* inspect closely for double charges—*it can* happen to anyone. Even after several decades of handloading, I still discovered one not too long ago. Nobody *does everything right* all the time. No matter how sure we are of ourselves, we still have to double-check.

When propellant charges are relatively large for the case volume (as in full-charge loads in .380 and .45 ACP), spillage is a distinct probability. Merely setting the filled loading block down a bit hard on the bench can cause propellant to slop or splash out of the cases. This is especially likely to happen with ball-type propellants which slosh around almost like a liquid. In fact, they're worse than a liquid which has surface tension to help hold it in place. I've been frustrated many times when an inadvertent jar slopped propellant out of half the cases in the loading block. Once the stuff is spilled, you have no choice but to dump everything and recharge all the cases. There's no accurate way to determine just which cases lost propellant, or how much. This same characteristic of ball powders which can be so frustrating in one respect constitutes a distinct advantage when charging the cases. It flows through the measure much more smoothly and uniformly than flake or rod propellants. Given equal performance for the job at hand, you'll be wise to select ball over the other type propellants because of its smooth flowing and metering characteristics. In my ex-

When charging cases with propellant, each and every case should be carefully inspected to insure that the charge bulks up approximately the same in all. Any case appearing to contain significantly more or less than the norm should be emptied and recharged. When a single case is being charged, followed immediately by starting the bullet, the propellant charge should be inspected before the bullet is put in place.

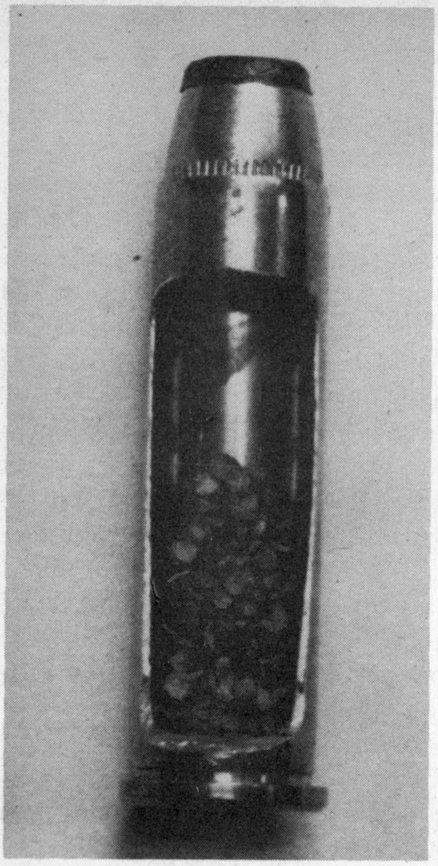

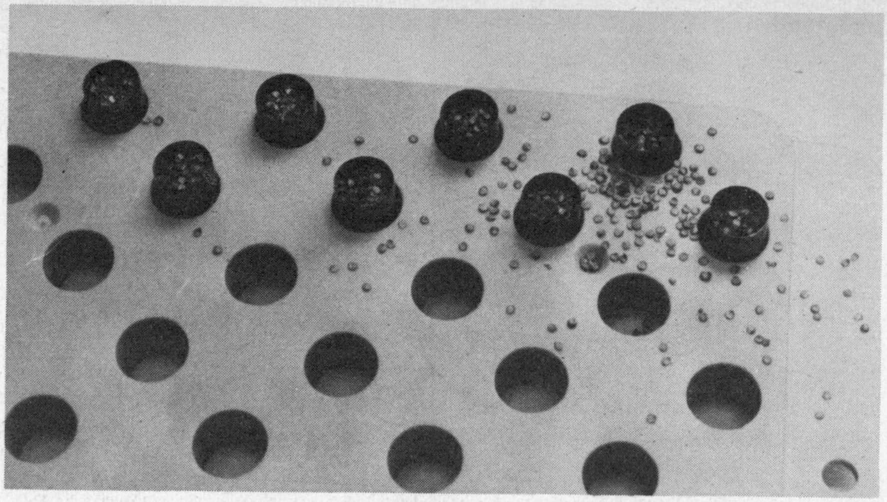

Above — When the required amount of propellant bulks high in the case, spillage becomes probable. If even reasonably good performance is to be expected from these loads, *all* the cases should be emptied and recharged.

Left — This .357 Magnum has had its wall cut away between web and bullet base to show how little space is occupied by a .38 Special-class propellant charge; a full-charge magnum load would require enough propellant to fill the case almost completely. The real problem here is that the case will hold *two* light charges (maybe three) without overflowing; *careful* inspection is needed to avoid such a dangerous condition.

perience, the pistol varieties of ball powder simply *never* bridge or clog or hang up in the drop tube because of static electricity. Incidentally, static electricity can cause problems in just that fashion, especially with the finest flake powders like Bullseye. Anti-static devices are sold by photo supply shops for use on negatives and should work equally well on the plastic drop tubes and reservoirs of powder measures. Grounding the measure is probably your best bet if static electricity is a problem in your area and climate.

BULLET SEATING AND CRIMPING

With the cases charged with propellant, we're coming down to the wire and approaching very closely the time when you can pluck a fully-loaded cartridge out of the shell holder, all ready to be fired. As already mentioned, bullets should be started in the cases immediately after the entire block of cases is charged with propellant. This is to prevent spillage or contamination, but aside from that, best results will be obtained when the bullet is forced tightly into the case by hand and aligned with the axis of the case as accurately as possible before being placed in the die. Bullets that are rather long for their diameter, such as the typical big-bore rifle type, will generally be straightened up in the case by the die and the seating punch without any harm. Since most handgun bullets are short and stubby, they are particularly susceptible to being seated cock-eyed if they aren't started straight. Simply setting the bullet on the case mouth and ramming both into the die won't do the job. Press the bullet tightly into the case mouth and give it a bit of a twist to center and align it; then take a look, and if it isn't straight with the case, apply a bit of corrective pressure.

Screw the seating die into the press until it is about 1/16-inch from the top of the shell holder; then, unlock the seating screw and back it out most of the way. Run an empty case into the die, screw the die down until it contacts the case and then back it off 1/8-turn and lock it temporarily in that position. Now, insert an empty case with bullet in the shell holder and run it into the die and turn the seating screw down until it contacts the bullet. Back the cartridge out and turn the screw in a bit deeper. Keep repeating this until running the case all the way into the die results in the bullet being seated to the correct depth.

Above — In this bullet's-eye view you can see the concave working surface of the bullet-seating stem or screw; this cavity should fit the bullet nose *very* closely.

Right — To speed up the bullet-seating operation and to protect the propellant charges from contamination and/or spillage, bullets should be started in the cases immediately after charging. Bullets should be aligned as accurately as possible in the case, however, the tendency is to tip a few. A slight tipping probably won't cause any trouble, but in this photo several are tipped far enough that they may not straighten up while entering the seating die, and may result in a bulged case or shaved bullet.

Seating Depth

What is the correct depth? Well, some loading data will specify either a loaded cartridge length or a seating depth, the latter meaning the distance to which the bullet enters the case; these two values are easily measured with a vernier caliper. If that information isn't available, there are other ways to go. While lead bullets normally contain a crimping cannelure or groove near the front of the cylindrical bearing surface, jacketed bullets for revolvers normally contain only a knurled cannelure for crimping. With either, seat the bullet sufficiently deep so that the case mouth does everything but cover the groove or cannelure. Seat ungrooved bullets to produce the same overall length as factory loads using the same weight and shape bullet; or, until most of the cylindrical bearing surface is inside the case—just be sure the overall cartridge length isn't excessive. In reality, bullet-seating depth is seldom critical in handgun loads except where overall length is important to functioning or the case is completely filled with propellant and compression of the powder is not desirable.

Once this sample round has the bullet seated to the correct depth, back out the seating screw about 1/8-inch. Then, by trial and error, turn the die deeper in the press until its crimping shoulder produces the desired degree of crimp. The amount of crimp can be important to accuracy and is critical to functional reliability with heavy loads in both revolver and autoloaders. With heavy loads it's usually best to apply a little more crimp than you really think is needed, and with target loads a little less. Once the desired degree of crimp has been obtained, secure the die with its lock ring as already described, and with the cartridge home in the die, turn the seating screw down until it makes solid contact with the bullet. Then, secure that adjustment.

The die is now set up to simultaneously seat bullets and crimp the case upon them. It's a good idea to make a habit of guiding the bullet with your thumb and finger as it enters the die; even though you started the bullets into the cases by hand, it's still wise to guide them into the seating die. Jacketed types, in particular, might be jostled out of alignment by simple handling—they should be guided all the way. If you don't guide the bullet, it may tip enough to strike the edge of the die mouth before you can stop your stroke, thus ruining both the case and bullet. Examine

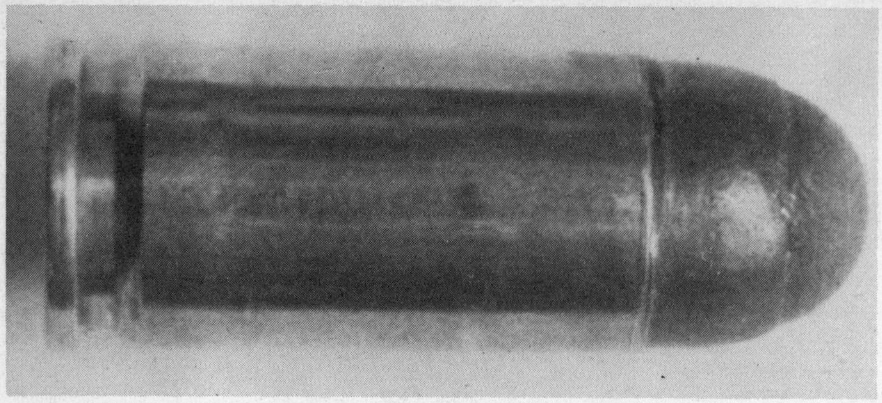

The ring around the nose of this bullet offers a classic example of what happens when a lead or soft jacketed bullet is seated with a seating stem which does not fit the bullet nose correctly. Here a semi-wadcutter cavity was used to seat a round-nose bullet and deformed it as shown. In this particular instance, accuracy probably will not be affected because deformation is both slight and concentric. Nevertheless, an improperly-matched bullet-seating stem certainly *can* influence accuracy and/or feeding.

the first few completed cartridges very closely around the bullet/case junction. There's always the possibility that the case mouth might not be flared quite enough to prevent the shaving of, or digging into, the particular bullet used. In this respect, it's better to have too much than not enough flare even though we've already pointed out the disadvantage of over-flaring. The appearance of minor slivers of lead scraped up by the case mouth isn't terribly important except for ultra-accurate target loads. If a large amount is piled up, the ridge of lead might interfere with the chambering, and the cases require more flaring.

Case Length

Uniform case length has a distinct affect upon the degree of crimp that will be formed by the die, and a similar affect upon the flaring produced by the expansion die. Flaring the crimping can be uniform *only* when case length is uniform. With a given die adjustment, cases longer than the one upon which the die adjustment was based will be flared or crimped more, and, shorter cases less. Generally this doesn't present any particular problem in good-condition cases all from the same manufacturer and/or lot. However, due to different manufacturing tolerances, when makes or types are mixed, length may vary as much as .040-inch or more—and that can play hob with both flare and crimp.

Since the difference between a roll crimp and a taper crimp is difficult to present clearly in a photo, this sketch serves the purpose by emphasizing that difference. At left is a roll crimp showing the bullet and case mouth in section and the inward-turning of the entire perimeter of the case into the bullet. At right is a taper crimp shown in the same fashion, with the taper beginning about one-third of the way up from the bullet base.

Case-length problems can be corrected or accommodated either by trimming all of a given batch of cases to the length of the shortest example, or by simply sorting cases into batches of approximately the same length. Sorting avoids the cost of a case trimmer and the effort of trimming, but it requires readjustment of the seating die to accommodate the different batches. In addition, batches might become mixed after firing and require sorting all over again. Where accuracy is of utmost importance, it is therefore a good idea to carefully select enough cases for the job, then trim them to uniform length, and take steps to insure that they retain their identity thereafter. This will produce the most uniform crimp; and, as long as you keep that lot of cases together, no sorting will be required.

Crimping

What we've said thus far refers specifically to a *roll-type crimp* in which a convex shoulder in the die turns the case mouth into the bullet, forming a smooth radius. This is the type of crimp generally applied to revolver cartridges and only occasionally to autoloader calibers. It is also the type of crimp that's most difficult to apply uniformly, but it is the type that must be applied to heavy revolver loads.

The *taper crimp* is another form which is most applicable to light target loads in revolver calibers and to rimless autoloader cartridges which headspace upon the case mouth. Specific details of the differences in applying the two types of crimps and their effect upon accuracy and other factors are covered in Chapter 13 on accuracy and high-performance loads. The main difference between the two types is in their shape and the fact that the taper crimp must be applied as a separate operation with a separate die. The cavity of a taper-crimp die tapers very slightly inward toward the top, and when the case with fully-seated bullet is forced into it, the forward segment of the case mouth is squeezed inward on a straight taper. No bullet crimping groove is required and, in fact, the case is forced into the bullet surface to a slight degree. Thus, the case mouth remains square and can still serve its function to provide a headspacing surface.

A roll crimp may also be applied as a separate operation and while it isn't generally considered necessary, I prefer doing it this way. To do so, the die is first adjusted as described for seating the bullet and all cartridges are run through it. Then, the die is readjusted to crimp the case correctly, but unlike before, the seating screw is backed off (or removed) so that it cannot contact the bullet. Then, all the cases with bullets seated are run through it, and crimping is complete. As we mention elsewhere, four-die sets are available for straight handgun calibers containing separate seating and crimping dies, the latter available for either roll or taper crimp. For the few dollars extra such a set costs, one can crimp separately without continual readjustment of the seat/crimp die. However, the four-die set *does not* have the capability of seating and crimping simultaneously—the choice is yours.

Seating and crimping complete the mechanical processes of loading the cartridge. I suppose one might consider them ready for shooting—

This bullet has been properly seated in the cartridge case (which has been properly flared) and is now ready for crimping as a separate operation. Note the crimping groove in the bullet at right and the extent to which the case mouth covers it in preparation for crimping.

At *left*, a case inadequately flared so that the bullet cannot be started into the mouth manually; *center*, a case properly flared, and with the bullet accurately and correctly started for seating; right, a case excessively flared, but not sufficiently so that the mouth will be damaged in entering the seating die. It's wise to remember, though, that excessive flaring of the mouth will reduce case life.

This photo shows us two factors important to accurate seating of bullets. First, the case mouth is flared just barely enough to permit the bullet to enter without shaving; second, the bullet is accurately aligned with the case so that it will enter smoothly without bulging the case or receiving any damage of its own.

nevertheless, they should be carefully inspected for any defects produced by the reloading process or overlooked in previous component inspection. Then, particularly if they are lead-bullet loads, they should be cleaned to remove all traces of lubricant, lead shavings, finger prints, etc. Now's the time to wipe or tumble that newly made ammo in a solvent-moist towel.

If you're assembling a *new* load (meaning one you've not tried before) for a particular purpose, I'd recommend functional testing before depending upon it. This is especially true where autoloaders are concerned, but it also applies to revolvers. The best way to do this is to plan ahead for your functional testing by loading only a small batch for the test shooting. Then, if everything checks out okay, proceed to load whatever quantity you require. It would be a bit irritating to load 500 rounds and then discover by live-firing that they do not function as intended. You're then stuck with a large quanity of ammunition that may be acceptable for some purposes, but not the one for which it was loaded.

PACKAGING AND LABELING

The last reloading step I consider necessary is most often overlooked—that of carefully packaging and labeling your handloads so that their identity is clear and the ammo itself is protected from dust, dirt, grit, moisture or whatever outside influences might exist. Any kind of label which identifies the case, propellant, bullet type and weight will be satisfactory. Some people also like to include the number of times the case has been fired, muzzle velocity, the particular gun for which it might be loaded and, perhaps the primer type as well. The ideal container, particularly if the ammunition is to be carried around in an automobile for a while, is a molded, plastic, compartmented cartridge box for the particular caliber. It protects the cartridges from one another and, to a large degree, from any outside impact. While these boxes are not completely dust or moisture-proof unless sealed with tape, they are far better than the typical cardboard/plastic factory-load carton. If you can't justify the cost (which isn't really all that great for the quantity you need to carry ammunition afield) ordinary, transparent, zip-lock plastic bags large enough to hold 50 rounds or so will do very nicely. Dump in the cartridges, slip the label inside the bag, seal it up and your ammunition is ready for storage or use.

BASIC RELOADING OF MILITARY CARTRIDGES

Early in this chapter we indicated that fired military cases should be set aside for special treatment. Generally, such ammunition has a ring crimp applied around the primer after it is seated. You might be lucky enough to run into brass which has not been treated in this fashion, but don't count on it. In autoloader cartridges this crimp is applied to prevent the primer from backing out when fired in unlocked-breech automatic weapons (normally submachine guns or machine pistols). Excess chamber and bore wear, excess headspace and incorrect interchange of parts in the field, can cause primers to back out upon firing, perhaps falling into the gun's mechanism causing a malfunction. The ring crimp prevents this to a large degree and is generally specified in military contracts. The reason for this crimp is clear in autoloader cartridges, but I've yet to find a valid reason for crimping the primers in revolver cases. Nevertheless, most military revolver ammunition does have the crimp. I suspect it is applied simply because the basic specifications for military small arms ammunition call for it, and the fact that it can serve no useful purpose in a revolver is academic as long as the specification is in effect.

The ring crimp we've just described is almost invariably used in U.S. military ammunition and

Case and bullet should be held in alignment between thumb and finger in this manner and guided into the mouth of the bullet-seating die. If this isn't done, the bullet may tip slightly and snag the die mouth, or the entire assembly may shift in the shell holder and cause the same problem.

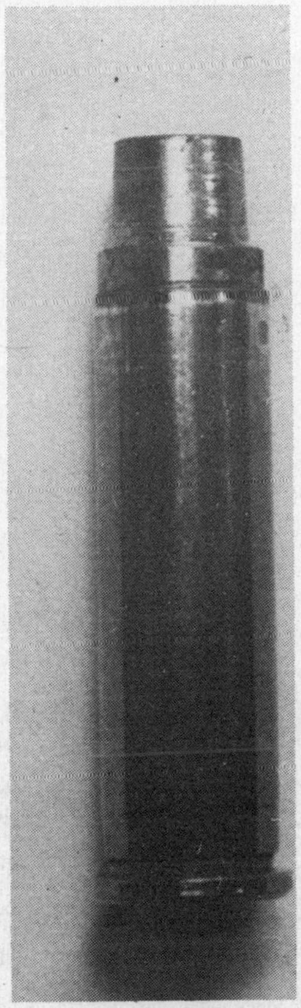

Left — From the preceeding photographs, this case has been seated and properly crimped upon the bullet. If you've followed the bullet seating routine as outlined, this is the type of round your efforts will produce — no bulges, no wrinkles, etc.

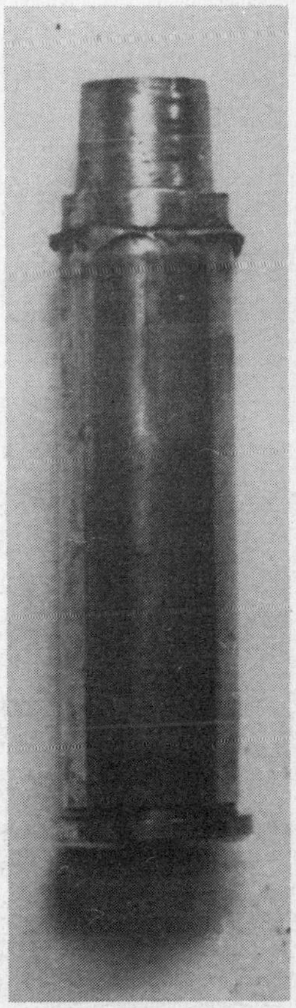

Right — If your results look like this, it probably comes from attempting to seat the bullet in a case that has not been adequately flared. Note that the case mouth sheared off a ring of lead as it was forced over the bullet, and that lead now prevents chambering of the cartridge. The protruding lead can be peeled off easily enough, but accuracy will normally suffer because of the bullet deformation it produced.

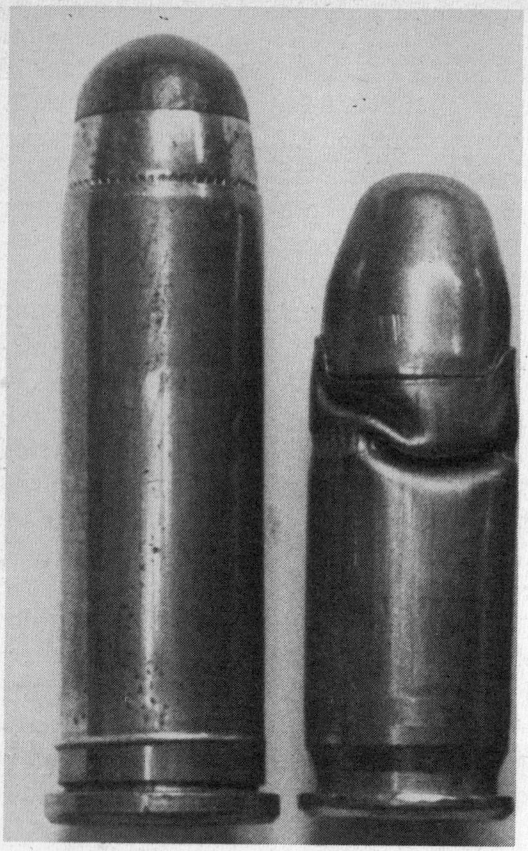

Left — The .38 Special case at left was damaged — note the ring of brass ahead of the rim — because dirt in the rear of the shell holder recess prevented it from seating fully home. In this instance, the ridge of brass is not quite sufficient to prevent the round from being chambered and fired safely. The second cartridge is a .38 Super loaded with a 90-grain JHP bullet which was not properly started and overlapped the edge of the mouth sufficiently to collapse the case as seen. It's unsafe to fire and should be discarded immediately. The alert handloader should constantly be on the lookout for these types of problems during the final inspection of his reloads.

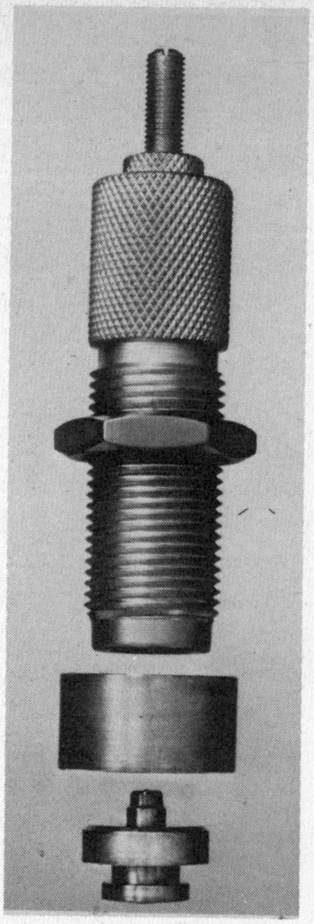

Right — This is the primer-pocket swaging tool made by RCBS, for use in a conventional loading press, to displace (rather than remove) the military primer crimp. In this author's experience, this particular setup produces the most consistent results when military crimp must be removed to permit reloading of fired cases.

will be encountered to a lesser degree in that of foreign manufacture. More common on foreign ammunition is a stab crimp. It is formed by triangular punches striking the case head just outside the primer pocket mouth and turning a very narrow lip over the edge of the primer. Usually there will be three stabs, though other numbers may be encountered.

Removing crimped primers will often require some additional decapping effort, but not enough to present any problem. After decapping, the crimp must be removed or forced back out of the way before a fresh primer can be seated correctly. Attempting to seat a new primer without removing the crimp can result in a crumpled primer, fractured pellet, or in severe instances, detonation of the primer to the hazard of the operator. The crimp can be cut away with a very simple and economical "primer pocket reamer" made in both standard primer diameters by Lyman. As an alternative a sharp, slender knifepoint can be used to peel the entire crimp out in a single, thin shaving of brass. Faster and more convenient to use is a primer pocket swage, typified by the unit supplied by RCBS. It consists of a die-like holder which screws into the press and contains a rod which enters and supports the case. This rod, coupled with a small button swage which enters the primer pocket, forces the crimp back into the case head and smoothly radiuses the mouth of the pocket. Cases may be processed very rapidly on an ordinary loading press with this tool, and it does a more uniform job than either of the cutting methods. It has the added advantage of making it impossible to enlarge the mouth of the pocket so much that it doesn't support the primer correctly. Incorrect use of a reamer or knife can cause that problem. Once the crimp has been removed, by whatever method, military cases are primed in the conventional manner.

The foregoing constitute the basic operations of handloading. Naturally, as you get deeper into the subject there will be variations and extensions of them, and even some exceptions and perversions. As we cover the assembly of various types of loads in this book, you'll find more detailed descriptions and instructions for the variations. Even so, you need to understand the basic operations. Any problems you might encounter later may well be resolved by going back to the basics.

CHAPTER 7

Propellant Selection

YOU'RE PROBABLY passably familiar with automobiles. At least you are familiar enough with them to know that they require fuel in the form of gasoline and that whether you drive one of the most recent mileage-stretching economy cars or one of the fire-breathing muscle wagons of the '60's, the fuel must be matched to the engine and drive train. Fill the tank of your '67 muscle car with today's government/environmentalist-dictated, no-lead regular gasoline, and it will barely stutter and stumble to life. Even worse, it can damage that howling 389 or 409 engine, the likes of which will never be seen again. Turn that around and inadvertently fill the tank of your little four-banger Chevette with fast-disappearing leaded premium, and everything will go to hell in a handbasket. First, the catalytic converters will go, and after a while the engine will get progressively sicker, and in no way could it deliver that wonderful gas mileage those big four-color ads brag about.

As we have described elsewhere in this book, your handgun is basically a heat engine, just like the engine in any automobile. Consequently, it must burn fuel in order to do work, and that fuel is the nitrocellulose compound known technically as "small arms propellant" which we most commonly (and erroneously) refer to as "powder" or "gunpowder." It would be both more technically correct and descriptive if we referred to this fuel as "propellant" or "propellant powder," but the habits of many generations aren't overcome that easily.

In any event, to achieve efficient fuse of the propellant and to produce optimum ballistic performance, the propellant for a given load must be carefully selected with consideration being given to many factors. The most important of those factors are: bullet type and weight; the velocity (and therefore energy) levels desired; the maximum allowable pressures for the particular gun/cartridge combination; the volume or capacity of the cartridge case as it relates to diameter and weight of bullet; the type of handgun involved; and the length of the barrel. Less important but nevertheless worthwhile considerations are: muzzle flash, blast-and-report; the amount of residue (ash) remaining after combustion; propellant cost per shot; ease of ignition; the manner in which the propellant flows through volumetric measuring devices and its effect upon uniformity of charge weight. It should, at this

In a pinch, virtually all levels of handgun loads except full-charge magnums can be assembled with these two propellents: Hercules Bullseye or Unique. At some loss of efficiency in the smallest calibers, the job could be done with Unique alone.

point, be mentioned that while there are many *brands* of ''powder,'' each brand uses *one* of three *types* of propellant: *rod, flake* or *ball*. ''Rod'' type propellant has an extruded, rod-like shape; ''flake'' propellant usually has the appearance of paper-thin, small circular discs while ''ball'' propellant is round, resembling tiny BB's. Of all three types, the ball powder flows through a powder measure the best — almost like water. Obviously, some *very* uniform charges can be thrown with ball propellant.

As a practical matter, most of the more popular handgun-cartridge propellants are of the double-base type, and they differ only very slightly in energy content. Put another way, they all give off pretty much the same amount of energy per unit of weight in the form of heat and gases when burned. Incidentally, a double-base propellant is one which consists of nitrocellulose but which has a specified amount of nitroglycerine added; single-base propellants consist of nitrocellulose alone, without nitroglycerine. Of course, other additives exist in very small quantities, but generally the purposes they accomplish are not of particular interest to the handloader. Technically, double-base propellants do have a slightly higher energy content than the single-base types. Chemically, all single-base types are virtually identical except for the minute amounts of additives; double-base types differ mainly in the percentage of nitroglycerine.

The single most important factor in choosing a propellant is its *burning rate*. Burning rate is not an absolute value, but rather a comparison of different propellants. Thus it may be said propellant B burns less rapidly than propellant A, and propellant C burns less rapidly than B, and so on through the entire list.

The term ''relative quickness'' has been applied by some writers rather than ''burning rate.'' Actually, this seems to me a more descriptive term, and its use might make understanding the problem of propellant selection a bit easier. Since there are many propellant manufacturers and each develops propellants to suit particular needs (and performs its own tests), there is no single listing which places all propellants in precise order of burning rates. Various individuals have from time to time constructed lists, but none are entirely complete nor do they necessarily agree because of different experiences and data

Relative Quickness of Propellants

Rank	Powder
1	"Bullseye" Hercules (1)
2	#230P WW (1)
3	Super M Balistite (3)
4	#6 Du Pont* (1)
5	AL-101 Alcan* (3)
6	"Acapania"* (3)
7	"Nike"*(3)
8	#450-LS WW (3)
9	"MB"* (3)
10	AL-120 Alcan (3)
11	"Top Mark" Hodgdon (2)
12	"Red Dot" Hercules (2)
13	"Trap No. 14" Hodgdon (3)
14	"Hi-Skor" Shotgun & Pistol Du Pont* (2)
15	"Hi-Skor" #700X Du Pont (Double Base) (2)
16	"Green Dot" Hercules (3)
17	P-5066 Du Pont* (1)
18	PB ("PB-6") Du Pont (2)
19	"Bulk" Du Pont* (3)
20	"Unique" Hercules (2)
21	SR 7625 Du Pont (3)
22	AL-5 Alcan (2)
23	#500-HS WW (3)
24	HS5 Hodgdon (2)
25	AL-7 Alcan (2)
26	#540-MS WW (2)
27	HS6 Hodgdon (2)
28	"Herco" Hercules (2)
29	SR 4756 Du Pont (2)
30	AL-8 Alcan (2)
31	R-23 Norma*
32	H240 Hodgdon* (2)
33	H110 Hodgdon (3)
34	"Blue Dot" Hercules (3)
35	#2400 Hercules (2)
36	#295 HP WW* (2)
37	SR 4759 Du Pont (3)
38	H4227 Hodgdon (3)
39	IMR-4227 Du Pont (3)
40	"Hornet" Nobel (3)
41	N-200 Norma (2)

Notes: The number following each powder designates: (1) Powders originally developed or intended for handgun use; (2) Powders made for other purposes but widely used in handgun loads over the past few decades; (3) Powders generally used for other purposes but do have *some* utility in certain types of handgun loads.

*Production discontinued

input. One such list was compiled a number of years ago by Dean Grennell for the now-defunct "American Reloaders Association," and is shown here. You'll note that it does not include the current series of W-W Ball Propellants but is otherwise generally complete insofar as those propellants suitable for handgun loads are concerned.

Take note that the foregoing table was originally much longer, but we have deleted those higher-numbered propellants which are not suitable for any practical handgun usage. Only those propellants designated (#1) were originally developed or intended for handgun use, while those designated (#2) were made for other purposes but have become widely used in handgun loads over the past few decades. Those with a (#3) are generally used for other purposes, but do have *some* utility in certain types of handgun loads.

The fastest-burning propellant shown on the table is Hercules Bullseye. Its principal use is for full-charge loads in small-capacity cases of small caliber (under .38/.357) at chamber pressures well under 20,000 CUP. Thus, it is virtually a standard propellant for full-charge loads in auto cartridges such as .25, .32, and .380 ACP and for full-charge loads in revolver cartridges up through .38 S&W caliber. Its secondary use is for *light* target or plinking-type loads with lead bullets in virtually *all* handgun calibers. In this role it is almost the universal choice among handloaders for super-accurate, mid-range loads in .38 Special and .45 ACP. It is definitely *not* for full-charge loads in the larger magnum (or near-magnum-type) cases with *heavier* bullets.

Full-charge loads in .38 Special (and heavier calibers) require a slower-burning propellant typified by Hercules Unique, No. 20 on the list. Exceptions to this would be a few of the old black powder, low-velocity, large-caliber loadings such as the .450 Boxer and .455 Enfield. Their established velocity levels, even with relatively large and heavy bullets, are so low that the use of the fastest-burning propellants is safe and acceptable.

The foregoing demonstrates the basic principle of propellant selection—that as bullet diameter and weight, case volume and velocity level increase, the optimum propellant burning rate decreases. Put another way, the smaller cases in smaller calibers with light bullets and short barrels require the fastest burning powders; and the

Hercules Unique powder is especially versatile in the hands of a perceptive handgun handloader. At left is the .32 S&W Long cartridge and at right the .44 Magnum; in between a small sample of extruded Unique which is sliced into thin, solid, round wafers from rods or strings. Unique allows the .32 to be safely loaded to performance levels above that of factory ammunition, and in the .44 it will safely produce loads falling only slightly below the full-house magnums employing much slower burning propellants.

largest calibers with heaviest bullets and high velocities in long barrels require the slowest-burning propellants.

While we can have an almost infinite variety of loads, it's obvious that we cannot have an infinite variety of propellant burning rates. Fortunately, the propellants that are available today (and there are over 40 that are suitable for one type or another of handgun load) are remarkably versatile. One can, in fact, produce a wide variety of full-charge or light loads in *all* handgun calibers (everything from .25 ACP through .44 Magnum), utilizing only three propellants: Bullseye, Unique, and 2400 (all from Hercules). These three powders cover the entire range though some loads in the magnum cases would be more efficient if a slightly faster-burning propellant (than 2400) were available for use with the lighter bullets. For over 40 years 2400 has been used quite extensively in the magnum cases with all types of bullets and has proven to be remarkably adaptable. However, it burns just a wee bit slower than is desirable for optimum efficiency with anything but the heaviest bullets in the magnums. Bullseye and Unique will equal or exceed factory-load performance levels in all the non-magnum calibers and will also perform quite well in light loads.

A number of years ago, Lee E. Jurras, then president of Super Vel Cartridge Corporation, extensively investigated the factors involved in selection of propellants for typical handgun loads. This investigation involved the firing of thousands of rounds for both pressure and velocity in a modern, well-equipped ballistics laboratory. An outgrowth of all of this work was a construction of a "Handgun Powder Selection Table" which indicates the most efficient propellants for some loads in the more commonly handloaded calibers. This table clearly indicates those propellants which are most efficient with particular bullet weights in specified velocity ranges, assuming standard test barrel length. The table is not entirely current, since 230P propellant has since been replaced by W-W 231, and W-W 296 performs somewhat better than H110 in the load range listed for it. Admittedly it would be ideal if this table could be thoroughly updated at this time, but to do so would require quite a few thousands of dollars be spent in the laboratory—needless to say, this scribe is not in a position to do that.

As we've indicated, *you can* cover the entire

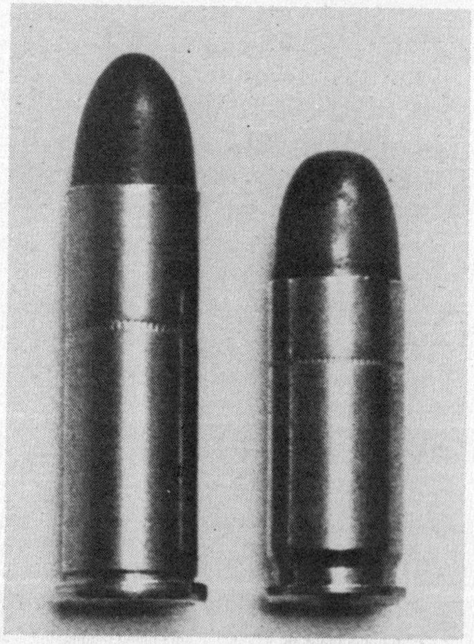

Here we have the typical effect of propellant upon cartridge design: Left is the .38 Special, alongside the .38 ACP. Both were designed at about the same time; the .38 Special to accommodate black powder (but with the capability of employing smokeless); the .38 ACP for smokeless only to deliver much higher velocities and chamber pressures. Note that the latter caliber employs a case of much less capacity due to the higher energy content of smokeless propellant.

HANDGUN POWDER SELECTION TABLE

Caliber	Bullseye	230P	Unique	H110	2400
.380 ACP (4″ barrel)	80-90 gr. bullets at 900-1000 fps	90-110 gr. bullets at 800-850 fps	110-125 gr. bullets at 700-775 fps	NA[3]	NA[3]
9mm Para. (4″ barrel)	80-90 gr. bullets at 1200-1400 fps	90-110 gr. bullets at 1300-1400 fps	110-130 gr. bullets at 1100-1300 fps 90-110 gr. bullets at 1300-1400 fps	NA[3]	NA[3]
.38 Super[4] (5″ barrel)	NA[2]	80-90 gr. bullets at 1400-1500 fps	90-110 gr. bullets at 1400-1500 fps 110-130 gr. bullets at 1200-1300 fps	NA[3]	NA[3]
.38 Special (6″ barrel)	145-160 gr. bullets at 700-800 fps[1]	95-110 gr. bullets at 900-1000 fps	95-110 gr. bullets at 1200-1350 fps 125-135 gr. bullets at 1100-1200 fps	NA[3]	NA[3]
.357 Magnum (8⅜″ barrel)	NA[2]	NA[2]	110-135 gr. bullets at 1400-1600 fps	125-160 gr. bullets at 1200-1600 fps	137-170 gr. bullets at 1200-1500 fps
.44 Magnum (8⅜″ barrel)	NA[2]	NA[2]	200-250 gr. bullets at 900-1000 fps	180-240 gr. bullets at 1400-1800 fps	240-260 gr. bullets at 1400-1500 fps 180-240 gr. bullets at 1400-1700 fps
.45 ACP (5″ barrel)	185-230 gr. bullets at 800-850 fps[1]	NA[2]	185-230 gr. bullets at 900-1050 fps	NA[3]	NA[3]
.45 Colt (6″ barrel)	NA[2]	NA[2]	240-260 gr. bullets at 900-1000 fps[5]	NA[3]	250-300 gr. bullets at 900-1000 fps[5]

NOTES:

NA Not Applicable.
1 Not full charge load; only to show use of fast Bullseye in mid-range target loads.
2 Powders burning too fast for practical curve, but may be used for reduced or mid-range target loads *only*.
3 Burning rate too slow to produce acceptably uniform velocities.
4 Data applicable to 9mm Bergmann-Bayard, and to 9mm Steyr when used in locked-breech guns.
5 For use only in Ruger SA revolver.

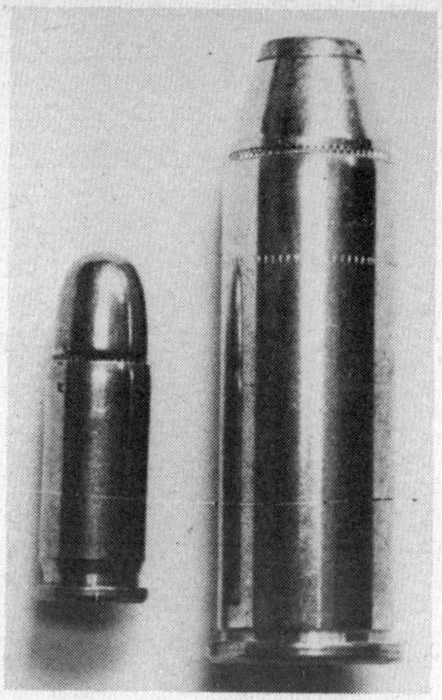

Left—If you're limited to just three propellants (Bullseye, Unique and 2400), you can actually load the entire range of handgun cartridges from the .25 ACP (left) up through the .44 Magnum.

Below, left—Comparison of early handgun propellants (still with us) and the black powder propellant which they replaced. Left to right, Hercules Bullseye, Hercules Unique, and FFFg Black.

Below—Left is Hercules Bullseye extruded into rods and sliced into very thin wafers; right is W-W 230, technically a ball propellant, but rolled into thin flakes. Bullseye is the fastest burning of the extruded handgun propellants; 230 burns fastest of the ball propellants.

range of handgun loading with only three propellants. From a study of this table, it becomes obvious that by substituting 231 for 230P and replacing H110 with 296, a total of five propellants will offer us an *optimum* choice for full-charge and high-performance loads in all standard calibers from .380 ACP upward. Even that doesn't tell the entire story, for the faster-burning propellants listed in the table perform equally well in those calibers beneath .380. Quite simply, it boils down to this—that unless you simply want to experiment, for your own edification and without particular hope of achieving better results, the most you'll ever need to cover all handgun calibers is five of the 40-odd propellants that are generally considered suitable. Further, if you confine yourself to those five propellants, your loads will most likely be more efficient than if you go picking and choosing among the others.

Note again that in the table the velocities quoted are based upon standard-length test barrels. When employing shorter barrels, particularly in the magnum calibers, velocities will be a good deal less. Probably the most outstanding example of gross velocity loss in shorter barrels is to be found in the .357 Magnum loaded with 2400 propellant. With the *heavier* bullets, this combination does perform efficiently in the long 8⅜-inch magnum barrel; however, cutting back to 6-inch barrel length produces quite a significant velocity loss, and further reduction to 4 inches produces performance not a great deal better than the .38 Special in the same length. Thus, 2400 (when used with the heavier bullets) would be probably the poorest choice of propellants for the .357 Magnum with a 4- or 6-inch barrel. This problem is overcome by going to a faster-burning propellant, specifically Unique. Unique will produce a bit less velocity (with heavier bullets) in the 8⅜-inch barrel (than 2400), "peaking out" in the 6-inch length. In fact, a given charge of Unique may produce *less* velocity in the long barrel than in the 6-inch tube. Where the faster Unique shines, in this caliber, is in producing relatively little velocity loss when dropping from a 6-inch to a 4-inch barrel. Keeping this in mind, it becomes apparent that 2400 is the best propellant to use for heavy bullets in the longest barrels; so, if that's all you'll be shooting, it's a good choice. But, if you want a general-purpose load which produces perfectly acceptable velocities in the long barrels but which will be used extensively in the shorter barrels, Unique is a far better choice. If you're one of those people who feels he must shoot a 2½-inch .357 Magnum, then even Unique burns a bit too slowly; and, "quicker" Green Dot would be a better choice for velocity. However, Green-Dot loads would then peak out in

Many people not thoroughly familiar with the subject seem to think that so-called "ball" propellants are truly spherical in form. Actually, it doesn't work out quite that way; at left is W-W 230 and at right is W-W 630P. Both are classed as "ball" propellants, and both are manufactured by the same methods. They differ in that 230 is manufactured in spherical form, but is then run between rollers which flatten the individual granules to irregular flakes. This gives the individual granules much more area and thus increases their burning rate, even though their basic energy content differs little. W-W 630 and other ball propellants are also run between rollers, but with a much wider spacing, so that only oversize granules are flattened slightly. As a result, the burning rate is slower.

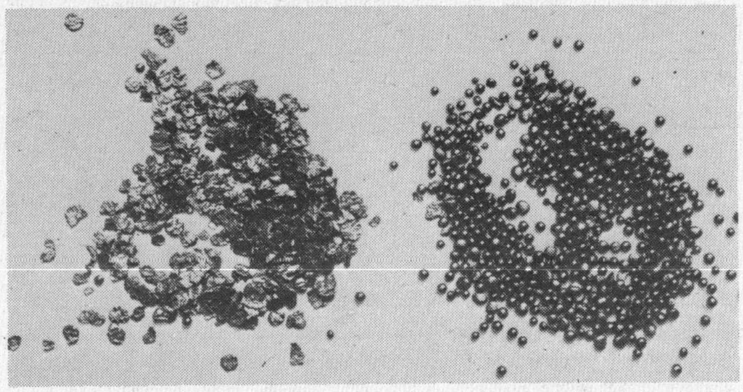

Long-obsolete, small capacity, low-velocity, black-powder cartridges such as (l to r) .450 Revolver, .455 Mark II and .455 Eley (Colt) are readily loaded to their full factory performance with our quickest burning propellants such as Bullseye and 230.

barrels less than 6 inches in length, and velocities would fall off substantially in the longest barrels.

The foregoing should make it clear then that Unique propellant is the best choice for use in the widest range of barrel lengths. It would not, however, be the best choice if your load is to be used primarily in very short or very long barrels. As a handloader, you have the option of selecting a propellant that best matches a particular use; on the other hand, the ammunition manufacturer does not have this option because his product will be used in all barrel lengths and therefore he must settle on a compromise propellant. That is just one more very good reason for handloading.

A similar situation exists in all calibers for which guns are available in widely-varying barrel lengths. Generally this is most applicable to revolver calibers inasmuch as the majority of autoloaders vary relatively little in barrel length. For example, of the many guns made in .45 and .38 ACP calibers, barrel length ranges nominally from 4 inches to 5 inches. On the other hand, most revolvers are offered in lengths ranging from 2 inches to at least 6 inches, and often more. Fortunately, the effect of barrel length on velocity is quantitatively a good deal less in other calibers than with the .357 Magnum mentioned. Even the .41 and .44 Magnums are less critical in this respect. The lower-intensity calibers such as .38 Special, .38 S&W and .32 S&W Long, are less affected by barrel length, therefore propellant burning rate is less important.

We've already indicated that as a practical matter you can get by with as few as three propellants in a pinch and that by expanding this to five, you can cover virtually any real or imagined need in loading handgun calibers. Yet, when you pick up any of the well-known loading manuals, you'll find as many as 12 different propellants listed for a single bullet weight and type in a single caliber. This doesn't mean that I am wrong or that the manual is wrong; it is simply the result of the manual publisher's desire to offer as many *safe* (if not necessarily efficient) loads as possible. At least partially, the publisher means to give the handloader a variety of alternate choices in the event he cannot obtain or for some reason doesn't want to use a "standard" propellant. After all, even though Unique is in my opinion the most efficient propellant for use in the .38 Super Auto, Herco performs very nearly as well with its only very slightly slower burning rate, and if it's available, there's no reason *not* to use it. If some of one's other loading operations cause him to keep Herco on hand and not Unique, then why not use it and avoid the cost and inconvenience of maintaining a stock of Unique at over $8 per pound? The Speer #9 manual lists

Above—This entire range of standard, conventional autoloader cartridges, smallest-and-weakest to largest-and-most-powerful (.25 ACP to .45 ACP), can be loaded safely and with reasonable efficiency using only Unique propellant. Bullseye will do nearly as well for all except for the .38 Super because it burns just a wee bit too fast.

Below—There really isn't any point in getting excited about chosing propellants. This typical page from a Speer loading manual lists more propellant types and charge weights for each bullet weight and type (adaptable to a particular caliber) than the average handloader can even begin to use. One will be far better served to select his propellants and charge weights from such published references, rather than looking elsewhere.

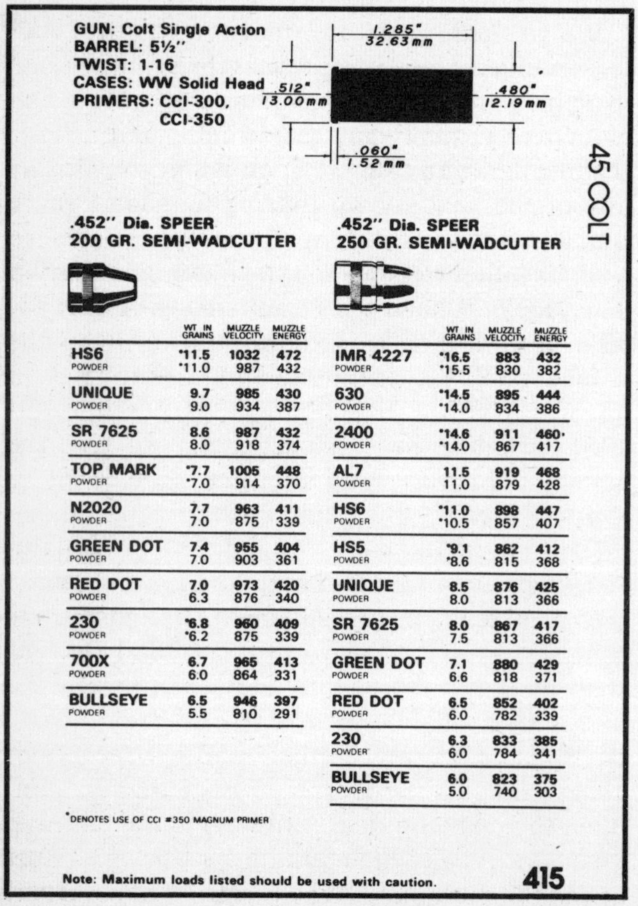

GUN: Colt Single Action
BARREL: 5½"
TWIST: 1-16
CASES: WW Solid Head
PRIMERS: CCI-300, CCI-350

45 COLT

.452" Dia. SPEER
200 GR. SEMI-WADCUTTER

	WT IN GRAINS	MUZZLE VELOCITY	MUZZLE ENERGY
HS6	*11.5	1032	472
POWDER	*11.0	987	432
UNIQUE	9.7	985	430
POWDER	9.0	934	387
SR 7625	8.6	987	432
POWDER	8.0	918	374
TOP MARK	*7.7	1005	448
POWDER	*7.0	914	370
N2020	7.7	963	411
POWDER	7.0	875	339
GREEN DOT	7.4	955	404
POWDER	7.0	903	361
RED DOT	7.0	973	420
POWDER	6.3	876	340
230	*6.8	960	409
POWDER	*6.2	875	339
700X	6.7	965	413
POWDER	6.0	864	331
BULLSEYE	6.5	946	397
POWDER	5.5	810	291

.452" Dia. SPEER
250 GR. SEMI-WADCUTTER

	WT IN GRAINS	MUZZLE VELOCITY	MUZZLE ENERGY
IMR 4227	*16.5	883	432
POWDER	*15.5	830	382
630	*14.5	895	444
POWDER	*14.0	834	386
2400	*14.6	911	460
POWDER	*14.0	867	417
AL7	11.5	919	468
POWDER	11.0	879	428
HS6	*11.0	898	447
POWDER	*10.5	857	407
HS5	*9.1	862	412
POWDER	*8.6	815	368
UNIQUE	8.5	876	425
POWDER	8.0	813	366
SR 7625	8.0	867	417
POWDER	7.5	813	366
GREEN DOT	7.1	880	429
POWDER	6.6	818	371
RED DOT	6.5	852	402
POWDER	6.0	782	339
230	6.3	833	385
POWDER	6.0	784	341
BULLSEYE	6.0	823	375
POWDER	5.0	740	303

*DENOTES USE OF CCI #350 MAGNUM PRIMER

Note: Maximum loads listed should be used with caution. 415

12 propellants for the .38 Super, thus offering a wide variety of alternatives. The slowest burning of those listed is IMR 4227 which is relatively inefficient. In addition a .38 Super case will not hold a sufficient quantity of this propellant to provide full-charge velocity. At the other end of the scale is W-W 230, which actually burns too fast for this caliber and reaches maximum chamber pressures *without* producing full-charge velocities. In between these two numbers—which are inefficient for the caliber—there are several which produce entirely satisfactory performance, even though they might not be quite as efficient in doing so as Unique.

In summary, propellant selection is *very* important to both performance and safety; yet on the other hand, it need not be nearly so difficult as is often believed. Assuming that you want to use a propellant which approaches optimum efficiency in a given caliber with a given bullet weight, then you need look no farther than our table. Select the particular propellant from the Handgun Powder Selection Table, then go to your several sources of loading data and select loads using that propellant. That's the simplest, easiest, and safest way to go.

CHAPTER 8

Bullet Selection

THE HANDGUN IS a wondrous instrument in many ways, possessed of numerous qualities of portability, convenience, speed and ease of handling. It also has great appeal as the most personal of firearms (and weapons) in general use today. However, the gun per se *does no work*. It may buck and roar in your hand and lance forth a great tongue of flame, and its appearance may cause visions of great deeds done by some pistolero ancestor to burst forth. But those things are all icing on the cake — the only part of the entire firearms system that does any really useful work is the bullet (the projectile) which streaks invisibly downrange, possessed of little or great kinetic energy which it transfers to the target upon impact.

In short, insofar as effect-on-target is concerned—and it's only for that effect that a shot is fired—the gun itself is simply a projectile-launching tool, as it were. Its sole purpose in life is to enclose the cartridge and allow it to give the bullet the velocity necessary for its travel and effect and, further, to give the bullet direction so that it will strike its target. Like all firearms, the handgun is, again, simply a launching device for the real star of the show, the *bullet*.

Inasmuch as only the bullet does any real work, it is the most vital part of the entire system. To be sure, we must have the gun to launch the bullet, we must have the primer to ignite the propellant, and we must have the propellant to give the bullet velocity, and the case to tie them all together; nevertheless, all the rest are useless without that leaden projectile which must strike the target to accomplish its mission. This means, quite simply, that for the best results—whether this be the closest possible 10-shot group on a piece of paper, or the instantaneous incapacitation of a large two- or four-legged animal—the bullet must be carefully chosen.

Many factors must be considered in selecting a bullet to meet a particular shooting need. The bullet must first of all be compatible with the type of gun in which it will be fired. For reasons already discussed, shapes and lengths and types of construction which will function perfectly in revolvers and single-shots are often totally unsuitable for autoloading pistols. Revolvers are totally independent of bullet shape, weight and construction insofar as functioning is concerned. This doesn't mean that you can't contrive a bullet which might not work quite right in a revolver,

Externally, the now-discontinued Super Vel, 180-grain, JHP bullet looks as if it could be expected to expand quite efficiently. However, it was originally designed with a thick jacket and hard lead core by Lee Jurras. It was made to penetrate deeply without a great deal of expansion for use on large game such as moose. This emphasizes the fact that you cannot simply presume that a bullet will have certain performance characteristics from its external appearance.

Expanding bullets for autoloaders and revolvers differ substantially. Left is a JHP 9mm number intended for autoloaders; the jacket extends fully forward and rolls over the core slightly so that no lead whatever can touch feeding surfaces in the gun and it has no cannelure. At right is a typical .38/.357 revolver bullet containing a deep cannelure, and a substantial exposure of lead forward of the jacket. Other factors being equal, the exposed lead promotes rapid expansion and the cannelure permits the application of a heavy roll crimp to secure the bullet against recoil forces. The revolver bullet will not normally feed well in autos because of the exposed lead.

but it does mean that all bullets currently available commercially, and all of those for which bullet molds are available, will function correctly in revolvers if of approximately correct diameter for a given caliber. On the other hand, maximum reliability of functioning in *production* autoloaders—and we refer to the broad range of guns, not special-purpose types—requires a bullet generally of round-nose, metal-jacketed type. The autoloader bullet must also be of sufficient length (and therefore a certain weight) to produce a loaded cartridge of the correct overall length for proper feeding. It must also be of sufficient weight to produce, at the velocity desired, a recoil impulse falling in the minimum and maximum range for proper functioning of the gun in question. If not of full-metal-jacketed design and construction, it must at least be made of a material hard enough to withstand feeding impacts without significant deformation.

The bullet must also be chosen for the target upon which it will be used and for the effect it must produce upon that target. Pure target shooting is generally (but not always) practiced on paper targets at formal ranges where adequate impact areas and safety zones are established and enforced. Consequently, a bullet for target use may be chosen without even considering factors of penetration, expansion, ricochet probability, energy transfer, etc. Over the years, the ideal paper target bullet for revolvers has proven to be of relatively soft unjacketed lead, lightly lubricated and of solid-base form when cast, hollow base when swaged. Except for the intrusion of

Back in the late 1800s this is the type of bullet you'd have been forced to use in any cartridge revolver. It is a soft-lead, hollow-base, heel-type bullet dipped in sticky lubricant after being seated in the case. The reduced diameter heel was expected to expand radically to fill the rifling by the action of expanding propellant gases inside the rather large base cavity.

Here we have the entire range of 9mm/.38/.357 bullet weights normally handloaded. L to R: 90-grain 9mm JHP; 110 grain cannelured JHP; and 170 grain case lead SWC. Lighter and heavier bullets have been used in this diameter, but they are not often encountered as they serve to fill limited purposes.

This is a recovered FMJ, military-type bullet which penetrated deeply in test medium but did not expand at all. Note that there is only very slight nose deformation, and that this is by no means enough to increase its lethality. A very poor choice for defense or hunting.

These three bullets all started out as 230-grain, JHP, .45 ACP slugs. They were fired into recover-medium at about 850 fps. Note that they did not expand significantly nor reliably at this velocity. Perhaps they would expand well if driven a good deal faster; but, in the gun for which they were intended, they could not be safely given such velocity.

lubricant-filled grooves, it is of cylindrical form and of the same diameter throughout its length. In autoloaders, a modified semi-wadcutter form has proven best, cast of a very hard lead alloy when made by the handloader, but generally fully jacketed and with a *soft* lead core when made by the factories. This *does not* mean that the jacketed bullet is more accurate than the plain lead bullet, but simply that given current manufacturing technologies, the factories find it *easier* to produce a target-quality bullet of jacketed construction than one of very hard lead alloy.

Bullets for paper target shooting are generally driven at the lowest velocity compatible with the desired degree of accuracy. On the other hand, bullets used for IPSC-type combat matches (and "bowling-pin" matches) are driven at much higher velocities. In IPSC shooting the higher velocity is required (albeit indirectly) because only full-power loads for the caliber are allowed by the rules. In the bowling-pin shoot, there is no stated requirement for high levels of velocity or energy, however, maximum energy transfer is necessary in order for the bullet to reliably knock the target over which is necessary to score. Consequently, while the same shape and weight bullets (as for paper targets) are acceptable, they must withstand higher velocities, and this necessitates *harder* bullets in revolvers; autoloader bullets are sufficiently hard, a requirement for reliable functioning already mentioned.

Long-range metallic silhouette shooting with handguns requires quite a different type of bullet. The basic requirement is that the bullet possess

Bullet weight (and, consequently, length) have an effect upon an autoloader's ability to feed cartridges, especially the .38 Colt Super. Left is the standard 130-grain FMJ service load at a length of nominally 1.28 inches. Next is a shorter cartridge assembled with the discontinued 112-grain JSP, Super Vel bullet (which feeds fairly well); last is the same case loaded with a 90-grain JHP bullet which travels over 1600 fps and is by far the most lethal of the three. Unfortunately this last round is too short for reliable feeding. While it might seem practical to seat shorter bullets to a greater cartridge length and improve feeding, it really isn't; without adequate seating depth, bullet pull values drop and become inconsistent, and both accuracy and velocity fall off.

maximum energy clear out to a 200-meter range and, secondarily, that it transfer *all* of that energy to the impenetrable, thick, steel-plate target in order to knock the target over. A hit alone, no matter how evident, does not score unless the target is knocked off its base. Empirically, it has been determined that bullets of "soft" construction, whether lead or jacketed, are not as effective in transferring energy and knocking over the target (other factors being equal) as a "hard" bullet made with a stiff, thick jacket of open-base type. Further, a moderate round-nose shape and relatively heavy weight for the diameter has proven to provide the best combination of retained velocity and energy, and energy transfer. Only a few such bullets are commercially available and only in .357 and .44 calibers. Components for making such bullets are not generally available; however, some rifle-type bullets may be substituted or modified, and jacketed .45 service bullets can be reduced to .44 caliber and perform quite well.

Plinking and informal target shooting loads permit the widest latitude in bullet selection. Though the various types of jacketed bullets will perform satisfactorily in this role, their cost cannot be justified. Lead bullets reign supreme, and no particular hardness level is required except as

Because of their relatively low velocities, bullets for autoloader cartridges are generally made with nose cavities as large in diameter as feeding will allow. Here on the left we have the Remington 185-grain load showing the short notches that are spaced around the cavity to assist in low-velocity expansion. The .38 Super bullet on the right contains no such notches, but has less need of them because it is generally driven about 400 fps faster than the .45.

At right is the conventional round nose, FMJ, bullet for the 9mm Parabellum cartridge. At left is the highly effective Super-Vel bullet first offered in this caliber in the 1960's. It is of jacketed, hollow-point design, 90-grain weight, and of truncated-cone shape. When the 9mm case is loaded with the lighter bullet to achieve greater velocity, the completed cartridge is of necessity a good deal shorter — this can cause feeding problems because of insufficient cartridge length.

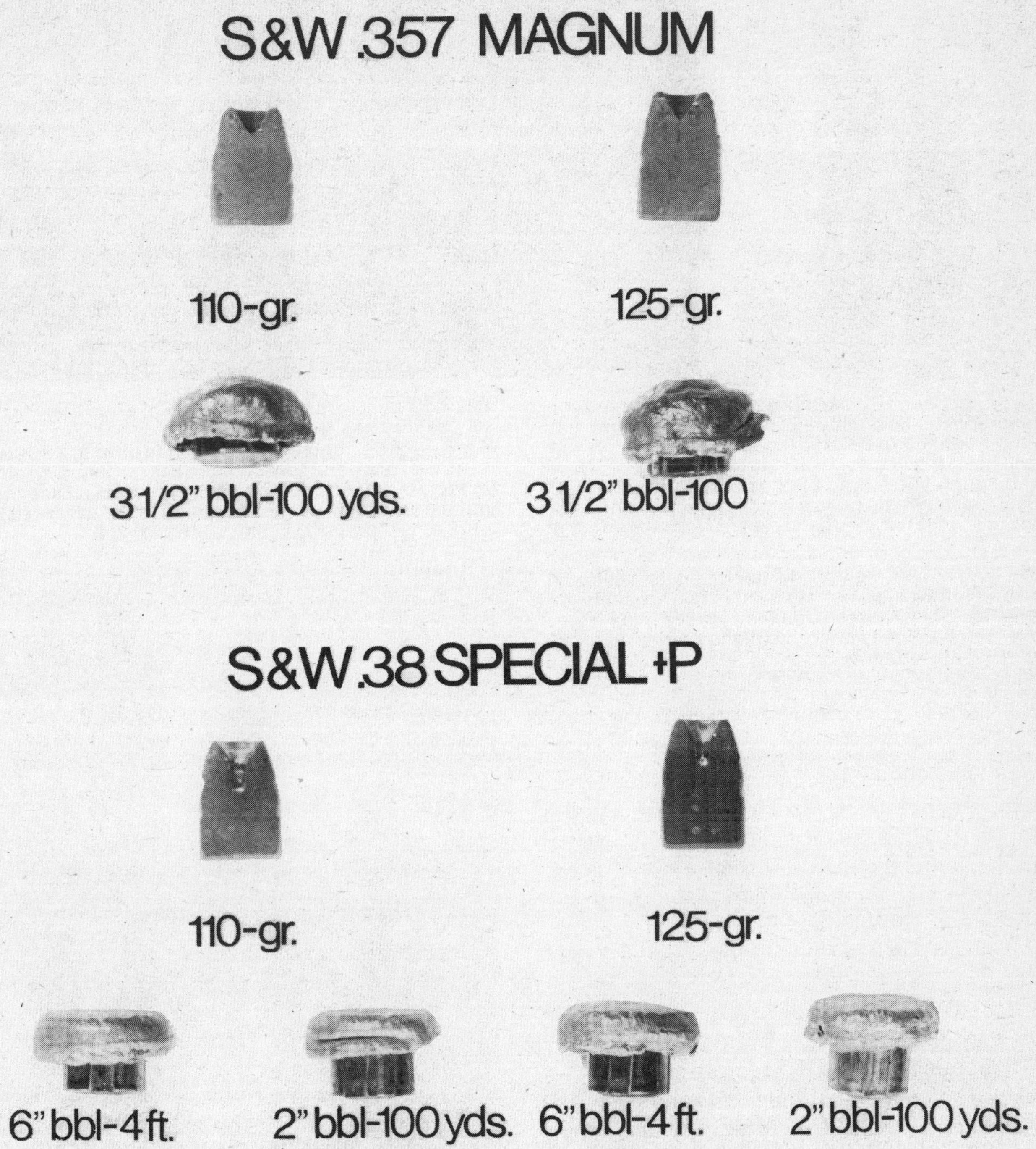

This factory-supplied photo of sectioned and recovered S&W bullets shows classic "mushroom" expansion produced in ballistic test medium intended to approximate the resistance offered by animal tissue. Results such as these are common enough in laboratory tests where conditions are closely controlled, but occur far less often in the field. Only field-testing on game will show what a bullet can really do — and even then the amount of testing one can do is limited and the results will vary a great deal.

Left — At top is the "closed base" design required for JSP and JHP bullets, allowing the core to be inserted from the front and to be exposed so that expansion will develop quickly when the bullet strikes the target. Below it is the "open base" type of construction used in full-jacketed, military-type bullets which require the nose to be entirely covered by the jacket. That requirement makes it necessary that the base be open so that the core may be inserted from the rear.

Right — Photos such as this are often used to emphasize the power of a handgun load, but can be very misleading. Impact on water and thin pieces of wood makes for an impressive photo, but doesn't really tell much about how a bullet will perform in animal tissue and bone.

Right — While most lead bullets handloaders use carry lubricating grooves, some employed in factory loads do not. This 158-grain lead SWC bullet is loaded into Federal's latest high-velocity police load and carries all of its lubricant on the surface.

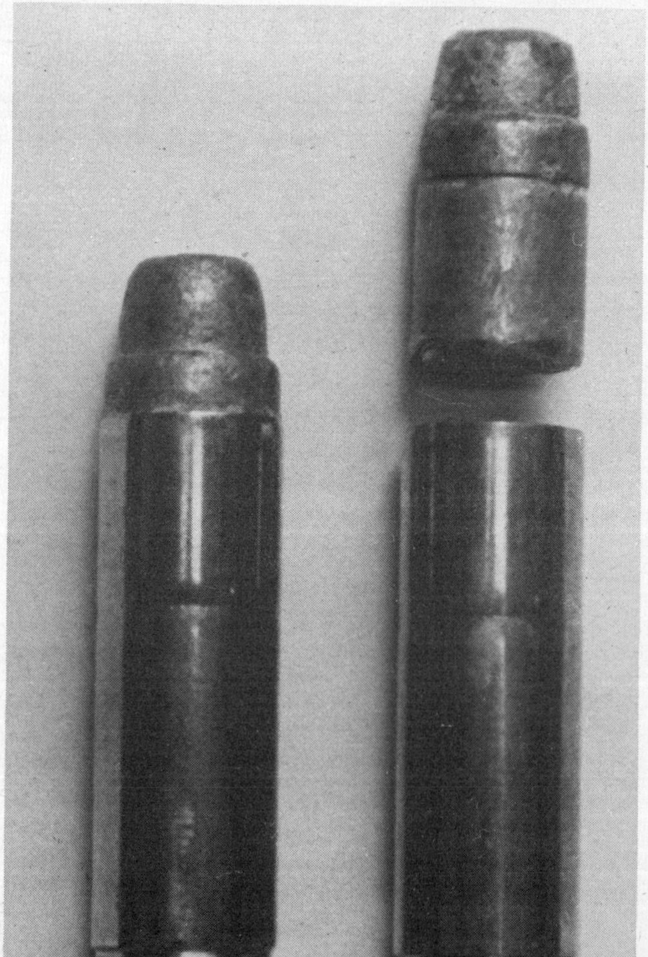

dictated for reliable functioning in autoloaders. Plinking requires no particular degree of expansion or penetration, so the simplest and cheapest forms of cast lead bullets are the wisest choice. Shape and weight are not important except as they relate to autoloader functioning, so the semi-wadcutter in light-to-medium weights is the most common and effective choice.

Particularly in the western states, many pistoleros engage in long-range forms of plinking or casual shooting, and this requires higher velocities and heavier bullets than easterners use for barnyard and creek-bed shooting. For this type of work, the home-cast semi-wadcutter of standard weight for the caliber and driven at 1000-1200 fps is still the best choice. Again, the cost of jacketed bullets can't really be justified, and since factory-swaged lead bullets are generally quite soft (in addition to being expensive) the home-cast projectile of adequate hardness is the best.

Aside from assorted special-purpose loads discussed farther along in this tome, this leaves us with powerful hunting and defense loads for which bullets must be chosen carefully. Actually, hunting loads are not so easily defined because the target may be edible small-game at short ranges, smallish varmints at long ranges, or big-game taken at the shortest range the stalking skill of the hunter will allow. I think we can dispense with bullets for small-game by stating quite simply that bullets best suited for plinking are very good for putting squirrel, rabbit and grouse in the cookpot. No great amount of energy transfer is required nor is deep penetration needed or desired. The cast-lead, semi-wadcutter projectile at moderate velocities works out exceedingly well.

Varmint hunting bullets are required to expand violently on small, light, animal targets such as prairie dogs and woodchucks at relatively long ranges. Generally, this means we should choose a bullet as light as is practical for the cartridge concerned, in order that maximum velocity may be obtained. This will give us the flattest possible trajectory and highest striking velocity. Under these conditions, obtaining maximum accuracy requires a jacketed bullet of soft construction; and, obtaining maximum expansion also requires a very soft construction utilizing a pure lead core and the thinnest and softest jacket practical. For revolvers and single shots, the hollow/soft-point type bullet containing a large nose cavity and exposing a moderate amount of lead core will produce moderate expansion. Autoloaders don't

These obsolete foreign revolver cartridges illustrate the type of bullets employed around the turn of the century. What we use today bears very little resemblance to those old designs. Bullet on left is paper-patched, while the center cartridge has the bullet enclosed entirely within the case mouth. The latter is the 7.62mm Nagant cartridge employed in the Russian service revolver of that era and still sees limited use in superbly accurate target revolvers among international shooters of the Soviet block.

Left — Here we have the entire range of bullet shapes that have been employed successfully in autoloading pistols. Each profile shown has proven its ability to feed with satisfactory reliability so long as loaded cartridge length and recoil impulse are compatible with the gun in question. Any significant deviations from these shapes may cause feeding problems.

Left — The Remington 185-grain, jacketed, hollow-point, high-performance bullet in .45 caliber is one that feeds very well in most .45 autoloaders. It is of essentially round-nose shape, with a large-diameter meplat formed by the nose cavity, and the jacket is turned over the mouth of the cavity so that no soft lead core can come in contact with the gun's feeding surfaces.

Above — Most factory-made autoloader bullets intended for handloading do not carry a cannelure, as at left. However, a few do (right) for those individuals who either want to apply extra crimp or perhaps to use the same bullet in revolver loads where a stiffer crimp is needed.

feed well with this type, so the jacketed hollow-point with *no* lead exposure offers better functional reliability at some cost in expansion.

Big-game bullet requirements are quite different. In them we need maximum striking energy combined with a good balance of penetrating ability and expansion. To achieve all this, we need a moderately heavy bullet of jacketed construction that is strong enough to allow reasonably deep penetration and, in particular, to permit breaking or penetration of heavy shoulder bones without bullet fragmentation. In addition it must still allow a moderate degree of expansion when bones are not struck. Actually, it's impossible to design and produce a bullet which will perform ideally in all of these respects; a compromise is mandatory. Note now that we're talking about a bullet for game bigger and/or tougher than the typical whitetail deer. That takes in wild boar, all of the bears, big western mule deer, elk, etc. In other countries, it would include virtually all species except the smaller hoofed plains animals weighing less than about 150 pounds. Insofar as revolver loads are concerned, this limits us to .45 bullets of 190-220 grains, .44 bullets of 180-200 grains, .41's of 170-200 grains; and, in .38/.357 caliber, 125-150 grains. With the exception of the discontinued Auto-Mag and the as-yet-to-be-introduced Wildey design, autoloaders simply can't produce those levels of performance and are therefore not generally considered suitable for the larger species of game. In the Auto-Mag and Wildey calibers, bullet choice should be made as for revolvers; however, in the other and lesser calibers, it will be necessary to use a slightly lighter bullet in order to achieve an effectively high striking velocity. This will mean in 9mm/.38 calibers that a weight of 110-115 grains is a better choice, and in .45, a weight of 180-190 grains is essential. These less potent loads are not so effective on the bigger game, no matter how good the individual bullets might be, and must be considered marginal for the bigger species; whitetail and smaller animals can be hunted quite successfully with the big-game bullets just described. Nevertheless, bullets offering more expansion and less penetration are quite acceptable and in some instances even preferable. Since these are generally the bullets best suited for defensive purposes, we'll refer you to the next

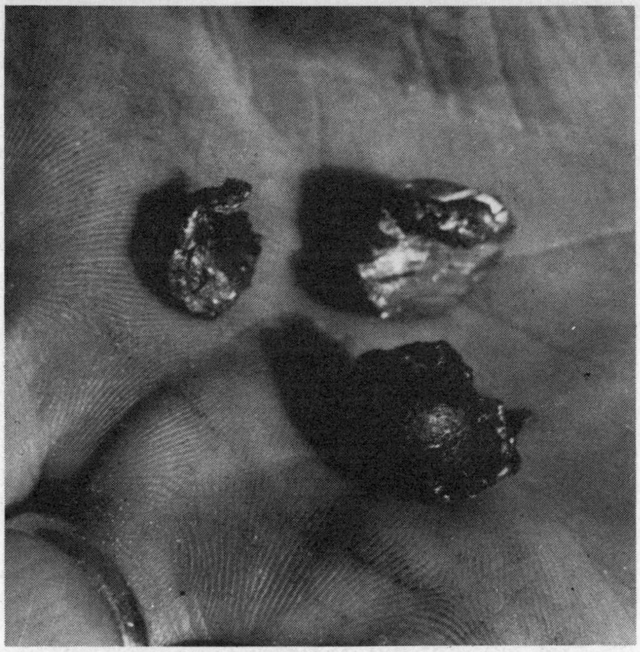

Even lead bullets will expand well in game when conditions are right. These fragments represent most of the weight of a single lead bullet that killed a large wild boar. The bullet broke up on a shoulder bone.

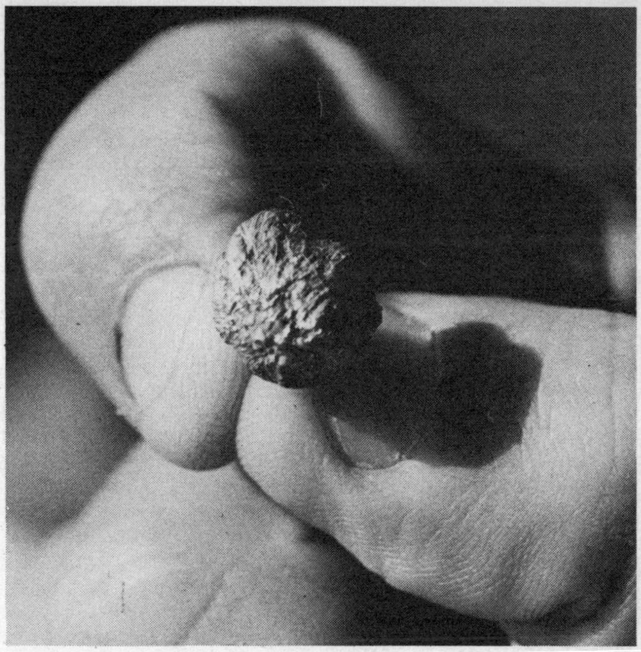

This began as a 137-grain, .357, JSP bullet. Driven at about 2000 fps, it struck a South African Blesbok at about 85 yards and expanded to nearly twice its original diameter. Recovered weight was 122 grains, showing very little weight loss and no breakup.

Handgun bullets *can* be designed and made to expand well in game, even at long ranges. This antelope was killed by Lee Jurras at 217 yards with a 137-grain, .357-inch diameter bullet he designed and loaded to over 2000 fps in the Auto-Mag pistol shown. The author witnessed the kill and checked the range.

paragraphs devoted to that type.

It might be said that the ultimate use of the handgun is defensive use against aggressive human beings in order to preserve one's life and property. This is the use to which the law enforcement officer puts his handgun, and it is the use (at least in the back of the mind) of almost every handloading pistolero. In fact, defensive use seems so much in the forefront of almost all handgun discussions today that we might even say there is *too much* preoccupation with it. The human animal is certainly by far the most dangerous of all, and therefore it requires the most destructive, and most lethal projectile that is practical for handgun use. Intensive study and development over the past couple decades has by now established quite well that maximum effectiveness against human targets is obtained by a bullet which produces maximum *wound cavity volume* (simply put, the biggest possible hole *inside* the target) while transferring the maximum portion of its energy to the target. Aside from

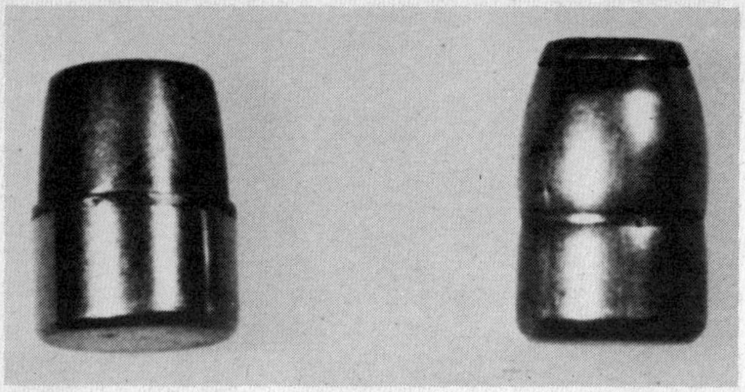

Here we see the significant difference between a half-jacket bullet and the long-jacket type. At left can be seen the small amount of full-diameter lead core just ahead of the jacket mouth. This alone may not cause leading, but the large-diameter, soft-lead core beyond it will expand radically to scrub the bore as initial acceleration causes it to upset and foreshorten. On the right, the jacket extends forward and prevents lead from touching the bore and causing leading.

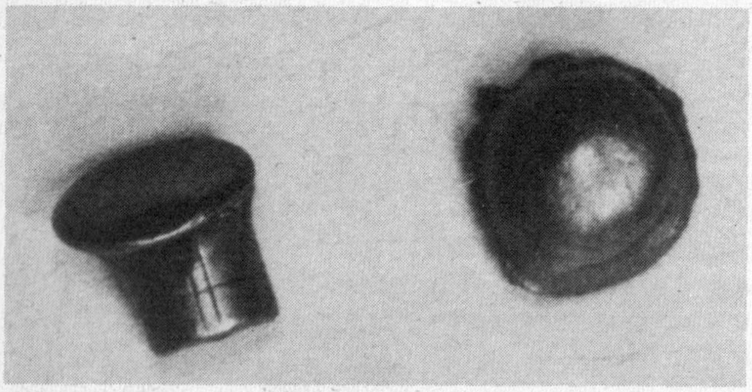

Both of these JSP bullets were fired and recovered under exactly the same conditions; one expanded quite well as can be seen on the right, while the other simply flattened out a bit at an angle. Naturally, the fully-expanded slug would have done the most damage, but even so, the one on the left would have been far more effective than a hard lead or FMJ type which simply punched through without significant deformation.

That jacketed bullets fired from autoloaders will expand reliably is evident in this photo. The bullet on the right, fired at a velocity slightly over 1000 fps, expands consistently in duxseal as shown on the left; and it expands almost as much in animal tissue.

rare hits upon critical nerve centers (which are instantly incapacitating or fatal with any type projectile) nothing less will immediately *stop* an aggressive human opponent. In addition to this clearly defined target effect, it is *desirable* that the defensive bullet present as little ricochet probability as possible in order that bullets missing or passing through the target present a minimum hazard to bystanders and property.

All of this requires a bullet which is highly deformable upon impact and produces maximum expansion, even fragmentation, upon entry into animal tissue; and, these requirements in and of themselves make the highest practical velocity necessary. This requirement in turn dictates a bullet very light in weight for its diameter in order that adequate velocity be achieved within the mechanical limitations of the gun and within the recoil-absorbing and recovery capabilities of the shooter. As a practical matter, all these requirements are met best by bullets of the Super Vel type first introduced by Lee E. Jurras in the early 1960's. In autoloaders they are of light weight, in the range of 80 grains in .380 ACP, 90-100 grains in 9mm/.38 ACP, and 185-190 grains in .45 ACP. In revolvers the range is 95-125 grains in .38/.357, 170 grains in .41, 180 grains in .44, and 190-200 grains in .45 caliber. They are of very soft construction with quite thin copper jackets and pure lead cores. They are slightly undersized by conventional standards, measuring slightly less than barrel groove diameter, and of hollow-point and/or soft-point design. The hollow point produces best results in autoloaders, and in the lower velocity ranges is most effective from revolvers, while the soft point is normally more effective from the higher velocities in the longer-barrel revolvers. These bullets are loaded to velocities approaching 1100 fps in .380 and .45 autoloaders, and to 1400-1650 fps in 9mm/.38 autos. In revolvers of .38/.357 up through .44 caliber, they are driven at velocities ranging from 1400 fps up to 1800 fps. Of course, the higher values cannot be achieved except in the overlong barrels of "magnum" six-guns which are generally considered impractical for defensive use.

The *original* Super Vel bullets no longer exist, but most other manufacturers offer similar types today. Unfortunately, though they all *look* much the same, some do not possess the soft construction necessary to insure maximum expansion, nor the undersized diameter in combination

When you look down into the cavity of the Hydra-Shok defense-load bullet it resembles nothing so much as the hollow-base wadcutter inverted in the case and made with an integral stem rising from the center of that cavity.

While the typical handloader uses cast bullets, more and more factory-swaged numbers of the target type are available. It is difficult to cast match-quality hollow-base bullets, so the swaged type might be a better choice for serious competitive use.

therewith which allows maximum velocities to be reached without excessive pressures. Thus, it isn't practical to generalize as we have up to this point and say simply that, "a 90-grain JHP is best in 9mm Parabellum and .38 Super." There is a wide range of effectiveness to be found in different makes and models of bullets of this "high-performance" type which appear to be similar and are offered for the same purpose. With frequent new models being introduced and with invisible changes being made for reasons of manufacturing ease and economy, it's almost impossible to say that a particular bullet is the best of the crop at any particular time. All of those offered by the major makers perform well, but where maximum effectiveness can be essential for survival, it may well be in your best interests to experiment with a variety, testing them for expansion in a suitable medium. Remember, other factors being equal, the bullet which produces the largest wound cavity *inside* the target offers the highest probability of instantaneous incapacitation with a single solid hit.

In this respect, we should not overlook the fact that just recently so-called "explosive" handgun bullets have entered the market. As of this writing they are not available as reloading components but only in loaded cartridges. While it appears at first glance that they offer greater effectiveness than pure expanding bullets, we simply don't have sufficient data at this time to pass judgement upon their reliability, consistency, safety or effectiveness.

So, for the foreseeable future at least, your defensive loads will have to be assembled with conventional high-performance type bullets. There is plenty of the type available today, though barely 20 years ago none were offered except in small quantities by a few independent custom makers. In this respect, we are many times better off than in the early 1950's when the only practical way to improve bullet effectiveness for defensive use was to simply drive it faster and hope for the best. You can still do this, of course, with either lead or jacketed bullets if you are so inclined. However, insofar as I have been able to determine—and that determination includes the examination of countless reports of gun fights—the soft, jacketed-expanding, undersized, high-performance bullet of relatively light

Above — Here we have the nose and tail of swaged, lead, hollow-base wadcutter bullets as employed in factory .38 caliber match ammunition and also supplied by independent makers to handloaders. Left is the nose, flat except for a slight projection rising from its center; right is the base, showing its wide, deep cavity which serves to shift the center of mass well forward and (secondarily) to reduce loading density of the assembled round.

Right — The unusual Glaser frangible bullet has, in this instance, struck a block of ballistic gelatin. Fragments of the envelope and the many shot pellets are shown clearly in this x-ray view of the block. Individual pellets scatter and act as secondary projectiles, thus destroying far more tissue than a conventional one-piece bullet.

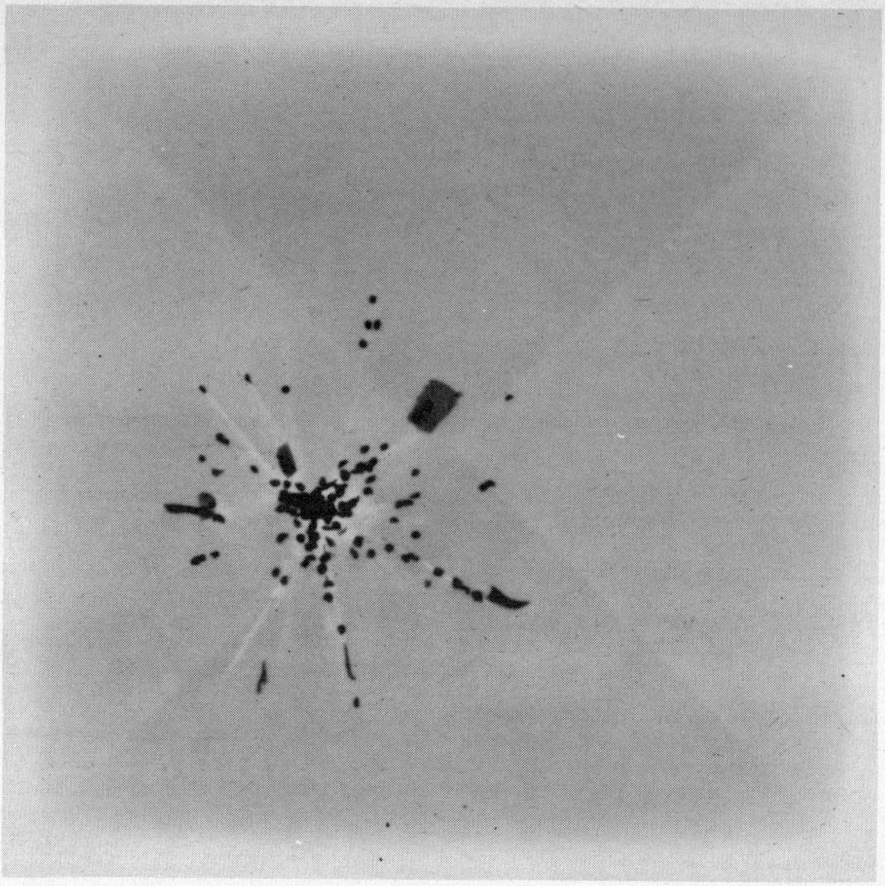

weight is by far the most effective when *all* factors are considered.

There are other bullet or projectile selections you might be called upon to make. Invariably they involve special-purpose loads—types of loads that are covered in detail in other chapters of this book. Inasmuch as the essential projectile characteristics are discussed in detail under those appropriate headings, there's not much point in going into them here.

Do keep in mind that even the most meticulous selection and preparation of all the other components can be negated to a very large extent by choosing a poorly-suited bullet or by simply using what happens to be at hand. As we've said in the beginning of this chapter, only the bullet does any real work. To achieve optimum effectiveness in *any* load, the bullet must be chosen to meet the shooting need. Remember, ''handsome is as handsome does'' and only results count. Choose the right bullet, and you'll obtain the right results.

CHAPTER 9

Casting Lead Bullets

TRADITIONALLY, handgun bullets have been made of lead. Lead's advantages for the purpose were and are many: being very dense and heavy; soft and generally nonabrasive; easily worked; dimensionally stable; relatively free from corrosion damage when in storage; reasonably priced; chemically inert; and—of particular importance to the handloader—readily formed into almost any shape by casting at low temperatures with simple equipment. Until the 1890s, virtually all handgun ammunition was loaded with plain, lubricated, lead bullets, though some use had been made of a paper patch insulating the bearing surface from the bore. While since the 1870s and 1880s almost all commercial production of lead bullets has been by swaging under great pressure, the same basic casting methods and tools first employed during the muzzle-loading era are still being used by most handloaders today—with those tools and techniques suitably improved and refined, of course.

SOURCES OF LEAD FOR CASTING

Casting a handgun bullet consists essentially of pouring molten lead alloy into a metal cavity made in the shape of the bullet desired and then removing the formed bullet after the lead has hardened or "frozen."

So, to cast a lead bullet, one needs three things: lead, a means of melting that lead, and a mold containing a bullet-shaped cavity into which the lead is poured. The sources of lead for casting are many but not all are practical for the average handloader, because of the lack of availability and cost. A century ago pig and sheet lead for casting bullets was a common item of trade and could be obtained almost anywhere at a reasonable price (for the time). Today lead is still available, usually in the form of 5-pound ingots, but only at plumbing supply houses and the better hardware stores. It is by no means as widely distributed or as readily available as it once was and is costly. Lead alloys suitable for casting bullets are commercially available, but usually only at handloading supply houses and retail shops doing an extensive handloading business. Handloading shops also offer extruded lead wire (intended for swaging bullets) at a fairly reasonable price, and it is often easier to obtain than pig lead or bullet alloy. Lead shot for loading shotgun ammunition is usually much easier to ob-

Left—Used wheel weights from your service station are the simplest and cheapest source of *good* bullet-casting material. The steel clips, which hold them to the wheel, float to the top as the lead melts, and are easily removed. You will occasionally encounter weights made of zinc alloy, and they make themselves known by refusing to melt at ordinary lead-melting temperatures and by floating to the top of the melt. When that happens, discard them or the entire batch of lead would be ruined by zinc contamination.

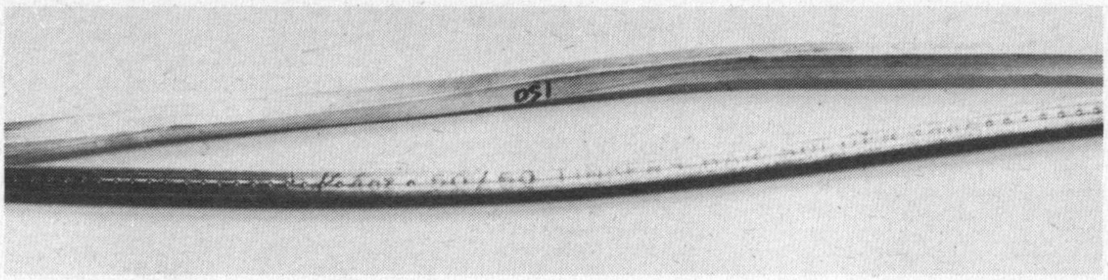

Right—The simplest and quickest way to obtain tin for hardening scrap lead is to use this 50-50 solder available from many hardware stores. Unfortunately, it is *expensive,* these ¼-lb. bars costing $1.50 each.

tain than either; in fact, it can be purchased for less than you'll usually pay for pig lead.

However, our highly mobile and mechanized society has provided us with a ready supply of low-cost lead that can be used for making bullets—the common, ordinary, automobile wheel weight which is usually discarded every time a tire is changed. Tire stores amass them in tremendous quantities, and even the lowliest service station will accumulate 40 or 50 pounds of them in a short period of time. I think I'd be safe in saying that more handgun bullets are cast of wheel-weight metal than all the other types mentioned combined.

The best way I know to assure yourself of a continuing supply of lead suitable for handgun bullets is to give your favorite service station operator a fifth of good Scotch for Christmas, then immediately let him know you'd like him to throw all of his cast-off wheel weights in a 5-gallon grease bucket placed in the corner of the shop. Naturally, you'll be glad to carry the bucket out each time it gets comfortably heavy. If some other handloader has beat you to it, consider transferring your customary "fifth" to another station that will accommodate you. Failing in that, locate a good-sized tire store and ask the manager to call you whenever he's ready to haul his accumulated wheel weights to the scrap yard. He'd much rather have you come and pick them up than have to send an employee to the yard with them and will doubtless make you a good price. The last weights I bought cost me 15¢ per pound, but if you're a customer of the house, you can easily pay nothing for all you can carry home.

Scrap lead in other forms can also be scrounged or purchased. When old houses are being wrecked, lead pipe will often be found among the plumbing; underground cables were at one time almost always enclosed in lead sheathing; and discarded auto batteries contain lead plates.*

Lead purchased in pigs, as shot or as wire, will be fresh and clean. On the other hand wheel weights and other scrap will be almost indescribably dirty and greasy, normally mixed up with other debris and foreign material. Wheel weights will also contain steel clips (for attachment to the wheel) which must also be removed. While larger pieces of scrap lead can be washed or hosed off to some degree, the most practical way of cleaning the scrap is to melt it in fairly large batches, flux it well, and then skim off the dirt and debris as it floats to the top.

*Beware when melting down battery plates. Pockets of trapped acid within the plates are quite apt to erupt and splatter hot lead. If you do want to melt down battery plates, flush them thoroughly with water and dry them off before melting.

This SAECO furnace has been on the market for well over 20 years now and has achieved an excellent reputation for durability and reliability. This author used one to cast over 1½ *million* bullets.

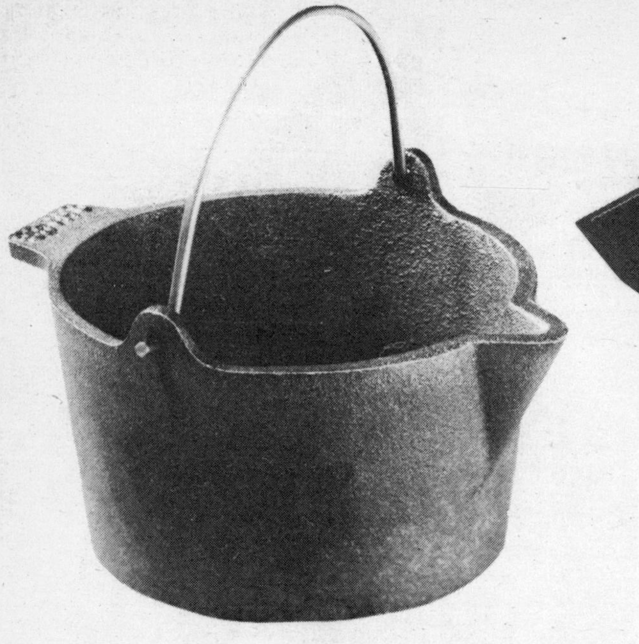

One need not spend a bundle for an electric melting pot; this simple, cast-iron pot may be used to melt bullet-metal on a gas range, single gas burner, electric range or hot-plate, or even on a hot bed of charcoal coals.

One or more ingot molds are a must. There is no other really practical way to store lead that has been prepared for use.

Over the years, I have found that wheel-weight lead is hard enough for most handgun bullet applications as it comes from the pot after cleaning. In fact, the only need that it occasionally might not meet is for magnum bullets driven at velocities over 1400 fps. Pig lead will be far too soft for anything but low-velocity, target loads; and, as a practical matter, the same may be said of shot and lead wire. Shot will usually be the hardest of the three, but even it will require a bit more hardening for high-velocity loads. If you obtain some linotype metal, it will be relatively hard, harder than many people prefer for light and target loads, but quite suitable for the highest velocities for which you'll want to drive lead bullets in handguns. Of the lot, by virtue of its composition and intended use, *fresh* linotype metal generally casts better bullets than any of the other lead types mentioned. If money were no problem and linotype metal were readily available, I'd be most happy to use it for *all* cast bullets. Even though it is a bit harder than some people prefer for some uses, it still casts beautiful bullets and can be driven over the entire practical velocity range without leading. In addition bullets made of linotype feed better in autoloaders than the softer alloys, and will produce superb accuracy in mid-range target loads as well as at the higher velocities. The only condition is that to do all this it must be *fresh* linotype metal, not burned-out, scrap metal ready for discard or reconstitution after extensive use.

Nevertheless, most handloaders will use scrap lead to the maximum extent possible, with very few of them shelling out cold, hard cash for pig lead, linotype, or the commercial bullet alloys. If you will clean and blend scrap lead properly first, you'll encounter far fewer problems in casting perfect bullets later. Assuming you have several types of scrap accumulated—let's say wheel weights, bullets recovered from the backstop or bullet trap, and some old, lead pipe—there is a proper method of cleaning and combining the lot to produce a large quantity of reasonably uniform alloy. While this can be done with the usual 12- to 20-pound electric casting furnace, the job takes forever and consumes a great deal of electric power. An old gasoline plumber's furnace (a plumber friend probably has one in his garage you can borrow) will usually handle about 100 pounds of lead at a time and is perfect for the job. Lacking that, any heavy, metal vessel holding at least 50 pounds of lead will do. A large, cast-iron, pipe cap will do, so will a section of large-diameter pipe with a steel plate bottom welded on it. If no other heating source is available, a deep,

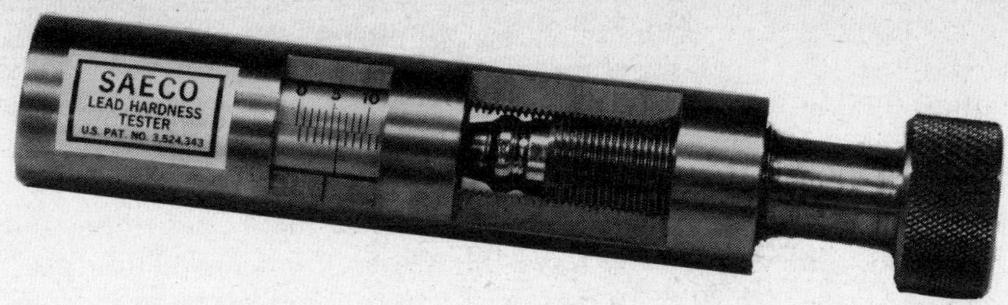

Left—If you are the kind of handloader who wants to know *exactly* how hard a particular bullet alloy is, this SAECO hardness tester will do the trick, accurately.

Right—A typical pyrometer which is a good investment for casting bullets, regardless of what type of melting unit you use. With it you can always tell at a glance what the exact temperature might be and can further record the best temperatures for different alloys.

hot, charcoal fire (those same briquets over which you grill steak in the summertime) will do fairly well; but, an old-fashioned, gas burner is easier to control. In addition to this, you'll need some kind of ladle for dipping the molten lead, something to skim off the dirt and debris and an ingot mold. For the latter, I prefer the multiple ingot molds supplied by the makers of the electric furnaces, but an old seamless, muffin tin will turn out nice round ingots easily, and you can scratch a mark inside each individual cup and fill just to the mark so that the ingots are of reasonably uniform weight. For a ladle, an old, metal, kitchen ladle will do nicely, or even a new one from the discount store providing the handle is sturdy enough that it won't bend under 5 or 6 pounds of lead in the bowl. Raid the kitchen again for an old slotted or perforated spoon to do the skimming.

Segregate the different types of scrap; the wheel weights in one pile, recovered bullets in another, and lead pipe in yet another. Each type should be separately melted, fluxed, skimmed, and cast into ingots and kept separated according to type for the time being.

Fill the pot with small scrap and turn on the heat until it melts. Depending on your heat source, this may take 30 minutes to an hour—even more with a large amount of scrap. When the mess becomes fluid, stir it vigorously to help loosen dirt and debris and those wheel-weight clips if they're present. Take care not to splash any out and burn yourself. When the mix is completely fluid, take the skimmer and get rid of as much of the *coarse* debris floating on top as possible. In removing this "coarse debris," don't include any of the *gray* scum that may also be floating amidst the garbage. That "scum," in all probability, is tin—the secret to your alloy's hardness. Fluxing will recombine the tin with the molten lead. Have an old bucket or other metal container handy to dump this very hot junk into and make certain the container isn't sitting on something flammable. Best put it on a couple of bricks.

Now it's time to flux the mix. Simply put, fluxing keeps the alloy thoroughly mixed and floats out the smaller particles of dirt and debris. Ordinary grease, beef tallow, rosin, bullet lubricant and many other materials will flux the melt reasonably well, but all will stink, smoke, and flame. I strongly suggest that you obtain a chemical flux compounded specifically for lead. The best one I've found is Marvelux, and it does a beautiful job without stench, smoke or flame. In using Marvelux, approximately one heaping teaspoonful for each 20 pounds of lead is all you'll need; but if you're using one of the other fluxing materials, about twice that amount will be required. As the flux melts, stir the mix *thoroughly* (remember how badly splashed lead can burn), scraping the bottom and side of the pot to loosen any foreign material that is clinging there. All the junk (it's called "dross") will float to the top, so skim off all that you can with the slotted spoon. Any small particles that remain can be carefully skimmed with an ordinary serving spoon.

Set out your ingot mold, whatever you're using, making sure it's not resting on anything flammable. As with the trash bucket, a couple bricks under it is a good idea. Make *absolutely certain* that there is no water, or even moisture of any kind, in the ingot mold—if there is, it will flash to steam the instant the lead hits it, causing an explosion of molten lead. I learned this to my sorrow over a quarter-century ago, and the white scars from the burns received are still visible. Carefully ladle the molten lead into the ingot mold, using either as many molds as you have, or enough to empty the pot. Leave about 1/2-inch of molten lead in the bottom of the pot (you probably won't be able to get it all out with a ladle anyway) to speed up melting of the next batch.

Once all of your scrap has been cleaned, fluxed and cast into ingots, count the number of ingots you have of each type. Blending is usually ignored but is essential if your stock of salvaged bullet metal is to be uniform ingot-to-ingot. Let's say your pot produced 20 ingots per filling and that you have 100 ingots of wheel-weight metal, 20 of lead pipe and 10 of salvaged bullets. Combine the three alloys into 10 separate piles, each pile containing 10 percent of each type. That will mean one recovered bullet ingot, two ingots of pipe lead, and 10 ingots of wheel weight for each pile. I know, each pile may represent less than the full capacity of the pot you are using. I also didn't figure out the numbers so it would come out even—and it usually won't with your salvaged metal, either.

Now, put one of those piles back in the pot, remelt it, flux it, stir it well and recast it into ingots. Repeat for each pile, and when all is done, you'll have 130 ingots all of which contain essentially the same alloy. It's a good idea to record the proportions of the three different kinds of scrap, and then the next time you have a big batch of them, you can use the same formula and come up with an alloy that is very close to this batch.

In this particular instance, the recovered bullet metal and the lead pipe will have reduced the

Left—Virtually all makes of double- and single-cavity molds employ interchangeable handles such as these from RCBS. This saves money, for a single set of handles will take care of any number of molds.

Above—Almost any old stick of wood will do for breaking the sprue cutter free, but if you want to be fancy, you can buy this polished walnut mallet made just for the purpose. Author used an old hickory hammer handle.

Left—These inner faces of a typical set of single-cavity mold blocks have diagonally-opposed alignment pins in one block and matching holes in the other. The extremely fine grooves across inner faces of the blocks are to allow air to escape from the cavity as lead enters and thus prevent voids or failure of lead to fill all the nooks and crannies of the mold cavity.

hardness somewhat of the wheel-weight metal. However, the reduction won't be much, and you can still treat the resulting alloy as if it were simply wheel-weight metal. The important thing about this blending operation is that it gives you a large supply of bullet material that is uniform throughout. This means that the last batch of bullets you cast from that several hundred pounds of lead will be of essentially the same hardness as the first.

If by chance your salvaged metal is mainly lead pipe or pig lead—both of which are essentially pure lead without significant hardening components—the blended result will be too soft for high-velocity loads, and perhaps too soft for proper feeding in autoloaders. Antimony is the ideal hardening agent, but it is difficult to obtain and even more difficult to alloy properly with the lead without fairly sophisticated equipment. You'll be far better off to use tin instead, and while it can be purchased in ingot form as "block tin" from handloading supply specialists, it will probably be more convenient to trot down to the local hardware store or plumbing-supply house and pick up a few pounds of "50/50 solder." This material is sold in bars or strips, usually clearly marked, and consists of one-half tin and one-half lead. Assuming your basic melt is essentially pure lead, calculate the amount of tin you'll need in a single potful to produce the alloy you want; 1:10, 1:15, 1:20, etc. For convenience sake, let's say you'll be working with 20-pound batches. For a 1:10 mix, that means 2 pounds of tin and 18 pounds of lead; since the solder is only half tin, you'll need 4 pounds of it to produce the 2 pounds of tin, so add 16 pounds of lead. Flux and stir well, then cast into ingots. You'll save confusion later if you'll scratch or mark on each ingot the identity of the alloy.

I'll admit most people don't go to all that trouble. It probably isn't necessary for 90 percent of the shooting we do, but your bullets will be better and perform more consistently if careful attention is paid to preparing the alloy before casting begins. Still, as I said earlier in this chapter, straight, clean, wheel-weight metal will serve for almost all purposes, and if you can get a quantity of it cheaply enough, why bother? I know one avid handgunner who is a pretty good scrounger of scrap lead from house wreckers, so he usually just trades off his lead pipe to other handloaders, and muzzle-loading shooters in particular (they are always hungry for pure lead), for wheel weights. His bullets perform beautifully, and he does nothing at all to that metal except clean it thoroughly and flux it well.

Right—A four-cavity mold *can* be manipulated freehand, but it's big chore and the probability of spilled lead is high. It is far better to use a mold guide such as this unit made by Lyman to fit the Lyman casting furnace. It is adjustable for height and fore-and-aft position to suit almost any make of mold. It speeds up the process and makes it much safer.

Below—A variation of the spouted dipper introduced by Lyman many years ago, this type of unit (now supplied by RCBS) works extremely well for filling single-cavity molds by hand.

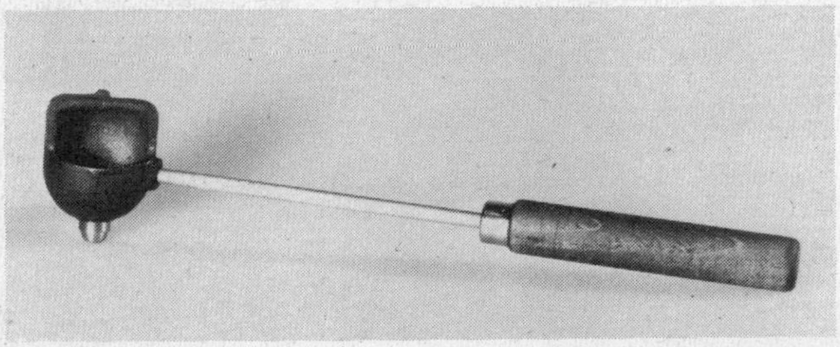

CASTING BULLETS

Equipment

Having solved the problem of a good supply of uniform lead alloy, we should now work our way into the actual casting. First of all, contrary to popular belief, you don't have to have a $100, thermostatically-controlled, electric, casting furnace to turn out perfect bullets. Good bullets can be cast using any suitable container that will hold 15 to 20 pounds of molten lead and that has a controllable heat source which will allow you to keep the lead at a fairly constant temperature. I've seen perfect bullets cast from lead melted in a cast-iron skillet on the kitchen range; and, I've also seen it done with the lead melted in a cut-off, 105mm shell case heated on a bed of glowing charcoal. The electric furnace *is* the most convenient; and it's also the cheapest in the long run, unless you can use gas.

More important than the container and the heat source is an *accurate* way of determining the temperature of the alloy. A pyrometer (a dial-indicating thermometer with a long stem that can be inserted in the molten lead) looks like an overgrown meat thermometer and can be purchased for less than the price of a box of factory-loaded .44 Magnum ammunition. With a thermometer in the lead, you can not only keep the temperature constant throughout an entire casting session, but you can also make note of the temperature that casts best with particular alloys or molds. Be sure to log these temperature readings so you can use them in the future to avoid frustration. Some older handloading references suggest that the lead has reached the proper casting temperature when it will char a pine splinter without causing an open flame; at best, that is a very imprecise method. (If you are using a thermostatically controlled electric pot, set the thermometer for about 650 degrees and adjust the heat up or down in, say, 25-degree increments, as needed. If you are using the simple pot-and-fire method, use the pyrometer to determine the temperature.)

If your melting pot does not have a bottom-discharge valve, or if you simply don't want to use it, you'll need a ladle for transferring lead from the pot to the mold. I particularly like the stainless-steel ladle offered by RCBS, but there are others, including the old, cast-iron, Lyman ladle which has been available continuously

When filling any mold from a bottom-draw electric furnace, play it safe. Note ingot mold to catch overflow. GLOVES are a vital safety device for this, but we have left them out of the photo for the sake of clarity. ALWAYS use them when casting bullets or processing molten lead for further use.

The same casting steps as in the previous photo viewed from a different angle. Note that a cheap plastic hammer is kept in the off hand for knocking the sprue cutter around.

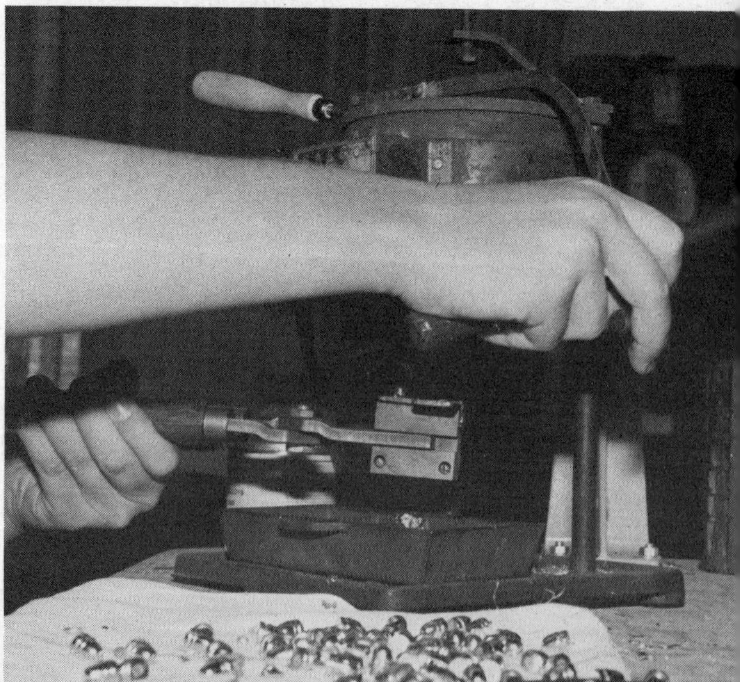

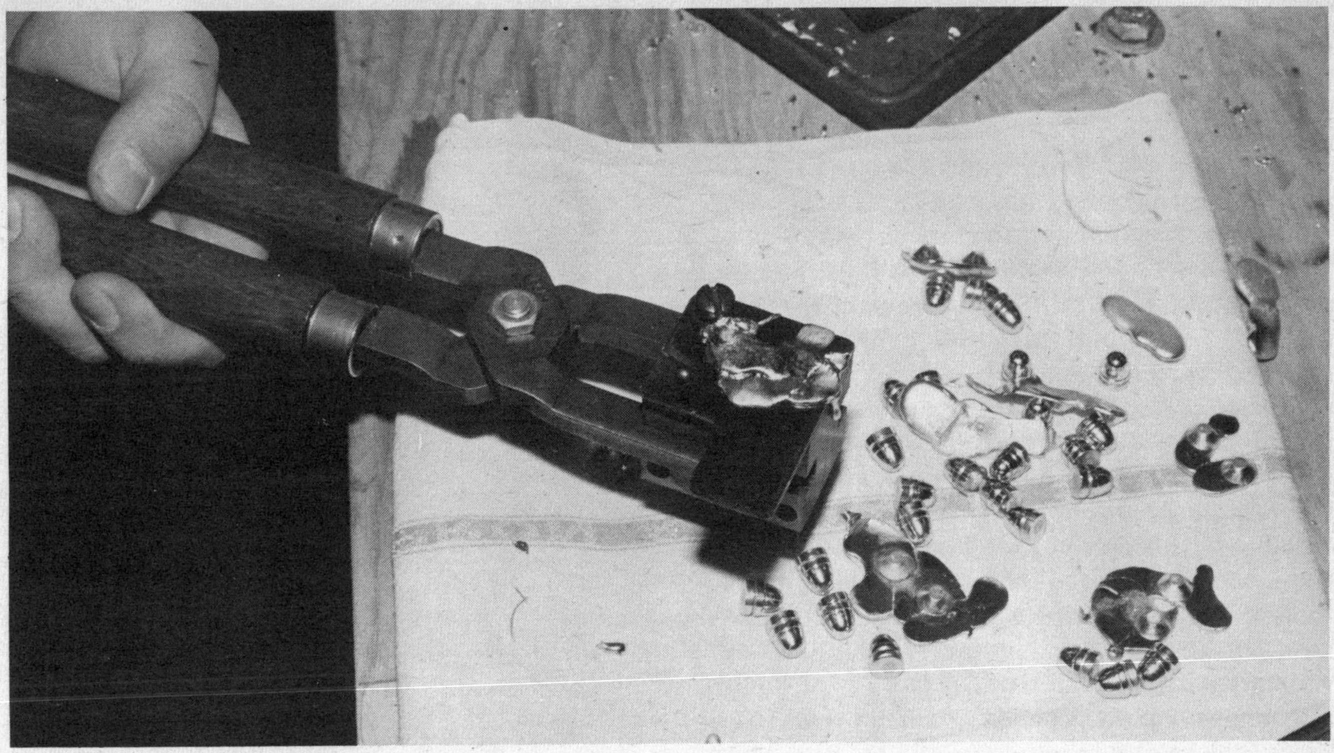

This sprue is larger than it should be, but purposely so in order that the silvery sheen of still-molten metal will show in the photo. As long as the sprue looks "shiny," it isn't ready to be cut off. Ordinarily this much lead in the sprue may lock the blocks together when it hardens and cause a few cuss words.

These two sprues show clearly the grainy, frosted look that indicates that they are hard enough to knock the sprue-cutter over and open the blocks.

since the last century. In a pinch, though, a large serving spoon will do.

Of course, the heart of any bullet-casting operation is the bullet mold. Modern molds are all patterned generally after the interchangeable-block type introduced many years ago by Lyman. Two, rectangular, cast-iron blocks are machined to fit together closely, then fitted with alignment pins so that they will always go together exactly the same. Then, the cavity for the bullet is machined—half in each block—by a cutter shaped and dimensioned to duplicate the bullet; this is done by moving the two blocks together from opposite sides of the rotating cutter in a powered fixture on a milling or drilling machine.

The blocks are mounted loosely in a set of plier-type handles, and the alignment pins and holes are given a "final fit" so that when the handles are closed, the blocks are brought together with the two cavity halves precisely matched. On the top of one block (usually the left one), is a pivoting flat plate. This is called the *sprue plate* or *cutoff plate,* which lays across the assembly and contains a tapered hole (sprue hole) centered over the cavity. Molten lead is poured through this sprue hole to fill the cavity, with the surplus lead remaining in the hole to harden; then, the sprue plate is struck with a non-metallic instrument (hammer *handle* or wood/plastic mallet). As the plate is struck, it pivots, shearing off the excess metal (known as "the sprue") leaving a smooth bullet base. Opening the handles then separates the blocks, and if everything has gone right, the shiny new bullet falls clear.

In the old days, the mating faces of the mold blocks were smooth, resulting in air often being trapped by the incoming, molten metal. This produced a depression on the surface of the bullet, or an invisible void inside. Old-time bulletcasters got around this by very meticulously "venting" the offending portion of the mold by filing a microscopic groove from the cavity to the exterior to allow air to escape. Present-day handloaders don't have that worry, because for many years now, the mating surfaces of all makes of molds have carried a parallel pattern of microscopically shallow grooves which allow air to escape. The grooves are so shallow that lead won't run through them.

No matter how nice it looks as it comes from the mold, the bullet is not ready for use except in certain rare instances we'll cover farther along. As cast, it is slightly larger than the required diameter and must be sized. When bullet sizing takes place, this "excess" bullet diameter is trued up and brought to a precise size. (Molds could not be made to cast exactly the same size bullet except at prohibitive cost.) Once sized, the bullet should be lubricated. Most, if not all cast bullets feature "grease grooves" for this purpose; and, when filled with the proper lubricant, those grooves help eliminate, or reduce, barrel leading. These operations can be performed separately, but there are highly efficient lubricator/sizer tools with which both operations are accomplished simultaneously at the rate of several hundred bullets per hour. This tool contains a smoothly polished, cylindrical die of exactly the right inside diameter. It features a handle/linkage/punch that enables the operator to force the bullet into and back out of the die—or in one design, straight through it. This type of bullet sizer also contains a lubricant reservoir connected to the die by passages. When pressure is applied to the lube reservoir by means of a ratchet and/or spring, the lube flows downward through holes in the die walls to fill the grooves in the bullet. Bullets come out of the lubricator/sizer perfectly round and properly lubricated, ready for loading into the cartridge case.

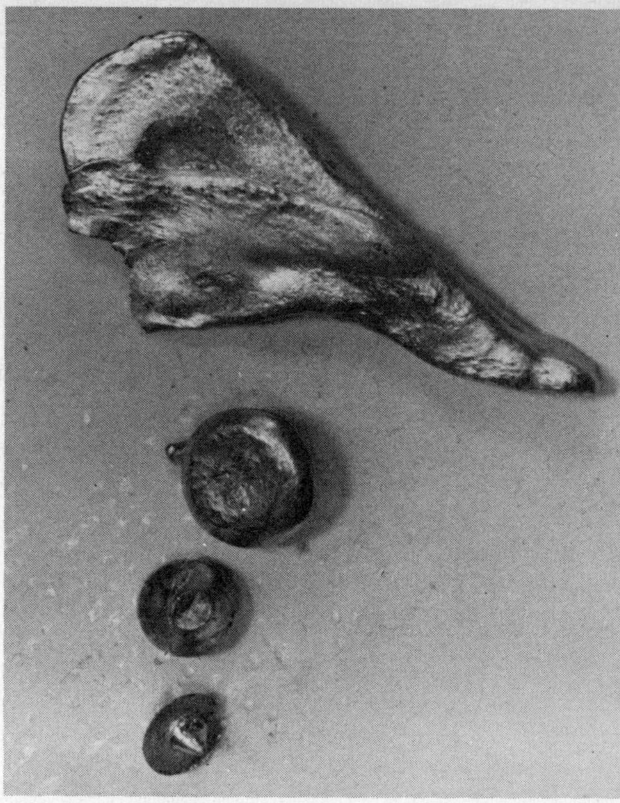

Typical cutoff sprues. The one at top is much larger than need be, and the next two are just about right; the bottom one is really too small, especially for a large-caliber bullet since it may not have contained enough molten lead to compensate for shrinkage, thus causing a void in the bullet base.

Casting Procedures

The actual casting of a perfect bullet is mechanically simple, but fraught with problems generated by variations that can be controlled only with constant attention. We'll point them out as we go along.

First, the bullet metal must be melted and brought up to the temperature that will produce bullets that *fully* fill every nook and cranny of the mold. Molds themselves are somewhat individualistic in this respect. While the metal is being melted, the mold should be thoroughly cleaned internally so that the slightest trace of oil or grease doesn't remain in the cavity; if it does, vapors will be produced when it is struck by the hot lead, and these will prevent bullets from filling out completely. Any solvent which does not leave a solid or greasy residue will serve for this purpose. (I've always obtained excellent results from acetone.) After the mold is cleaned, it should be placed near the heat source so that it, too, is brought up to operating temperature before casting starts. When an electric furnace is being used, it is common practice to heat up the mold before it's used by laying the mold across the top of the pot where the blocks will normally reach about the same temperature as the melted lead. When the lead is completely melted, drop in the proper amount of flux (as already mentioned) and stir it thoroughly. When the flux is completely melted, there will be a layer of dross floating on top; and, if casting is to be done with a ladle, the dross must be skimmed off to avoid being picked up and carried into the mold. If a bottom-draw pot is being used, the dross will float on the surface and do no harm. When the exposed surface of the lead achieves a very shiny, glossy appearance—almost silver in color, with perhaps just a trace of blue—it is probably just a bit hotter than it needs to be. The heat can be reduced slightly, and casting begun. As mentioned earlier, there's a better way to determine the alloy's temperature, and that is through the use of a pyrometer or metal thermometer. If this is available, start with a metal temperature of about 650 degrees, and as you gain experience, that can be varied. Remember, if the melt isn't clean, and/or the temperature incorrect, you won't cast a good bullet.

Make certain the blocks of the mold meet perfectly, and be sure to exert sufficient pressure on the handles so that the incoming lead won't force them apart. Also, make sure the sprue plate is tight against its stop pin. With a bottom-draw pot, hold the mold about ¼-inch below the spout and operate the valve handle briefly so the lead runs directly into the sprue hole. You'll have to practice this a bit, because you must learn to time the opening and closing of the valve so that the mold is completely filled, with the sprue hole also filled, but without a splashing and splattering overflow. A brick or other solid support which holds the mold at about the right height will be a big help here, and so will a "mold guide" sold by some suppliers of electric furnaces.

If difficulty is encountered in filling the mold in this fashion, try raising it so that the valve spout makes firm contact in the sprue hole. They are designed to work in this fashion, but I have always found that I obtain better results doing it the other way. Anyway, with the sprue plate and spout in contact, open the valve, then after a very short delay, simultaneously close the valve and remove the mold so that enough additional lead flows to leave the sprue hole filled as the mold moves away.

Leaving excess, molten metal in the sprue hole is essential because as the lead in the mold solidifies, shrinkage causes a cavity to form in the center and the excess metal from the sprue hole automatically fills this cavity as it forms. If there is not enough sprue left behind, there will be a hole in the base of the finished bullet. That ain't good.

If you'll be using a ladle to fill the mold, take the mold in your off hand and the ladle in the other. Dip the ladle full of molten lead (preheating of the ladle is necessary) then with the mold and ladle held horizontally, place the ladle spout in firm contact with the sprue hole. Holding the two solidly together, rotate them 90 degrees toward the hand holding the mold, allowing the molten lead to run into the cavity. Pause momentarily to allow the excess lead in the ladle to force lead into all the nooks and crannies of the mold, then "roll" the two apart, striving to leave a small puddle of lead in the sprue hole. This puddle of excess is just as essential with this manner of filling the mold as with the others, and, for the same reason.

If both the mold and the lead are at about the right temperature, the puddle in the sprue hole will remain silvery bright and fluid for a couple of seconds, then as you watch it, the texture will turn frosty and the color dull. The sprue hardens last, and this is an indication that the bullet beneath has solidified—don't get the idea that it's completely hard, because even though technically "frozen," it is still a bit soft and can be easily deformed.

If the puddle in the sprue hole is hard as soon as the lead quits flowing, either the lead or the mold is not hot enough; and, if it takes 4 or 5 seconds or more to harden, then either or both is too hot. I've always obtained best results when the sprue takes about 3 seconds to harden.

What we've said up to now implies that you'll be using a mold which casts only one bullet, i.e., a single-cavity mold. Actually, most molds supplied today contain two cavities; and, some companies no longer supply single-cavity molds in anything except the largest calibers. I prefer the double-cavity mold, because it takes hardly

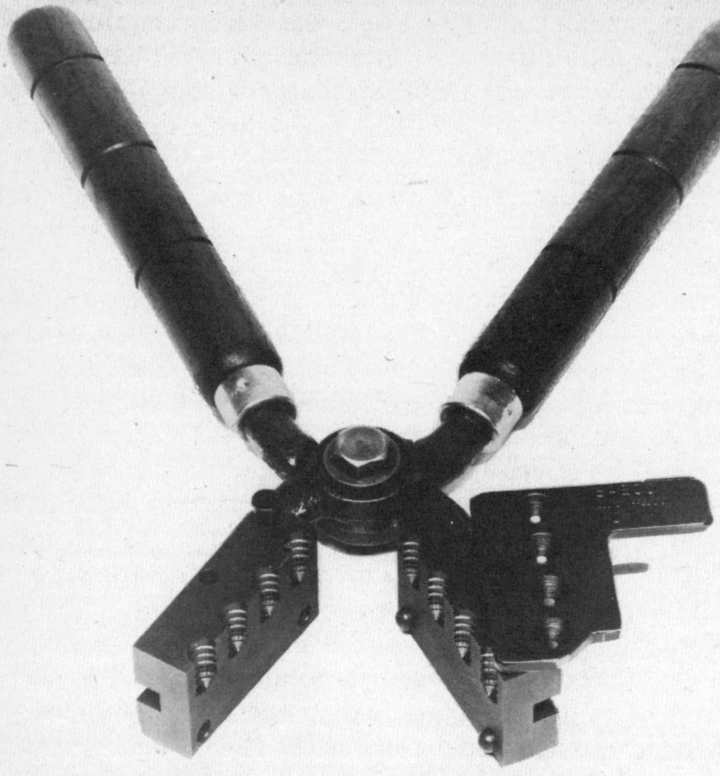

SAECO, which took over the old Cramer mold many years ago, makes this four-cavity number. While production from such large molds is high, they are more difficult to handle and to use. Many people can turn out bullets just as fast with a two-cavity unit because it is lighter and easier to use.

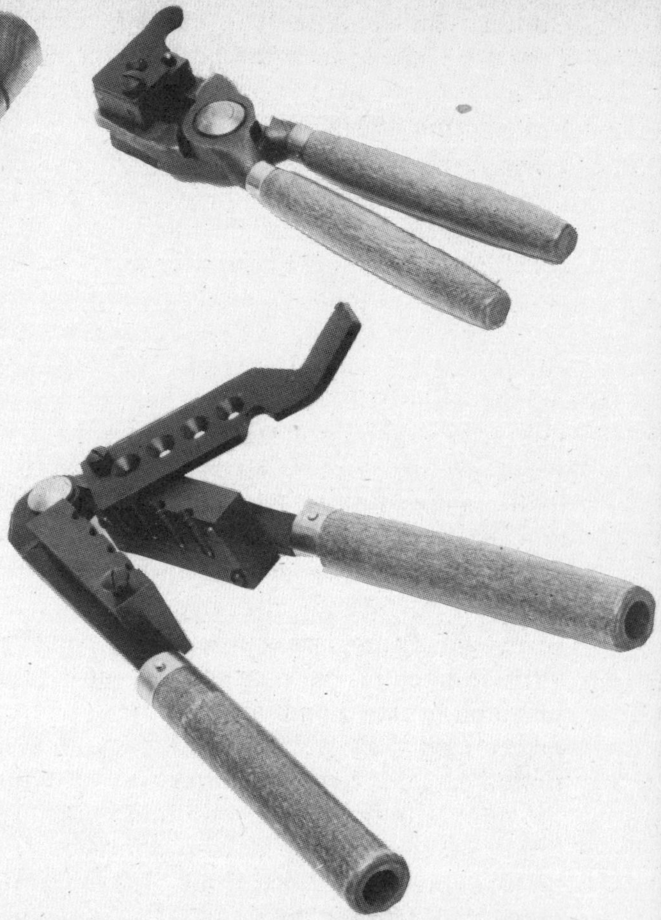

Lyman four-cavity mold, (above), compared with the same company's standard one-cavity unit (top). Note the big mold is hinged forward of the blocks, rather than behind as in some other makes and in all smaller molds.

any additional time to run through one cycle, yet it makes double the number of bullets and seems no more prone to problems than the single-cavity. The second cavity is filled exactly as the first, but it must be done quickly. This can get a little tricky when using a ladle; it will be easier at first if you let the sprue harden in the first cavity, then redip the ladle to fill the second.

Even larger molds, containing from four to 10 cavities are available, but they aren't adaptable to the needs of the average handloader.

As soon as the sprue has solidified, pick up the stick or mallet to be used, and strike the extension of the cutoff plate a forward blow, just smart enough to shear off the sprue and cause the plate to rotate clear of the cavity of its own momentum. To function in this fashion, the plate must not be held too tightly to the block by its screw and (usually) spring washer. This screw should be adjusted so that the plate can be moved freely through its full travel by hand, but without displaying any sloppiness. As the sprue is cut off, it should pop up out of its hole, but if it doesn't don't worry about it now. Open the blocks, simultaneously rolling the mold over, and the bullet should fall free. Often it won't, and when this occurs, rap on the handle hinge joint with your stick or mallet; *don't strike the blocks or the cutoff plate*. The latter might be bent, and the former might be banged together and dented or burred—you may even ruin the mold completely. Don't dump fresh-cast bullets from the mold onto a solid surface. As mentioned already, though solidified, the bullet is still quite hot and a bit soft, and if it strikes a hard surface, it will be badly deformed. The simplest and most practical method of preventing this is to fold up an old towel into a multi-layered pad and drop the bullets on it. After 5 or 6 minutes, the bullets will be hard enough that they can be pushed off the pad without damage. I've known some casters who dropped bullets directly from the mold into a bucket of water, thus avoiding both the problem of deformation and handling them while hot.

Once the bullet is out of the mold, close the blocks, and tap the cutoff plate back into position. This will usually jar the sprue out if it's still in place. If it still hangs in there, pick it out *carefully* (don't gouge the sprue plate) with a scriber or some other instrument. But, be sure you work on the sprue while the plate's *open*—one bad slip with a sharp tool can ruin the mold.

Usually, the first few bullets will be imperfect, either wrinkled and misshapen, or not fully filled out. This most often is the result of the mold being not quite hot enough yet, or of faint traces of oil. If you start without preheating the mold, bullets are guaranteed to be a wrinkled, misshapen mass, and the lead will probably freeze solid in the sprue hole before the cavity can even be completely filled. In either event, continuing to cast will bring the mold up to temperature, after which, the bullets should be okay. However, deformed bullets are not necessarily the fault of the mold as lead that isn't quite hot enough will cause the same problem. If bullets don't improve after a half-dozen or so, increase lead temperature slightly. This is where that thermometer can help a great deal, enabling you to start casting at a known lead temperature each and every time.

Mechanically, that's all there is to casting bullets. However, casting perfect bullets is more of an art than a science. Two apparently identical molds may handle the same alloy differently, one working best at a given temperature, the other with more or less heat. One mold may fill well from a bottom-draw pot with the spout in the sprue hole, another may not. One mold may produce its best bullets when used very rapidly and kept quite hot, another may require casting only half as fast so that it remains cooler. Some molds will cast nearly pure lead quite well, while others won't and do their best work with alloys containing a high percentage of tin. Some molds may work better with a particular flux, while others perform the same with almost any suitable material. In short, it may be necessary to vary the alloy, the flux, the metal temperature the method of filling the mold and the rate of casting in order to obtain the best bullets from a particular mold.

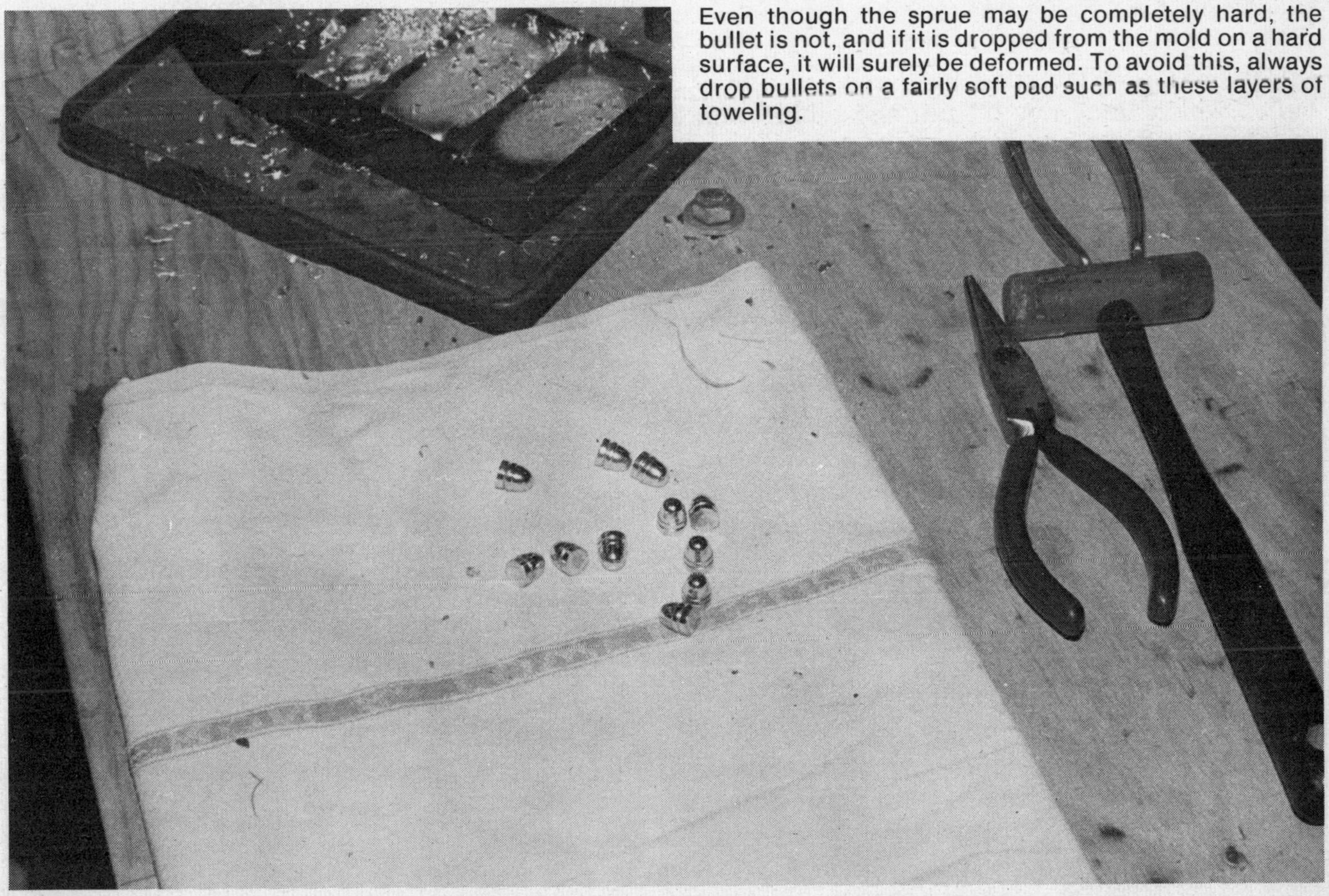

Even though the sprue may be completely hard, the bullet is not, and if it is dropped from the mold on a hard surface, it will surely be deformed. To avoid this, always drop bullets on a fairly soft pad such as these layers of toweling.

Left—This bullet isn't filled out well at all—this might result from oil in the mold, a cold mold, cold lead or any combination thereof.

Right—This bullet is filled out sharply in all areas except for the front edge of the front band. Actually, so small a defect in this area may not affect accuracy seriously, but it should be avoided if at all possible.

Diagnosing Cast Bullet Defects

Many different defects may be encountered in cast bullets. Probably the most obvious is the bullet which does not fill out the mold completely, therefore, the edges of its base, lubricant grooves, etc., are heavily and irregularly rounded. This may be caused by traces of grease in the cavity, or by vent grooves clogged with foreign material. It may also occur if the relative temperatures of mold and lead aren't quite right, or by insufficient pressure on the lead as it runs into the mold. More pressure can be applied with a ladle by holding it in position longer; and, with a bottom-draw pot, by keeping the valve open longer with the cutoff plate up against the spout. If the rest of the bullet fills out properly, but the base remains rounded, it's usually the result of air trapped under the cutoff plate. This ''air-trapping'' may be caused by the plate being too tight, or as a result of grease resting between the plate and the blocks, thus filling the gap through which air would ordinarily escape. In some instances, the sprue may not shear off flush with the base, leaving a slightly beveled protrusion — this is usually the result of the cutoff plate being loose or the inner edge of the sprue hole being damaged or dull. It can also be caused by the cutoff plate being slightly bent or warped. At some point in your casting hobby you may discover lead virtually soldered to the top of the blocks or on the underside of the cutoff plate. This can be the result of striking the plate too soon while the lead is still quite soft—by allowing a few seconds more for the sprue to harden you will cure the problem. This ''soldering'' problem may also result from the mold or lead being too hot. If no other explanation can be found, try *lightly* lubricating the underside of the plate with a *tiny* dab of beeswax—this will usually do the job. Of course, the smeared lead must be removed first, and it can be peeled or scraped off with a new razor blade.

Bullets may look right on the outside, and yet vary widely in weight. This is normally caused by internal voids, and these can be the result of dirt in the alloy, or by irregular splashing of the lead as it fills the mold, trapping bubbles of air. Varying the rate at which the mold is filled will usually help, and so may varying temperatures.

Bullets may come from the mold looking perfect except for slight fins around the mold parting line. If these fins are only very tiny and fragile

Above—The base of this cast bullet shows a central void caused by a lack of molten lead in the sprue. It also shows failure of the base to fill out sharply and some vertical misalignment of the mold blocks.

Left—Look closely at the parting line of the two blocks where the lube grooves meet and you'll see that this mold has a bad case of vertical misalignment. The blocks were also held apart by some foreign material on their inner faces so that lead flowed between them to cause thick fins. This bullet is suited only for tossing back into the pot.

teats of lead, they're simply the result of the molten metal running into the vent grooves and won't cause any other trouble. In fact, if you have a mold/temperature/flux combination which produces this, remember it—when lead flows that well, it should produce perfect bullets easily. If there's a continuous fin, then it's caused by the blocks not being held tightly together or from foreign material (usually flecks of hardened lead) on the internal mating faces of the blocks. If you clean the blocks and hold the handles tight, you'll have no more trouble.

If the mold has been dropping perfect bullets and then gradually the bullets become less well filled-out, either the mold or the metal is not being kept up to temperature. Maybe both. If the mold is at fault, casting a bit faster will cure the problem; and, if the lead is too cool, corrective action is obvious. At the other end of the scale, bullets may start coming from the mold with a frosty, crystalline surface. This is the result of the mold or the lead being too hot. Slow down casting and/or cut the heat a bit on the lead.

As the level of lead in the pot goes down, it should be replenished frequently or it will be chilled when a large quantity is added. The best procedure is to cast until the level has dropped just enough to accommodate one more ingot. Adding a single, 1-pound ingot will not generally chill a potful of lead enough to affect casting. If you let the level run too low before adding more metal, it may even solidify and you'll be stuck for the length of time it takes to remelt and come up to temperature.

Casting Hollow-Point/Hollow-Base Bullets

Hollow-point and hollow-base bullets require a special type of mold, and casting techniques are a bit different as well. Hollow-point cavities are cut nose-down in the blocks, while those for hollow-base bullets are cut with the base down. In both, a channel is cut between the blocks, centered on the cavity, and a shaped pin or punch fits in this channel, intruding into the bullet cavity so that when the molten lead fills the cavity it flows around the shaped end of the punch and forms the hollow needed. This "nose pin" or "base pin" is held in place by a flange formed on it, rotated underneath a screw or lug that's attached to the bottom of one block. In preparation for casting, the blocks are closed, the pin inserted and rotated to lock into position; the blocks are

Left—The holes in the lubricator/sizer die body allow lubricant to be forced into any bullet grooves which are aligned with them when pressure is applied to the lubricant reservoir. The smaller punch protruding at the top actually fits into the lubricator/sizer frame to force the bullet into the die and its lower end must fit the bullet nose closely to prevent tipping. The bottom punch serves to eject the bullet upward after being sized and lubricated, and it also seals off the holes to prevent lubricant from being forced in under the bullet.

Right—On the left is a typical sizing die as made by SAECO—the "3080" designation indicates the dimension (.3080) to which a bullet will be sized when this die is used. The unit to the right is a "top punch" which is internally shaped to accommodate the forward portion of the projectile during sizing.

then clamped more tightly together by handle pressure, and the mold filled as usual. When the sprue has hardened, the operator's grip on the handles should be slightly relaxed. The punch is then rotated to unlock, and withdrawn—opening the blocks allows the bullet to fall out in the ordinary way. These additional operations naturally require extra time, and as a result, it may be necessary to increase the lead temperature in order to keep the mold hot enough to cast good bullets. Hollow-points don't usually present any serious casting problems, but the relatively thin wall at the rear around the cavity of the *hollow-base* type bullet will not fill out properly unless all the factors are really right.

Generally speaking, the larger bullets cast the easiest. Very small bullets—.25 and .32 caliber—contain so little lead that it is sometimes difficult to keep the mold at proper operating temperature. This may require increasing the lead temperature and casting quite rapidly to keep the mold hot. Just the opposite problem may be encountered with the larger bullets, since they contain so much more lead in relation to the mass of the mold blocks. A double-cavity mold for the 255-grain, .45 Colt bullet can reach excessive heat rather quickly if you cast fast.

Casting Safety

Molds normally have wood handles, or rather, metal handles with wood grips. With a lot of use, the wood sometimes chars, the handles become loose, and may even split. *Never* work with a mold that has loose handles, or you may wind up with bad burns. Always wear gloves when casting, just in case some lead splashes; wherever it hits skin, it sticks and, even if the burn isn't deep, the skin comes off with the lead. Always wear safety glasses or goggles when casting, because even the tiniest particle of molten lead can destroy the sight of an eye. Don't wear boots with your pants tucked inside them; an exposed boot top is a natural funnel for splashed or spilled lead. A long-sleeved cotton or wool shirt buttoned all the way will protect your upper body from lead burns, but if you think just any old pair of pants will preserve your manhood in the case of a below-the-waist spill, you're wrong—molten lead will burn right through polyester and some other synthetics and you won't like the consequences at all. Ever since the first blister up high on my leg, I've worn a heavy shop apron for all bullet casting. It's an excellent investment, but loose, heavy cotton trousers will suffice.

BULLET SIZING

Some people seem to have the idea that relatively minor imperfections in the cast bullet can be ignored if the sizing/lubricating operations smooths them out. Nothing could be farther from the truth. Driving bands or bases not fully filled out may look clean and sharp after sizing, but more often than not, they'll result in the bullet being made eccentric (improperly shaped) as it's forced into the die. Nothing less than perfectly-cast bullets will come out of the lubricator-sizer right.

Make certain the lubricator/sizer is bolted, clamped, or screwed securely to the bench; make equally certain that the die is clamped securely in the frame. It needs to be tight to prevent movement during operation and also to prevent leakage of lubricant. Check the nose punch; the cavity in its lower end must fit the bullet nose closely

If you're going to cast bullets, you're going to have to size 'em. This lubricator/sizer, as made by SAECO, is a rugged, durable unit designed for years of use.

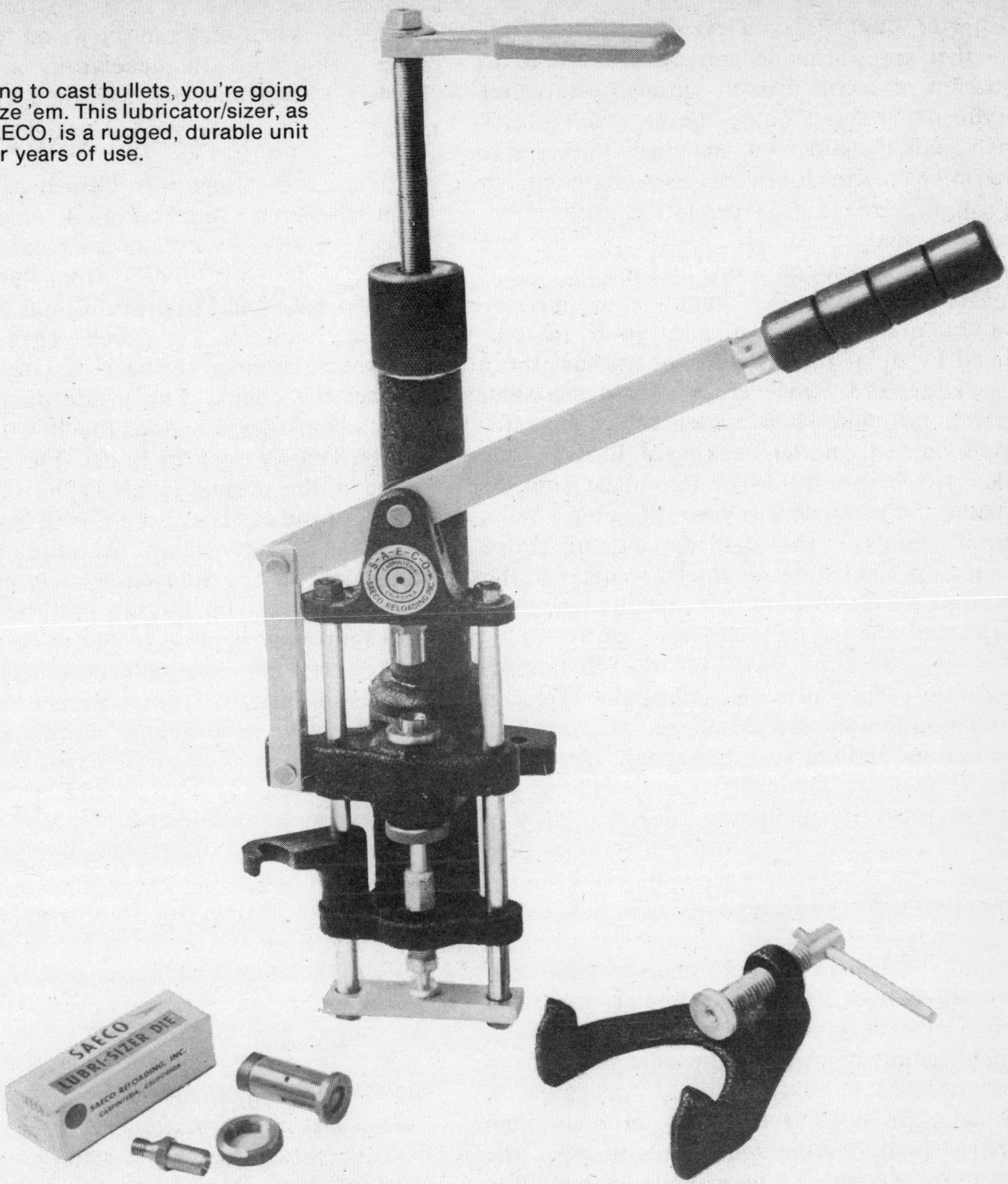

to avoid deformation; and, the punch must be held tightly in the sizer in accurate alignment with the bore of the sizing die. If it's a bit off center, the bullet will be canted as it is forced into the die, and both the die and the punch will be worn excessively at the point where they rub together. In order to be sized concentrically—or at least as near concentrically as possible—bullets must be held perfectly straight as they are forced into the die, and this is one of the functions of the nose punch. If it doesn't do the job, the bullet will come out of the die eccentric, and it will not shoot as accurately as it should. Modern sizing dies are of two diameters, joined by a smooth taper. This allows the bullet to be started in the larger diameter, then forced through the taper and into the smaller. This reduces the bullet without removing any metal, and produces a more concentric sizing job than the older step-type die. Don't use the older die; the sharp shoulder joining the two diameters inside can shear lead off one side or the other of the bullet, making it grossly eccentric and inaccurate.

Most lubricator-sizers are adjusted by running

the bottom punch (and bullet) well up into the die. You then apply a small amount of pressure to the lubricant reservoir (usually through a threaded handle or ''ratchet'' atop the reservoir itself). Then, back the punch out and check the bullet to determine whether lubricant has been forced into any of the grease grooves. If the grooves are partially filled, or completely void of lube, you lower the base punch a bit, and the process is repeated. With most dies and bullet designs, adjustment in this fashion will allow lubricant to be placed in all of the grooves, or at least those grooves desired. However, you might encounter a particular bullet shape which will require (for proper lubrication) the blocking of the die's lube holes with lead shot (driven into place from the outside). Or you may find yourself sizing a bullet which requires the drilling out of those previously-filled lube channels. Whatever the situation, be sure the groove to be filled overlaps one of the lube passages at least slightly.

Once the die and punches have been adjusted, the handle of the tool is raised, the bullet is placed in the mouth of the die and aligned as accurately as possible. The handle is then brought firmly but smoothly down against the stop. Then, the ratchet or lever is tightened ever so slightly to apply pressure on the lubricant column—just enough to fill the lubricant groove(s). If too much pressure is used, lubricant may squirt in between the bullet base and the bottom punch, messing up not only the adjustment, but requiring cleaning of the bottom punch before loading. It may also allow lubricant to pack up on the top of the top punch, and that prevents the next bullet from going deep enough into the die. If easing off on the pressure doesn't cure this problem then either readjust the die and punch, or block the hole in the die which is improperly feeding lube. A buildup of excess lube on the face of the bottom punch can easily occur if the upper edges of the punch are rounded. In that case, grind the punch down until the edges are sharp and square.

Once the tool is properly adjusted, and you've learned to ''sense'' the correct amount of lube pressure necessary to do the job, you will be able to—with a modest amount of practice—process several hundred bullets per hour through a conventional in-and-out tool. If you're using the more expensive, straight-through Star or Phelps tool, that rate can be doubled. I once had a young lady lubeing and sizing .38 wadcutter bullets for me; and, when really in the mood for work (she was being paid on piecework), she could do nearly 1500 bullets per hour on a Star tool.

GAS CHECKING

There is yet one more aspect of cast bullets that we haven't touched upon. This is the ''gas check,'' a shallow, copper cup placed on the base of the bullet to insulate it from the heat of the powder gases and to prevent bore leading. Bullets intended for gas checks have a reduced-diameter section at the base, just deep enough to accept the check. The inside diameter of the check is just large enough to allow it to be hand-pressed easily over the bullet. The bullet is then sized in the normal fashion. During sizing, the edges of the gas check are forced into the bullet, securing it tightly. Doing this requires a bit more pressure on the lubricator-sizer handle than when regular sizing is done; but, the operation is still simple and easy. It is necessary, however, to make certain that the bullet bottoms in the cup; if it doesn't, the effect is that of an angled base, and that destroys accuracy. While the gas check is usually just pressed on the base of the bullet by hand before sizing, I've found that more uniform results can be obtained if it is tapped solidly in place with a light plastic hammer. If there's full contact between the check and the bullet base before sizing, then you can be sure the check won't be tipped.

Gas checks aren't as popular as they once were, and they aren't really necessary if the bullet alloy used is hard enough. However, if you want to drive soft bullets fast, the gas check is the most certain way to insure against leading and its deleterious effect on accuracy.

As we've said elsewhere in this book, there are very few handgunning needs that cannot be met with cast, lead bullets. By choosing the proper design and alloy, they can be made to give superb accuracy at velocities of only a few hundred feet per second on up to over 1500 fps; and, about the only purposes they won't serve are for those requiring substantial bullet expansion at high velocities. To drive them at the upper limit without leading and with good accuracy, you will need hard alloys which won't expand well or consistently. The soft alloys which will expand fairly well will often lead bores seriously unless employed with gas checks, and even that doesn't always cure the problem.

CHAPTER 10

Swaging Jacketed Handgun Bullets

JACKETED BULLETS DIDN'T COME into being until the autoloading pistol, and then more from necessity to the gun than from any particular desire for the type. Until recent years, jacketed bullets were considered more a necessary evil than anything else among handgun cartridges. They were standard among autoloader cartridges, but appeared in revolver ammuntion only to give greater material penetration or in military loads to comply with the rules of so-called "civilized warfare." (It was determined that lead bullets were "uncivilized" because they just might expand or deform a bit and thus create more than a simple penetrating wound.)

However, in recent years it has been proven that a jacketed handgun bullet of proper design and construction can be made to expand quite well upon entering an animal target; this is contrary to pre-1960 experiences with early jacketed expanding types loaded by the various ammunition makers. Today we know that we can take a light-weight, lightly-constructed bullet and safely drive it fast enough from either autoloader or six-gun that it will expand to double or more its original diameter and thus do much greater and faster execution upon either an animal or human target. It has also been proven that when bullets are made toward those ends and are of uniform good quality they will normally be quite accurate—sometimes more so than even the best of lead bullets.

Because of all this, there is a plethora of good jacketed bullets available today from both the old-line and new independent makers. Unfortunately, like most other manufactured products, they are anything but cheap. The pistolero who is willing to invest in a moderate amount of the special equipment necessary can make perfectly good jacketed expanding bullets for substantially less money than it will take to buy comparable factory-made products. In fact, as in the lead bullet field, he can also make special-purpose bullets which simply aren't available from the big boys. While the cost of home-swaged bullets is by no means as low as those we make of lead—nor is the percentage of savings as great—the money saved is substantial and well worth the tool investment if one does much shooting at all.

As in some other handloading areas, there is a wide spread in the cost of equipment; for as little as $20 or so you can buy a simple set of dies which can be used in an ordinary heavy-duty loading

On the left is the Corbin swaging die designed for use in a standard (preferably "compound-type") press. On the right is a set of dies designed for use in the Corbin Mity Mite.

press to turn out entirely satisfactory bullets of the short-jacket, half-jacket, or gas-check type. At the other end of the scale, you can easily spend $200 or more for the best of dies or a special bullet-making press which will turn out bullets looking (and performing) just as good as the best of the factory products; in fact, it's no trick at all to make *better* bullets than some the factories brag about so much.

There is one area where the advantages of both swaging and casting can be combined to produce excellent bullets of uniform density and weight and, therefore, good accuracy. This is simply the process of swaging a bullet after it has been cast/sized/lubricated in the conventional fashion. The principal advantages are that even the slightest internal voids of surface irregularities are removed, and the bullet shape may be modified in the process. Even though voids and irregularities may be filled out by the swaging operation, this will not increase weight uniformity unless the swaging die used contains a bleed hole to permit excess lead to escape, and unless the die is adjusted to insure at least some bleed from *all* bullets.

The bullet is first cast, employing the methods and techniques we've discussed in Chapter 9, then sized and lubricated in the normal fashion. It is absolutely essential that all of the lubricating grooves be completely filled with lubricant. Even the smallest gap will cause the bullet to be thrown off-balance as lead flows under pressure to fill that gap. The lubricant is generally incompressible, so as long as the grooves are completely filled, they retain their shape and there is no loss of concentricity. The bullet is then placed in the swaging die which must be properly adjusted so that at least a minimum amount of bleed-off occurs. If the as-cast bullet shape is to be retained, then the nose punch of the die set must conform to that shape; however, if it does not conform, then the bullet will assume the shape of the punch cavity under pressure. Thus, if your mold drops a round-nose bullet and you want the SWC shape,

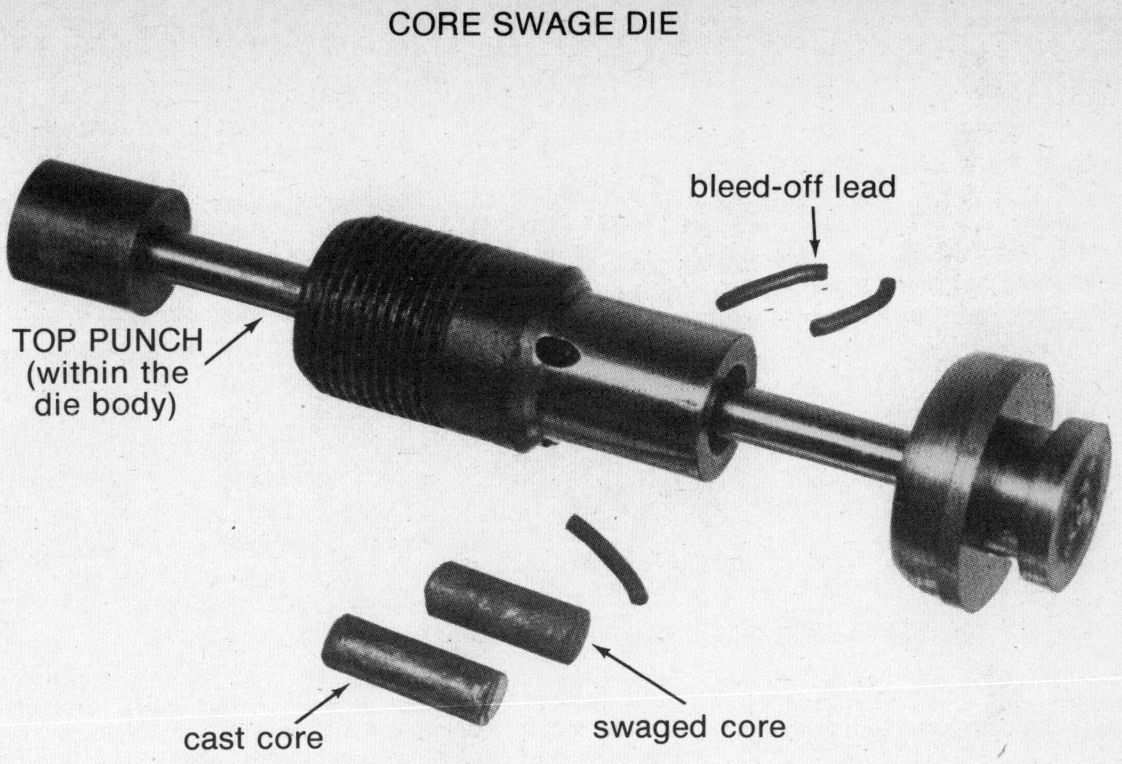

The handloader can even buy swaging dies designed to simply swage cores to proper length and diameter. The core swaging die shown here is surrounded by two cores (one cast, one swaged) and several wire-like pieces of lead. The wire-like lead is actually a secondary extrusion normally called "bleed," which is simply excess lead forced out of the die's bleed hole during sizing.

the change can be made in the nose punch. Likewise, if your mold drops a SWC bullet and you need a full wadcutter, the nose punch will do the reshaping. If you want to use a gas-check, it can be seated during the swaging operation, even though the mold did not produce the usual gas-check shoulder at the bullet base. In addition, if you should want a cap over the bullet nose such as that of the old "metal point" commercial bullet, then it is only necessary to place a half-jacket cup of the proper size over the nose of the bullet before swaging. The cup will be forced into the nose punch by the lead flowing into it, and the result will be a lubricated, lead bullet with a jacketed nose. I can't foresee any particular reason why you might want to do this, but Smith & Wesson once used this type bullet to insure reliable feeding in autoloaders while keeping bore wear down with the lubricated, lead, bearing surface. Ammunition makers once supplied this type to obtain greater penetration than an all-lead bullet would produce, yet avoiding subjecting the bore to a conventional bullet jacket.

This method of swaging cast bullets is the best way I know of to produce a hollow-base wadcutter of target quality. It's simple. Just cast/size/lube a conventional wadcutter bullet the best way you can, then run it through the swaging die using a hollow-base punch. Hollow-base bullets, as an example, are extremely difficult to cast perfectly; and, it's faster to turn out a good bullet through swaging than by mere casting. On the other hand, this is also an excellent way to produce a hollow-point, lead bullet. This type isn't particularly difficult to cast well, but the casting is quite slow and trying, so that forming the nose cavity after-the-fact in a swaging die has considerable appeal.

If you are of experimental bent, then swaging cast bullets opens up a whole new field in which to exercise your ingenuity.

The swaged, jacketed bullet, assuming proper design and materials, will give us the combination of accuracy and expansion at high velocity

Here we have a typical "half-jacket" slug from start to finish. *From right to left:* the jacket; core; core trimmed to proper size and inserted into the jacket—finally, the finished half-jacket bullet. Note that the entire bearing surface of the finished slug is *not* covered by the jacket—hence the name "half-jacket."

that is needed for defense and some hunting use. Essentially, we have two basic forms of jacketed bullets: those with a straight, cylindrical jacket which does not extend forward over the ogive, and which may or may not cover the entire bearing surface; the second type employs a longer jacket which extends forward beyond the bearing surface and is formed inward to cover all or part of the ogive. For convenience sake, we can call the first type "short-jacket," and when the entire bearing surface is not covered, we call it "half-jacket." The second type is best described by the term "long jacket," regardless of how far forward the jacket extends so long as it covers at least a substantial portion of the ogive.

Short-jacket bullets burst upon the scene nearly 20 years ago with the advent of the C-H Die Company's (now C-H Tool & Die Company) "Swage-O-Matic" bullet-making press with its very simple and low-cost dies. The type had been made before, but was not widely used until cheap equipment for making it became available. The dies were offered in a variety of bullet shapes, all of which embodied a cylindrical bearing surface joined to one of the several conventional nose shapes by a substantial sharp shoulder ahead of the jacket mouth. This shoulder was necessary in order to keep a fairly strong edge on the nose-forming cavity of the punch. By staying with this basic design, it was possible to use a cylindrical die body and low-cost punches to produce a functional bullet. This setup permitted completely forming the bullet in a single operation from a length of soft, lead wire, cut and placed in the copper-jacket cup. At first, these bullets were lauded to the skies and were proclaimed to be the greatest thing since sliced bread and free love. After a while, though, the disadvantages became apparent. Forming the bullet in a single operation without excessive effort in a low-cost press of simple design required that the core be virtually pure lead; in addition, it was necessary to have the jacket at least a little bit shorter than the total bearing surface. Often the jacket covered only about half of that surface (thus the name "half-jacket") and therein lay its major problem. At least a little bit of soft lead was exposed to contact the bore, and this often produced extensive leading. Because of the *soft* lead used with this half-length jacket, leading was worse than that encountered with cast bullets. In addition, even when care was taken to adjust the die to have the jacket cover as much of the bearing surface as possible, high-velocity loadings produced such great acceleration forces that the very soft nose of the bullet upset and flowed outward to contact the bore and caused leading anyway. This made the type much less suitable for high-velocity loads than it first appeared. At the other end of the scale, it was discovered that at low-to-medium, target-load velocities this type of bullet was not terribly accurate. Thus, rather than being the panacea, the short-jacket bullet found a smaller place in the sun, used for moderate velocity loads where bullet expansion was desired. It is still used for that purpose, and dies for making it in conventional loading presses are currently available, though the highly efficient Swage-O-Matic tool has been discontinued.

Above the core and jacket have been inserted into the die. Below the finished product is being ejected. During the swaging process the ram should be held at the top of the stroke for 2 to 3 seconds — this makes for a "stable," consistent series of swaged bullets.

MAKING SHORT-JACKETED BULLETS

Making short-jacket bullets is quite simple, requiring only a set of the C-H type dies, a loading press in which to use them, the proper copper jackets, lead wire for cores, and a core-cutting tool. The swaging die body, containing the base punch and ejector rod, is screwed into the press frame; the nose punch is secured to the top of the shell holder by the adapter provided. With the ram at the top of its stroke, the die body is screwed down until the nose punch enters the cavity a modest amount. A core is then cut from lead wire of the proper size and length so that it combines with the jacket to produce the bullet weight desired (keeping in mind that there is a minimum weight for each length of jacket). The core is then placed inside of the jacket, the assembly inverted, the protruding end of the core placed in the nose-punch cavity, and the ram raised gently until the jacket can be started manually into the die. Then, the ram is advanced until it either strikes its stop, or the jacket and core fill the die completely and prevent further ram movement. If the ram runs all the way up against its stop, then the die must be screwed deeper into the frame until the bullet is fully formed just as the ram comes up against its stop. Then, the die should be advanced just a tiny bit more, so that smooth, even movement of the ram to its stop produces just a small amount of bleed-off from the core. The die is then locked in position, and you're ready to swage bullets.

If the die is screwed too far into the press, ram travel will be halted by the fully-formed bullet before the stop is reached; in which instance, back the die out until a fully-formed bullet with just a small amount of bleed-off is produced as the ram reaches its stop.

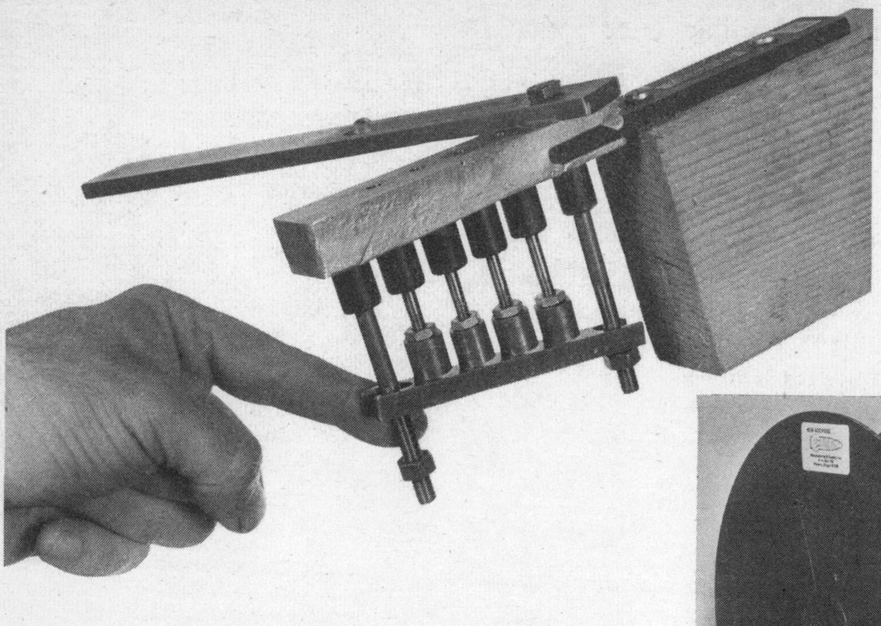

Left—Cores need not be cut from lead wire as you can actually use a core-casting mold as shown here.

Below—One of the benefits of swaged handgun bullets is that expansion can be controlled by the man who's making the slug. Here we have two rolls of dead-soft lead wire. When you combine this type of lead with a thin jacket (and a reasonable amount of velocity) expansion is almost guaranteed.

The core-cutter should be adjusted to produce cores just a couple grains over the exact weight desired. Cut all the cores you'll need, striving for as much uniformity as possible. Lay out a suitable number of jackets, and lubricate them ever so slightly; the simplest method is the one used for lubricating fired cases.

Place cores in the jackets, and run them into the die. Here is where some people go astray; they simply run the jacketed core fully into the die, immediately retract the ram, and eject the bullet. This will not produce bullets of uniform weight and density. A certain amount of time is required for the press and die to return to stable dimensions after the initial stress of swaging the bullet, and for the excess lead to bleed off. The ram should be held at the top of its stroke for a given, though short, period of time, like 2 or 3 seconds, to allow this all to take place. By then, stability has been achieved and the excess lead has been forced out through the bleed hole. The ram can be retracted, the bullet ejected, and far more uniform results will be obtained.

Some people object to the relatively high cost of soft, lead wire from which to form cores. Quite suitable cores can be cast in the special types of molds available from Corbin and, at one time, from Lyman. However, the lead used must be quite soft, or excessive force will be needed to produce a fully-formed bullet. Casting cores with the Corbin mold is unusual, inasmuch as the cavities are formed in a solid block of metal with adjustable stems or plugs which serve not only to vary core weight (by varying core length), but also to eject the solidified core from the mold.

The basic limitations of the short-jacket bullet already described are brought about by the inability of this particular type of die to do any shaping of the jacket. In order for the dies to be made most economically and to permit the bullet to be fully formed in one operation, the die cavity must be cylindrical. Then inasmuch as the die cannot do any shaping of the jacket, and the nose punch contains a shoulder around the mouth of its cavity, the jacket cannot be permitted to come in contact with the punch or it will be crumpled. To avoid this, the die must be adjusted so that a short section of naked lead is formed full diameter between the mouth of the jacket and the shoulder from which rises whatever point shape has been chosen. Even when quite short, this section of exposed lead can encourage or cause leading, and when half-jackets are used, the exposure is much greater and leading becomes much more serious.

Don't get the idea that there's anything basically wrong with *short-jacketed* swaged bullets. They were originally designed within certain

Above—Short-jacketed slugs (left) have their place; however, long-jacketed bullets (right) feature little or no exposed lead on the bearing surface—they can be driven at faster velocities, with little or no leading.

Below—Many shooters fail to realize that bullet expansion can be partially controlled by the jackets they use in swaging up a batch of bullets. The thin jacket on the left will permit much more expansion than the thick-walled jacket on the right.

specific parameters, and as long as they are employed within their limitations, they do rather well. I personally don't feel that they do the job significantly better than cast bullets, and they are more costly by virtue of the price you pay for the jackets, but they do appeal strongly to quite a few shooters. In any event, both in regard to the tools and equipment needed and the components, short-jacketed bullets are by a substantial margin the most economical jacketed projectile you can make.

MAKING LONG-JACKETED BULLETS

Long-jacketed bullets are at their best in high-velocity, high-performance loads. In fact, I can't think of a really good reason to use them any other way. Their principal reason for being is that they will give good expansion in animal tissue when properly loaded and that they completely eliminate leading while providing superb accuracy. There is also the lesser advantage that while swaging may be a good deal of work in itself, it is a cleaner and more comfortable job than casting and then sizing and lubricating lead

Handgun projectiles are designed to offer the shooter a selection of various types of bullets for varying purposes. Swaging dies allow the handgunner to do his own "selecting" at reduced cost. On the left is a long-jacketed H.P.; to the right is a long-jacketed soft point—the choice can be yours, if you swage your own.

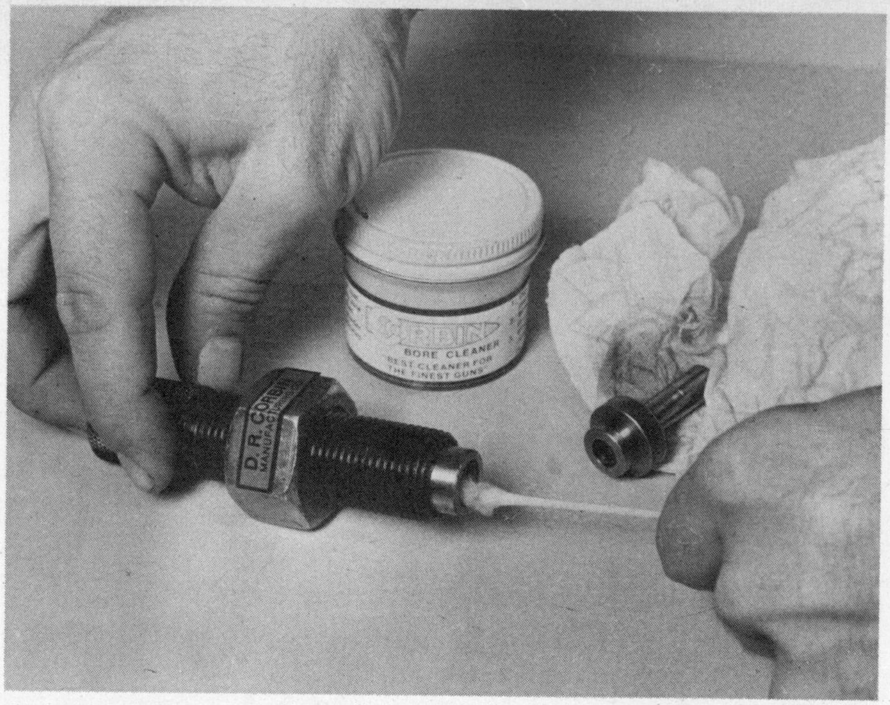

Swaging dies are precision instruments, and they should be treated accordingly. Keep them clean and they'll last a lifetime.

bullets. Their principal disadvantage is in cost—both the cost of the dies and other equipment to form them, and the cost of jackets and lead wire. To expand properly, long-jacket bullets must be made with a dead-soft core. Cores can be cast as already mentioned if you have a source of pure lead. Otherwise, commercial lead wire must be purchased. While the wire is by no means cheap, its use is so much more convenient than making and processing cast cores, that the cost seems insignificant unless you're in a real bind for shekels.

A wide variety of long-jacketed designs are available, and careful consideration to the type gun, barrel length, velocity range and intended use is necessary to make the proper selection. In revolvers, shape and core exposure is of no importance to gun functioning; but, in autoloaders, both factors can be critical. The shape must be one that experience has proven to feed well in a particular autoloader and the core material must not be exposed in such a fashion that it strikes the feed ramp or other part of the gun as the cartridge moves from magazine to chamber. A modified, round-nose shape with a relatively large flat tip (meplat) and the truncated-cone shape feed quite well providing the jacket extends all the way forward to the meplat and, preferably, turns over ever so slightly.

Also, it must be kept in mind that consistent expansion of this type bullet can generally be obtained only at velocities exceeding 900-1000 fps (and that's velocity upon *striking the target,* not muzzle velocity), so it is necessary to choose a bullet weight that can be driven at the requisite speed. Examples of the difference this can make is easily found in the .38 Special and .357 Magnum; the former can drive only a 100-grain bullet fast enough for good all-around expansion, while the latter can do the same with bullets of 130-140 grains. Bullet weight, as well as diameter and shape, must be tailored to the cartridge and to the gun if optimum expansion is to be achieved.

While pure lead is the best choice of core material for all long-jacketed bullets, more than one choice is available in jackets. A bullet intended for use on big-game at 1500-1800 fps from a long-barrel .44 Magnum revolver can be made to perform well with a fairly thick jacket. On the other hand, a bullet of the same weight to be used in the .44 Special for defensive purposes at velocities of 1200-1300 fps requires a much thinner and softer jacket to achieve maximum expansion. If you're making a bullet for use in the .45 auto pistol, the thinnest and softest jacket possible is demanded because maximum muzzle velocities, due to this gun's limitations, will be in the 1000-1100 fps range.

There is also the matter of bullet type, i.e., soft point or hollow-point. Other factors being equal,

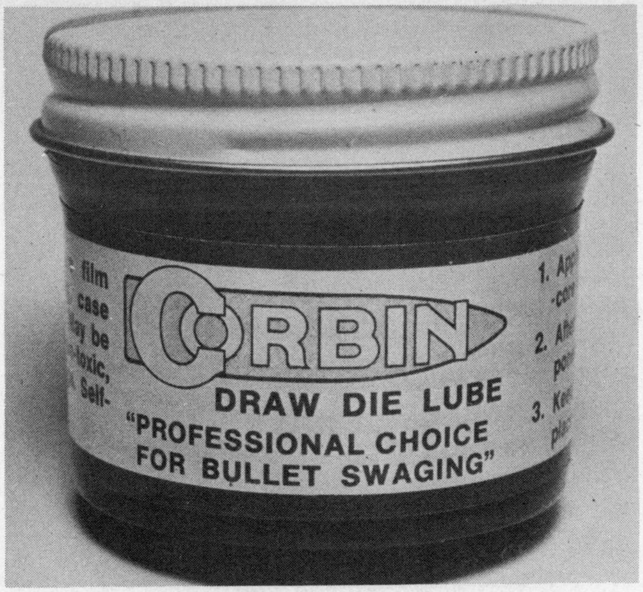

Before swaging, the jackets should be lubricated on the outside—Corbin's Draw Die Lube is a good choice.

the hollow-point will normally expand a bit more and do it a bit more quickly and penetrate less than the soft point. Thus, if the target is man, minimum penetration and maximum expansion is desired, so the hollow-point design will most often be best. On the other hand, if you plan on Indianing up close to an elk and busting him with your six-gun, you'll need more penetration, at the expense of expansion, and the soft point will be a better choice. If bullets are being made for a cartridge whose velocity potential is limited—such as the .380 and .45 ACP, then the hollow point is needed to insure expansion at all.

Most dies for making long-jacketed bullets can be adjusted to produce a wide range of weights. The ogive shape is determined by the die-cavity shape, so cannot be changed; but adjustment of the base punch will permit lengthening or shortening of the bearing surface to vary bullet weight. However, bullet length (and, therefore, weight) is controlled more by the jacket length than any other single factor. If we want an autoloading pistol bullet with the jacket extending all the way forward, then the bullet can be no longer nor shorter than the jacket. In this instance, then, it becomes necessary to obtain jackets of exactly the right length. Generally speaking, in this respect, you'll be limited to the jacket lengths that are available from the various suppliers; but, at the expense of extra time and money, you may buy overlong jackets and then trim them to whatever length suits you best. The best method I've found of doing this is using a collet to fit the jacket in a miniature lathe. Using an end mill in the tailstock, I advance it against the spinning jacket mouth to trim off the excess, running the tailstock against a stop on the way to establishing uniform length. I'm sure it wouldn't be too much trouble to build a similar rig for use on a drill press, somewhat like the powered case trimmers intended for drill-press operation. It would also seem that a conventional case trimmer could have its case holder replaced by a collet and employed manually for the same purpose.

Once you've chosen a particular design and weight of bullet, and jacket length, it's time to set up the tools. Begin as with the short-jacketed type, trimming a core so that it and the jacket combine to produce the weight you want. For use in revolvers, this doesn't usually present any particular problem, because the amount of exposed core can be varied substantially to achieve a particular weight; with autoloaders, though, you'll have to settle with the bullet weight that can be produced with a commercially-available jacket length unless you've equipped yourself to trim them. Actually, this doesn't present any really great problem for the simple reason that the manufacturers are smart enough to supply jackets of lengths suitable for the more popular bullet weights. Anyway, try as you might, you probably can't be absolutely certain of the finished bullet's weight until all die adjustments have been made and perfect bullets are formed. Minor variations in core length and weight are often necessary in final adjustment, particularly with autoloader bullets. Keep this in mind, and once you've succeeded in producing a bullet you like, record core weight and jacket length against the time in the future when you'll want to make up another batch. It can save a lot of fuss and bother if this information is readily available.

Once the proper core weight has been determined, a good quality core cutter, properly operated, will produce cores of sufficiently uniform weight for any ordinary use. However, if you want greater uniformity of bullet weight, (whether using wire or cast cores) the core can be swaged to consistent weight and density in a cylindrical die quite similar to the one used for making short-jacketed bullets. The die and punches are placed in the press, and the core in the die. The core is then compressed in the die between the punches and the excess lead is forced out through bleed holes. As long as the cores are held under pressure in the die for a sufficiently long and uniform time, they will vary remarkably little in weight. The pre-swaging of the core can be quite important in producing super-accurate rifle bullets, but I've not found it to be worth the effort involved for handgun use.

Bullets are best formed in two stages, using two separate dies. The first is the core-seating die, and it is adjusted in the press so that one punch supports the base of the jacket while the other forces the core into the jacket, enters the jacket itself, and upsets the core into intimate contact with the interior, at the same time expanding the jacket to *almost* finished-bullet diameter. The assembly thus produced is then run into the point-forming die whose cavity is shaped exactly as the bullet is intended, with a small-diameter ejector rod or punch centered in the narrowed end of the cavity. For soft-point types, this pin may have a relatively large head slightly cupped to carry out the nose radius of the bullet. In hollow-point types, the lower end of this rod will be shaped to form the cavity in the bullet nose as the lead core is forced up around it under swaging pressure.

In any event, the jacket with seated core is placed in the point forming die, and the base punch is run in by raising the ram. It contacts the jacket, forcing it into the narrowing portion of the die, which in turn squeezes the jacket inward to give it the ogive shape. As the jacket is squeezed, so the core is squeezed, and the lead moves forward either around the hollow-point nose punch or up against the soft-point punch. Vertical die position in the press is adjusted until the bullet is fully formed as the ram reaches the top of its stroke without requiring excessive pressure. Probably the most common failing in swaging bullets is in using too much beef on the press handle, thus applying too much pressure to the bullet. The minimum amount of pressure that will form the bullet fully in the die and expand the bearing surface diameter to that of the die cavity is the correct amount. When this is done correctly, the base of the bullet will be gently radiused, not sharp-edged as is often seen. Sufficient pressure to give the base sharp edges heavily overstresses both the die and the press. "Overstressing" can even cause the die to bulge permanently—possibly crack—where the bullet comes to rest, making it almost impossible to eject the finished bullet. Sharp-edged bases are by no means necessary or even desirable, and they are no more conducive to accuracy than bases uniformly radiused all the way around.

Several makes and types of swaging dies are manufactured for use in ordinary loading presses. This is a distinct economic advantage over a special bullet-making press, but the common open-front C-type press is not normally strong enough for swaging long-jacket bullets. This doesn't mean that it will break under the stress, but that it will distort temporarily under a heavy load, fouling up punch and die alignment, and causing excessive wear on both parts. By far the best type of press to use is the O-type, fitted with compound linkage, such as the RCBS Rock Chucker. The compound linkage means simply that the press possesses a great mechanical advantage and that relatively little effort will be required on the press handle to form the bullet correctly. This isn't terribly important when small quantities of bullets are being made, but if you're running several hundred in one batch, it will save sore muscles and a lot of sweat.

Bullets swaged for use in autoloading pistol cartridges can next be inspected and wiped free of lubricant and they are ready for loading.

Unfortunately, it's not quite that easy for revolvers. Since the main purpose of swaging jacketed bullets is to use them in high-performance loads, a heavy crimp is necessary to keep the bullet and case tied securely together under recoil. This requires a cannelure, and it is easily rolled into the surface of the bullet using the case and bullet-canneluring tools sold by Corbin and C-H. This little tool consists of a base supporting two, free-wheeling rollers upon which the bullet is laid, then a hand-cranked roller carrying a thin, serrated, rim is brought down on the bullet and held there under pressure while the crank is turned. The serrated rim rotates the bullet and impresses a typical cannelure into its surface. The location of the cannelure on the bullet is controlled by a stop screw against which the base of the bullet rests. The screw is adjusted to position the bullet correctly under the serrated rim, and once it is secured with its jam nut, subsequent bullets will all be cannelured at the same point. There's a tendency to make this cannelure

Left—If you want to make a full-metal-jacket handgun slug, you can do the job quite easily. On the left a lead core has been inserted into a small-caliber jacket. This is then placed into a large caliber jacket, lead core down (right) . . .

The entire assembly is then put into the press . . .

The ram is raised, bullet ejected . . .

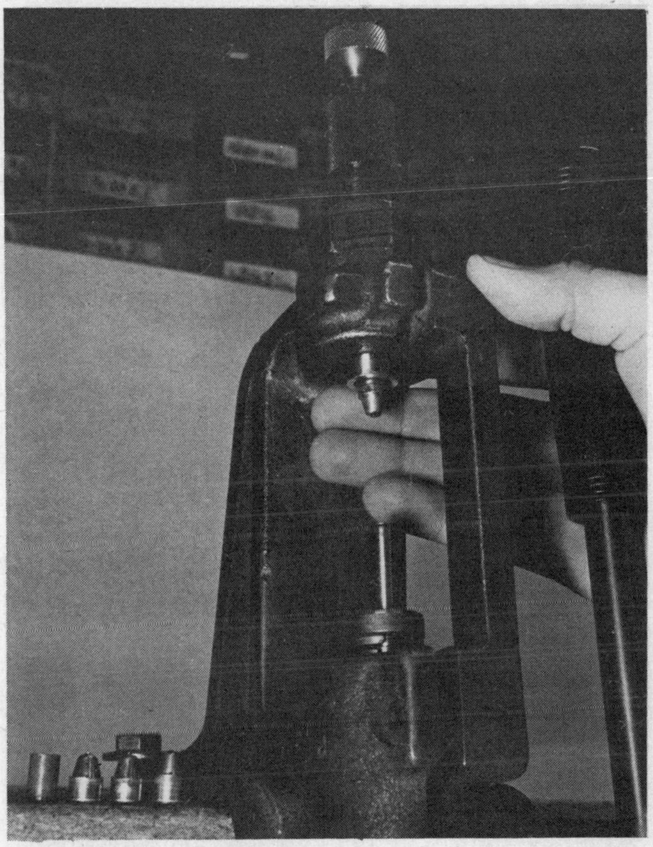

Right—And here's the finished product—a home-grown FMJ slug. This is but one example of the versatility bullet swaging offers the handloader.

Handgun hunting has increased dramatically over the past 15 years. Along with that increased interest has come a need for high performance loads and bullets. Here we have the start of what will end up as a "partition-type" handgun slug. From left to right: Large jacket and small segment of lead core over which is placed small jacket and large piece of lead core. When assembled and swaged you end up with the completed slug as shown—it's an excellent big game bullet.

too deep; this serves only to weaken the jacket and distort the bullet. The best thing you can do is attempt to copy the cannelure of a similar factory bullet and once this has been done, lock the adjustable stop provided in the base of the tool so that all subsequent bullets will be cannelured to the same depth.

CLEANING AND STORAGE OF SWAGING COMPONENTS

Bullet swaging dies and punches have their working surfaces very finely polished and hardened. Yet, they are highly susceptible to damage. The most common damage is scratching caused by grit carried into the die on the jacket or core. Cores and jackets should be thoroughly cleaned and kept dust free until actually placed in the die. Cores should be completely grease-free, and this is accomplished by washing them in acetone or similar solvent and avoiding any contamination thereafter until they are placed in the jacket. This is best done by washing the cores just before use, then covering them with a light cloth to keep off airborne dirt and grit, and handling them only with clean, cotton gloves. Jackets, on the other hand, must first be similarly cleaned, then very lightly lubricated. A number of lubricants will serve, particularly resizing lubricant, but a special bullet-swaging lube is offered by Corbin, and works extremely well. Before and after each swaging session, care must be taken to see that jackets don't pick up any grit going into the die.

As already mentioned, the die body, and to a lesser extent the punches, can be deformed by excessive swaging pressure. This can actually produce an enlargement up inside the die where the completed bullet rests, and a bulging of the base punch. By applying only sufficient pressure to form the bullet properly, a secondary damage caused by attempts to remove a stuck bullet is also avoided.

Before and after each swaging session the interior of the die body and the working portion of the punches should be thoroughly cleaned with acetone or similar solvent on a clean, cotton swab. The after-operation cleaning should be immediately followed by very light application of oil or preservative. The tiniest bit of rust will spoil the die, and as with other forms of damage, it can be removed only by lapping out the cavity with diamond dust to the depth of the defect. This means simply that if you somehow produce a .0005-inch deep scratch in the die, it will have to be lapped out that same amount to restore its pristine functional finish. Removing that .0005-

One of today's most popular swaging presses is the Corbin Mity Mite. It sells for around $70—a good investment for the handgun handloader.

inch all the way around means increasing the die diameter by .001 inch, and that has the effect of an equal increase in the diameter of all bullets that are made in it thereafter. That's not so good.

SWAGING SPECIAL-PURPOSE BULLETS

Swaging tools and methods and components may be adapted to produce unconventional bullets for special purposes. You might want a bullet with an extra-heavy jacket to resist deformation in penetrating tough material. One way of doing this is to place a smaller caliber jacket inside the one correct for the bullet being made, then place the core inside the second jacket. As this composite is swaged, the core forces the two jackets tightly together, producing a doubled-walled bullet of much stronger construction. You might also wish to make up full, metal-jacketed bullets to provide maximum penetration, or possibly for use in Handgun Metallic Silhouette Matches where they improve the transfer of energy to the steel-plate target. Conventional dies will not produce good FMJ bullets of the open-base design made by the factories, so another approach must be taken. Again we use two jackets, but this time the core is placed in the main jacket, then the next smaller caliber jacket is placed over the core and inside the first jacket. When this composite is swaged, it produces an FMJ bullet closed at both nose and base, with double jacket thickness where the two overlap. This is identical to the construction of the famous Winchester-Western "Silvertip" bullet, but with a full thickness jacket covering the nose rather than the very thin cap of the Silvertip.

There are other swaging operations that may be performed to produce special-purpose bullets. Handgun Metallic Silhouette competitors have found that heavily-jacketed FMJ bullets transmit more energy to the target than conventional lead or jacketed, sporting-type bullets. Lacking a suitable, commercial bullet of this type in .44 caliber, they have gone to swaging-down the .45 ACP service bullet to proper .44 Magnum diameter. Other jacketed bullets may also be reduced similarly in diameter if the need arises. Many years ago in Europe, I once made up a simple swage and punch to reduce lightweight, jacketed, 8mm rifle bullets to nominally .30 caliber for use in 7.63mm Mauser and 7.65mm Parabellum loads.

Swaging may also be employed to increase the diameter of a bullet. At one point in time, I needed jacketed, 9mm bullets of light weight, and having none available, I made a two-die, swaging set for use in a heavy bench vise which expanded

.30 carbine service bullets from .308-inch diameter up to .355-inch diameter. This is an extreme example and not really practical except in an emergency. Nevertheless, it did *work,* though the resulting bullets were not noted for superior accuracy. Jacketed bullets can easily be upset as much as .020-inch without too much difficulty by simply placing them in a swage of the proper diameter and applying pressure so that they expand radially (being foreshortened in the process) to fill the die. In this manner, I have swaged .44-40 bullets up to .45 caliber, and .38-40 bullets up to .41 caliber. Generally speaking, swaging bullets down or up is worthwhile *only* when there is no other possible way of obtaining what you need. If more than .015- to .020-inch change in diameter is involved, bullets tend to become eccentric and even asymmetrical unless the job is done in two or more small stages. Thus, upsetting a .44 bullet to .45 caliber *can* be done in a readily-available .45 swaging die, but to obtain best results, a specially-made, intermediate die should be used. That gets expensive, and when the cost of beginning with factory bullets is added in, you can see why it's pretty much a last-ditch proposition.

To a degree, you can have swaged, handgun bullets which expand rapidly on initial target impact, but then continue on to penetrate deeply after the fashion of the Nosler "partition jacket" rifle bullet. This done by taking a smaller-caliber jacket of short length, then seating a core in it that comes flush with the mouth, and at the same time expands the jacket so that it will be a relatively close fit inside a jacket of the proper caliber. Then, this assembly is placed in the main jacket inverted (mouth down) and a second short core is placed on top of it. The entire assembly (two jackets and two cores) is then swaged in the normal fashion, with the single exception that a base punch with a convex nose is used so that additional pressure is applied (through the base of the outer jacket) to the rear core and expands it radially without crumpling the edges of the inner jacket mouth. The result of all this is a double bullet, the front portion of which will expand upon impact, peel back or break off to create a massive entrance wound, after which the rear portion of the bullet will remain intact and penetrate just as if it were of FMJ wadcutter form.

There are some calibers for which jackets are not commercially available to swage bullets. Most prominent are the .25 ACP and the various .32 calibers. With special dies available from Corbin, .25 ACP jackets can be formed from shotshell primer battery cups, after which a reasonably effective JSP or JHP bullet can be swaged in the conventional manner. Jacketed bullets for the various .32s can be made by first cutting off a portion of the nose of a Speer Plinker or Hornady SJ .30 caliber bullet to produce the desired weight, then swaging the result in a conventional .32 die set.

Even odder swaging operations can be performed if you want to get deeply into the subject. For example, thin-wall, copper tubing can be cut to precise length and then formed into jackets in special jacket-making dies designed by Corbin. In this manner, using the extremely wide variety of tubing sizes available, jackets of almost any length or diameter can be produced, thus enabling some extremely unusual bullets to be made.

We really shouldn't leave this particular subject without mentioning at least briefly, special bullet-making presses. Most notable of the type is the Mity Mite originally designed by Ted Smith and now manufactured by Corbin. This is a heavy-duty, horizontal press which is very compact and easy to operate. It is superior to even the best loading press for almost any swaging operation, and anyone intending to make many bullets by this method might very well find it to be a good investment. From time-to-time, other presses of the same general type have been manufactured, and they may perform just as well—but this is the one with which I have a fair degree of familiarity, so I can recommend it.

The subject of swaging bullets is not one which can be covered in great detail in the space available here. A good deal has been written on the subject in various publications, particularly in different editions of the *Gun Digest* and in the *NRA Handloaders Guide;* however, the most current and extensive reference on the subject is the *Corbin Handbook And Catalog Of Bullet Swaging*. Being a major supplier of bullet-swaging tools and equipment, Corbin has followed the lead of component manufacturers who publish their own manual and has published this most excellent soft-cover book. If you'd like to know more about swaging bullets, I strongly recommend you obtain a copy.

CHAPTER 11

High Production Tools, Techniques

WITH ORDINARY TOOLS and accessories, carefully thought-out layout and procedures, plus a reasonable amount of experience, the average handloader can produce up to a couple hundred loaded cartridges per hour without sacrificing quality and without working terribly hard. That should be fast enough for almost anybody, providing he's loading only for himself. However, if you're also loading for friends or family, or if you're in serious preparation for a big match, you may well wish for a setup that will turn out more ammunition in a shorter period of time.

There are several ways to increase output, and all of them require additional equipment or help, or both. A good many years ago I was a member of a club and fairly large quantities of handloads were required for some of our activities, yet at the same time we couldn't afford to buy high-production loading equipment. With a little head-scratching, we came up with a fairly cheap operation that produced ammunition at quite a rapid rate. To accomplish this, we simply separated the loading operations, setting up in the club house one station for each operation, with a separate volunteer operating each station. One press we set up permanently for resizing, and a bit farther down the bench, an expanding and decapping station was set up. Then came separate stations and tools for priming and charging with powder, and at the end of the long bench was another press set up permanently for seating bullets. By putting a different club member on each station and having a couple more handy to fetch and tote, we could crank out a week's supply of ammunition in one evening of serious work. You might not have enough people available to work a setup quite as comprehensive as this, but a variation of it using mama and the older kids (you *can* get them to do it, can't you?) can speed things up a good bit. You might spend a couple evenings each month this way to replenish your reserve stock, and load by yourself in between.

HIGH PRODUCTION MOLDS AND SIZING TOOLS

No matter what you do to speed up the actual loading operations, casting and sizing/lubricating bullets can be a bottle-neck. At best, it's a hot and messy job, but it can be speeded up a good bit. Best-quality, four-cavity molds are the first step upward, and while I have used and prefer those made by Hensley and Gibbs, good ones are supplied by other makers. This type mold is much heavier than the usual single- or double-cavity model, but its large, heavy blocks contribute to uniform bullets, and it does throw *four* bullets with each filling. The fastest way of using these big molds is with a large-capacity, bottom-draw, electric furnace fitted with a mold guide or other form of rest that will hold the blocks directly under the spout. The mold is placed in the guide, positioned with the first sprue hole under the spout; the valve is opened to run lead into that cavity, and as it is filled, the mold is moved to place the next hole under the spout, without interruption in lead flow. This is repeated for all cavities, allowing excess lead to fill the groove connecting the holes in the sprue plate. The valve is then closed, the mold left in position for a few seconds until the entire sprue hardens, then withdrawn. Cutting off the sprue and dumping the bullets is as with the smaller molds. While I've never worked hard enough to reach a high production rate myself, I have had a couple people casting bullets for me who could turn out nearly 1,000 per hour with a four- or six-cavity mold.

A major problem in use of the large molds is in maintaining a sufficient supply of molten lead. Beginning with an electric furnace containing at least 20 pounds of lead, the temperature control should be set fairly high (adjusted as you gain experience), and the level should be kept up around 75 percent of capacity. This can be accomplished by adding 1- or 2-pound ingots after every few fillings; five to seven separate mold-fillings will consume a pound of lead, and a 1-pound ingot should be added at that time. If the molten-lead level is allowed to drop to 50 percent or so of capacity, then adding sufficient cold lead to fill the pot will literally "freeze" the contents, and you'll have to suspend casting until it comes up to temperature again. No matter how you do this, adding cold lead will cause the temperature of the melt to fluctuate. This may not cause any trouble, but some molds are sensitive to this sort of thing and won't do their best work. A solution to this problem is the use of two furnaces, one kept filled in reserve, while metal is dipped or run from it into the working furnace, thus maintaining a constant temperature. Molten lead can be transferred with a large ladle, but I know a couple people who have installed the reserve furnace above the working furnace in such a fashion that lead could be run directly from the former into the latter from the bottom-draw spout. This appeals strongly to me, but if I were to use a setup of this sort, I'd want a splatter shield to prevent the inevitable splashed lead from singeing my clothing or hide.

Sizing and lubricating more rapidly doesn't present all that great of a problem. The Star lubricator/sizer has been available for many years and is well proven. It differs from other units in that the bullet is simply forced into the die, then the top punch is retracted, another bullet is started in the die and forced into position; the second bullet pushes the first out the bottom of the die, so each single operation of the handle processes a bullet completely. An automatic, cam-operated plunger applies pressure to the lubricant, so one hand can be kept busy on the handle while the other places bullets in the die. Well over 1,000 bullets per hour can be processed with this tool.

Unusual Star lubricator-sizer is the fastest of all, each bullet being ejected downward from the die by the next one sized and lubed.

TURRET PRESSES

The next step up from using additional ordinary tools and helpers is the use of a turret-type loading press. A turret press resembles the C-type, but its upper arm is replaced by a heavy, circular turret pivoted on a sturdy, vertical axle. The turret contains a number of (four to 12) die stations, and all of those stations are easily rotated to position over the ram. Stations are held in alignment with the ram by a spring-loaded detent, and handles are provided for convenient rotation of the turret. Though the large turrets with eight to 12 die stations might seem to be particularly desirable, they are actually intended more for permanently setting up several sets of dies in different calibers than for greater speed of loading. A minimum of four stations is required for maximum speed of operation with a three-die set, but if you prefer the four-die set, then a six-station turret is the best choice. We need one station for each die, and if we are to avoid removing the case from the shell holder before it is completely processed, we need one additional station in which to install the powder measure. Actually, if the special expander die and powder measure made by C-H for its three-station press is used, a fourth station isn't necessary.

Typical turret-type loading press is this Lyman Spar-T with 3-die set and measure installed, along with automatic primer feed. Case is not removed from shell holder until fully loaded.

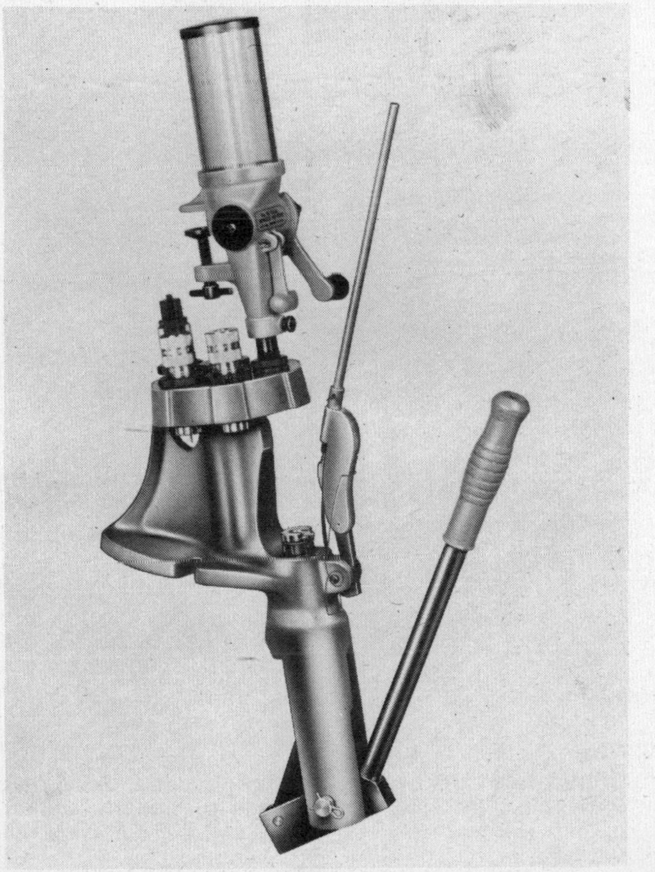

A turret press is set up by first making certain the turret moves freely, yet indexes positively. Once this is done, you should then determine which direction of turret movement will give you the most convenient operation. Then, arbitrarily pick a *number one* station and install and adjust the resizing die in it. Move to station two and install the expanding/decapping die. In station three, set up the powder measure with a threaded adapter, and, in some instances, the special drop tube supplied by the maker to permit this type of use. In installing the powder measure, make certain it is positioned so that there is plenty of clearance to operate the handle. Set up the bullet seating or seat/crimp die in station four, and if you'll be crimping as a separate operation, put the crimping die in station five. Any turret that contains five stations will have at least one more, so give thought to skipping a station in the vicinity of the powder measure to provide more clearance for operating it. I've skinned many a knuckle on sharply-knurled die bodies and protruding adjustment screws while working the measure. To make maximum use of a turret press, repriming will have to be done on the press, and it may be done either at the expanding/decapping station or, in some setups, at the powder measure station *before* the charge is dropped.

The loading process with such a turret setup goes like this: Insert a clean and lubricated case in the shell holder and position station one over the ram; run the case into the die, extract it, and index the turret to station two; run the case into the expanding die, at the same time picking up a primer with the off hand and transferring it to the primer punch; with the case in the die, activate the priming device, seat the primer, then extract the case from the die; index the turret to station three, and run the ram up *only* until the case mouth is in position under the drop tube to accept the powder charge; operate the measure handle as consistently as possible, and retract the ram; index the turret to station four, and place a bullet *straight* in the case mouth with the off hand; run the case into the seating die and extract it. With a three-die set, this produces a fully-loaded cartridge with a minimum number of operations and without removing the case from the shell holder. If you're crimping in a separate operation, all this will be followed by indexing to station five, running the case into the die and extracting it, and then removing it from the shell holder.

With quite a bit of practice and a layout which keeps excess hand movement to the minimum—plus a plentiful supply of properly-prepared components—it isn't too much trouble to load 250 rounds per hour. In addition to a higher production rate, the turret press allows permanent installation and adjustment of dies, and it avoids having quantities of cases in various stages of loading when you've shut down for the evening. Any time you decide to quit loading, no more than a few seconds is required to run out the case in the press at the time, leaving you with only a quantity of fully-loaded ammunition and the various components. This is better than working with a single press where you might spend most of the evening just getting cases ready to load and still not complete any shootable ammunition. With a turret press, complete cartridges begin coming out after only the first four or five strokes of the press handle.

One of the keys to maximum production with a turret press is careful layout and positioning of the components so that you won't waste a lot of time reaching and fumbling. I have found it convenient to use shallow trays for cases, bullets, and primers, blocked up so that they are angled toward the press. Set up this way, they cause the components to flow toward the lower edge so that you're always reaching to the same point to pick up the next item. This eliminates fumbling and saves time. Even paying careful attention to removing the loaded round from the press can save time; I place a towel-padded box or bucket directly beneath the press on the floor in which to carefully deposit completed cartridges. This is faster than attempting to transfer the round to a smaller container somewhere on the crowded bench top.

MULTIPLE-STATION PRESSES

A viable alternative—and normally a bit less costly—is the multiple-station fixed-head press typified by the C-H three-station, X333. This is an otherwise conventional H-type press with

Unusual (and expensive) Hollywood, manually-operated, progressive loader feeds propellant, primer, bullets automatically and can load 1,000-1,500 rounds per hour.

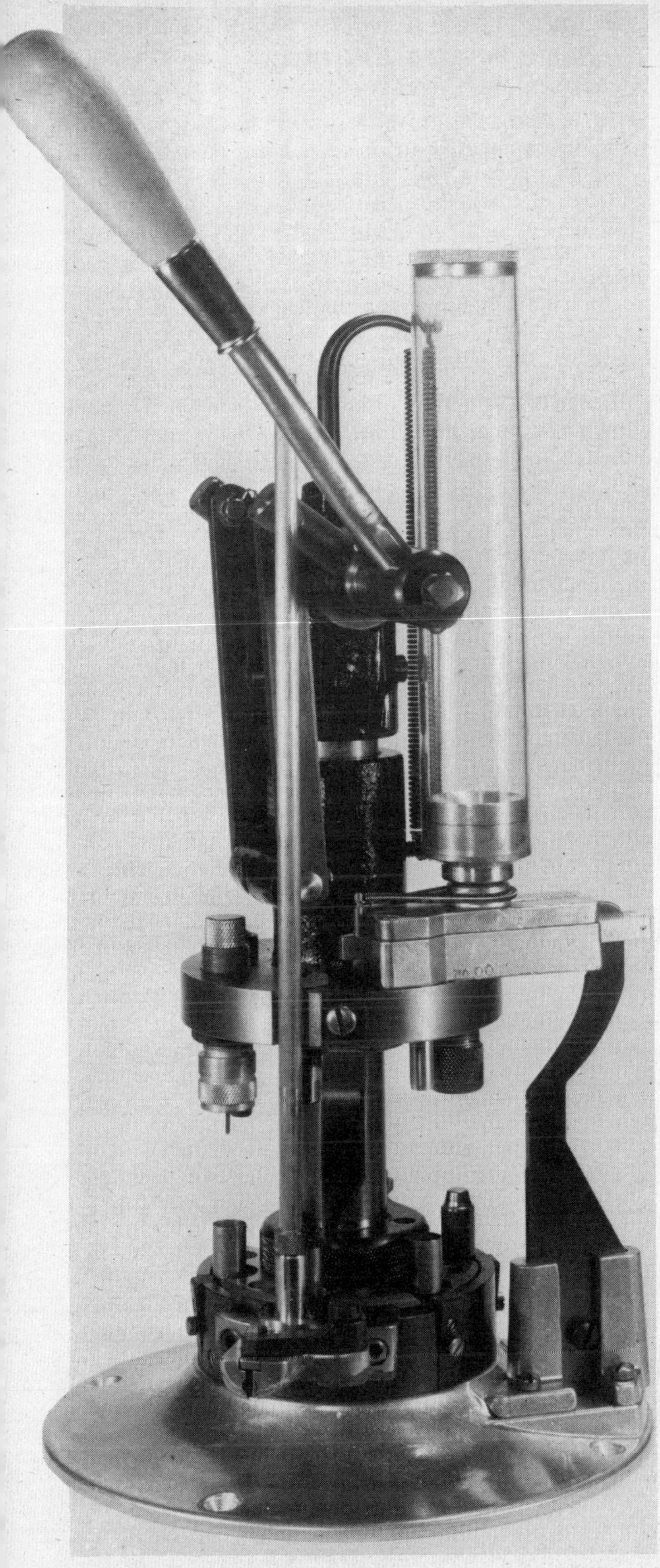

Star Progressive loader dates from well before WWII and is currently the most widely used high-volume loader made.

frame spacing, head size and configuration revised to permit the installation of three dies. The riser bar is also designed to contain three shell holders aligned beneath their respective dies. A single primer-seating punch is affixed to the base aligned with one of the shell holders. In addition, a sliding-bar, fixed-charge powder measure (with replaceable bushings) is designed for installation on this press. The three dies are installed and adjusted in the stations from left to right, expander die, resizing/decapping die, and seat/crimp die. This particular sequence of dies is required because the container for expelled primers is under the center station, and the primer-seating punch is under the left-hand station. The "push-button" powder measure is installed atop the expander die on the left station; it is held in place with its mounting bracket fitted over the die and secured by a pair of setscrews.

Operation is then quite simple. A case is placed in the center shell holder and run into the resizing die where it is also decapped, then it is extracted and transferred to the left-hand shell holder *after* a primer is placed in the primer-seating punch. The case is forced down over the primer; the case is run up over the expander and, while there, the powder-measure knob is pushed firmly forward, dropping the charge through the hollow expander. The case is now withdrawn and a bullet is immediately started in the case mouth; the complete assembly is then placed in the right-hand shell holder and run into the bullet-seating/crimping die (adjusted to crimp or not crimp as desired); when the cartridge is extracted from the die, all loading operations except inspection and cleaning or wiping have been completed.

With proper organization, a supply of components, a good loading-bench layout and a moderate amount of practice, a production rate of about 200 rounds per hour can be achieved and maintained on this C-H X333 multiple-station press. Other multiple-station presses have been manufactured, of course, but this is the principal onc with which we have had good experience. It's currently available and economically priced. While it is not terribly faster than an industriously-operated, single-station press, it does possess the particular advantage of completing all the operations on a given case in only a fraction of a minute. You don't have to spend a lot of time at the press before loaded cartridges are being produced—you are also not left with partially-processed cases when it comes time to shut down.

PROGRESSIVE LOADING MACHINES

The next step up from turret and multiple-station loading outfits is the "progressive" loading machine. For more years than I care to remember, the basic Star machine has been available and has been the first choice of those wishing to turn out large quantities of handloaded, handgun ammunition with a minimum of effort. Employing this loader without accessories, a single operator can turn out several hundred rounds per hour, with each stroke of the handle producing a completely loaded cartridge after all stations have been filled. In its basic form, primer feed is automatic from a vertical, tubular magazine, and powder charging is automatic by means of a fixed-charge, sliding-bar measure incorporated into the loader's construction. All other operations—placing the case in the loader, starting the bullet, indexing the rotary shell plate, removing the loaded cartridge (in the case of rimmed cases), and operating the handle—must be done manually. However, there are bolt-on accessories which eliminate the majority of these manual operations and which thus allow increased output. The first and most common of these is the Hulme case feeder. This bolts to the base of the tool, and consists of a cam-operated, sliding carrier combined with a vertical, tubular magazine. As the loader handle is brought down, a cam on the die head forces the carrier of the feeder outward, allowing one case to fall from the magazine (by gravity) into position in front of it; then, when the die head is raised, a spring compressed behind the carrier forces the carrier and case inward, positioning the case in the feed-way and on into its proper position in a case-holding recess in the shell plate. This eliminates the necessity for manually placing each case in the shell plate, leaving in this respect only the task of keeping the plastic-tube magazines filled with cases and exchanging them as they become exhausted. This produces a substantial increase in production rate. I have found that with a case feeder installed, it is possible to achieve a production rate of about 700 rounds per hour for short periods of time; a less hectic pace will usually produce completed cartridges at a rate of 400-500 rounds per hour. Another accessory which speeds up production is an automatic shell-plate indexer. In the manual mode, the plate must be rotated 1/6-turn after each handle cycle to move the cases successively under the proper dies. An automatic indexer accessory is available to perform this operation, actuated by a cam attached to the die head. Rotation occurs as the die head is elevated, thus no time is lost, and by occurring during the lightest part of the handle stroke, does not add significantly to the effort required. A case feeder may be employed alone, but it is not possible to employ an automatic indexer except in conjunction with a case feeder. The combination of an automatic indexer and case feeder is about as far as the typical Star loader user will want to go, and if a helper is available to keep an uninterrupted supply of cases, primers, and propellant flowing, a production rate of 800 rounds per hour is not particularly difficult to obtain.

Other accessories are available for the Star loader; however, they begin to get a bit costly. First is an automatic bullet feeder which functions much like the case feeder, transferring bullets from vertical, tubular magazines to a position over the case mouth where they are initially seated by a modified die. Automatic feed of bullets is a bit more dicey than other components simply because of the tackiness of bullet lubricant. It usually works well enough with dry-lubed bullets, but otherwise lubricant can rapidly accumulate in the magazine tubes and moving parts and cause malfunctions. The only way to avoid this problem is frequent and thorough cleaning before lubricant can build up to the point that it interferes with smooth operation.

At the top of the scale—both in complexity and cost—we have custom-conversions of the Star loader to power operation. Hydraulic power is the most common method of driving the loader, and I have personally performed such conversions utilizing surplus aircraft, hydraulic cylinders, control valves, pumps, etc. It is not a simple matter. However, if you are a competent machinist, you could probably design and complete such a conversion without a great deal of trouble. Custom conversions cost upwards of $1500, and this really places them out of the reach of anyone loading primarily for his own use. Only a custom-loading shop deriving substantial income from reloading could ever hope to justify the more than $2000 that would be invested in a

Star loader by the time it was fully equipped and converted to power-driven operation.

From time-to-time, other Star accessories are offered by various makers, including adjustable charge bars for the powder measure, automatic ejectors for completed rounds, stands or cabinets which facilitate operation and provide a container into which completed rounds drop automatically, large-capacity propellant reservoirs, etc.

Accessories aside, the basic Star loader is best described technically as of the "rotary-table progressive type." This means simply that upon its base rests a circular shell plate containing six carefully shaped and equally spaced notches or recesses around its perimeter (resembling and serving the function of shell holders) to accept the case heads. It is also provided with a strong ball detent which aligns the recesses beneath identically-spaced die stations in a "die head" which slides vertically on a strong, steel column above it. Reloading dies and powder measure are installed in this die head, which is connected to an operating handle by suitable linkage so that it may be lowered against the shell plate and raised several inches. Thus, when cases are seated in the shell plate, lowering the die head forces the dies and other operation stations over the cases. All operations on the case are performed in the die head with the exception of repriming; an automatic primer feed (fitted with a vertical, tubular magazine) and primer-seating punch are installed in the base of the loader and operated by a cam on the die head as it approaches the bottom of its stroke. The powder measure in the die head is likewise operated by a cam, but in this instance, the cam is installed on the base of the loader. The die head is keyed to the steel upright so that it does not rotate; the shell plate is rotated 1/6-turn after each handle cycle, placing each case successively under the next die or at the next operating station so that after one complete plate rotation has taken place, (assuming a correct supply of components) the first case in line has been fully loaded—and thereafter, each stroke of the handle and each 1/6-rotation of the plate produces a fully-loaded round. With practice and experience, all of the operations can be performed very quickly, and several hundred rounds can be loaded per hour.

The operations of indexing and inserting a fresh case can be combined (and should be) to speed up operation; removal of the loaded round is automatic (it falls through a hole in the base by gravity) with rimmed cases, but must be performed manually in rimless calibers; seating the bullet manually must be done separately, and can't be combined with any other operation nor can it be speeded up except by developing a high

The C-H "Auto-Champ" straight line, progressive loading tool incorporates automatic case feed, automatic indexing, automatic primer feed, automatic powder charging and automatic ejection of the completed round. This tool has more automatic features than will be found in the *standard* model of other progressive tools on the market.

1. (above) Here we see the left end of the C-H "Auto-Champ" loader with cases in the vertical clear-plastic magazine.

2. (below) Having begun with the handle (and die head) *down*, and with cases in the magazine, raising the handle raises the die head and advances the first case into position beneath the resizing/decapping die.

degree of digital dexterity.

I have owned and used Star loaders for over 20 years, both in their most basic form, and extensively modified to improve speed and ease of operation. The so-called "Progressive" model is the most desirable of the lot because of interchanging die heads, shell plates, and accessories it can be quickly and easily switched from one caliber to another. This model, in its basic form is probably the most practical for the individual who simply wants to load a lot of ammunition for his own use with a minimum of fuss and bother. When it comes to accessories, the case feeder is perhaps the most practical; and the cost is not prohibitive. Beyond this point, adding accessories becomes complex and costly, and increases the probability of malfunction. Consequently, I question the practicality of accessories beyond the case feeder for the average user. After all, with just the basic loader and case feeder, one's investment is well over $600, and the kit to change calibers adds a good deal to that. With just the case feeder installed, a couple hours spent loading one evening will produce more ammunition than you can possibly shoot over the weekend—and that seems more than a good investment to me. Even if you add several hundred dollars more in accessories, you'll still spend the better part of an evening loading the same amount of ammunition—so to me the added accessories aren't worth the extra cost.

The Star loader is really quite durable and long-lived. Twenty-odd years ago, I bought one new, and when it was finally disposed of a few years ago, it was still quite serviceable in spite of having loaded at least 1.2 million cartridges. Based just on that single personal experience and reports from many other owners, I feel safe in saying that if you buy a Star loader now, it will probably be just well broken-in by the time you turn it over to your son and will still be serviceable when your grandson's arthritis causes him to quit using it.

A much more recent high-volume loader is the Auto-Champ by C-H Tool & Die Company. This loader differs in almost all respects from the Star, basically because it is of the *straight-line* progressive type. Rather than the cases being carried around in a circle by a rotating plate, they are successively moved in a straight line from left to right under the various dies and work stations by a reciprocating shuttle bar. The Auto-Champ

3. (below) Lowering the handle all the way forces the die down over the first case, resizing, and decapping it and, at the same time, moving the primer-feed slide toward the operator to pick up the first primer from the magazine.

4. (above) Raising the handle again advances the first case to station two, and feeds a fresh fired case into station one; and, as the handle nears the upper limit of its stroke, the priming punch seats a fresh primer in the resized case at station two.

carries a case feeder as standard equipment, utilizing vertical, plastic-tube magazines. It also incorporates automatic primer seating (from a magazine) and automatic propellant charging. As the shuttle bar moves the case from left to right—with each pull of the handle—a die head carrying the dies and powder measure moves vertically on sturdy rods as the handle is rotated. Decapping is combined with resizing, and powder charging takes place through a hollow expander plug to further reduce the number of required stations. Additionally, there is a container for expelled primers to reduce loading mess.

The C-H Auto-Champ is a very advanced tool and it possesses the advantage of utilizing standard 7/8 x 14-threaded dies, thus avoiding the special dies required to fit the Star loader. At present the Auto-Champ seems to have reached its definitive form after several years of improvements. In this form, case feed, primer feed, powder charging, and case indexing are all automatic as the loader is shipped from the factory. Consequently, there is no real need for the extensive line of accessories that have developed over the years for the Star loader. The only thing one could possibly ask for on this loader is a bullet-feed system, and this really wouldn't speed up operation enough to justify its cost. In fact, if a bullet-feeding system were installed, there would be nothing for the operator's left hand to do, and that would be shameful.

In any event, while the C-H Auto-Champ does not have the many decades of public acceptance and satisfaction earned by the Star, it is a most excellent and advanced loader, offering 500 rounds or more per hour without difficulty and without costly additional accessories. It can be purchased for substantially less than the cost of a Star loader with the accessories necessary to give it the same capabilities. Its main disadvantage is that it's offered only in .38 Special and .45 ACP, while the Star can be had in many other calibers.

Regardless of the make and model of the progressive, high-volume loader one uses, there are a few rules to be followed if they are to give trouble-free service. First of all, it is absolutely essential that all dies be very carefully adjusted and locked and that automatic feeds such as case, primer, powder, etc, be double-checked and adjusted where necessary to the very "nth" degree. Minor maladjustments that would cause no particular problem with the ordinary manually-

5. (below) As the handle is lowered for the second time, the operations at station one are repeated, and at station two the hollow expander rod enters the case mouth to expand and flare it, and the powder measure slide moves toward the operator, dumping the powder charge through the hollow expander rod.

6. (above) As the handle is raised again, the first case is shifted over to station three, and the operations already described are repeated. When the die head is all the way up, the bullet is started manually in the case at station three, and the handle is brought down again to seat the bullet to the proper depth. The next handle cycle transfers its first case to station four, and applies the proper crimp. The fifth handle cycle pushes the freshly-crimped, completely loaded cartridge off the right end of the base, where it falls into a sheet metal chute which directs it off the rear of the tool.

operated press can be a continual source of trouble in these automatic mechanisms. The second and even more important rule is cleanliness; pure and simple cleanliness and neatness of the work area and the loader itself. Propellant spillage is a constant problem, especially when assembling loads in which the propellant charge bulks high in the case. And it becomes even more so with ball propellants. Anytime propellant is spilled, loading should stop immediately, and whatever steps necessary be taken to clean every granule of that propellant out of the working parts of the loader. If this is not done, propellant will combine with lubricant to gum up the works and make eventual cleaning much more difficult. Anything that interferes with smooth and orderly progression of cases will cause jerky parts movement which will in turn cause more propellant spillage, and create a real mess. All parts involved with the transfer of cases from station to station are critical to the correct functioning and must be kept clean and *properly* lubricated. Too little lubricant accelerates wear which causes its own problems, and too much lubricant simply collects airborne grit and also accelerates wear. Personally, I prefer very sparing appliction of molybdenum disulfide grease with the surplus thoroughly wiped off after the loader has been cycled a few times. This virtually eliminates lubrication problems, reduces wear, and makes cleanup of propellant spillages much easier. Occasionally a case feeder will malfunction, failing to position a case in the shell plate or shuttle. If you don't catch this the instant it happens, you'll suddenly be faced with a full charge of propellant dropped down into the works, thereby necessitating a disassembly and cleaning job. In addition, unless you are better at inspection than most people I know, a bent case will occasionally get into the feeder magazine and will be crumpled by the mouth of the resizing die. *Don't* try to save time by attempting to run the case on through. That will only give you headaches and may jam the loader up tight. Instead, as soon as the case is noted, stop loading, open up the machine, and remove the bad case. You may think that this takes more time, but believe me, it will save you time in the long run. Never run dirty or corroded cases through an automatic loader. Even though the carbide resizing die (with which such tools are usually fitted) will handle the cases well without damage, grit and dirt will be rubbed off the case by the die to accumulate on the track on which the cases move, forming a hard, gritty paste (with the lubricant) which greatly accelerates wear. It can, in fact, accumulate to the point where functioning is impared. You cannot avoid the hard gritty residue dumped into the case track by the decapping operation. Everytime a primer is punched out, some of that hard, black ash falls out of the primer pocket onto the track and is carried along by the case. Again, this material can accumulate and accelerate wear as well as produce malfunctions. It should be cleaned out periodically, and if you're using dry molybdenum disulfide lubricant, a jet of air does the job nicely every few minutes.

AUTO-PRIMING DEVICES

Automatic primer feeds are great—without them, we'd be hard put to build loaders that function nearly so rapidly. However, there is one potential danger inherent in their use. The moving parts must be made to fit closely in order to function correctly and align the primer properly for seating. Yet, you may still occasionally encounter (probably without knowing it) a primer with a high anvil or over-long cup that will rub slightly as it is transferred from magazine to seating position; in addition, perfectly normal operation of the loader produces a good deal of vibration which is transferred not only to the primer being positioned, but to the entire column of primers in the feeder magazine. These two factors combine to produce a phenomenon known in the trade as "dusting," the production of minute quantities of very finely-powdered priming compound. Even though the primer is sealed with a waterproof foil or paper disc over the priming-compound pellet, in any large number of primers there will also be a few exposed smears or traces of this compound which are eventually shaken loose and combined to form a highly-explosive dust. This dust will tend to accumulate in the feed mechanism and, if ignored entirely, will eventually accumulate to the point that excess friction or inadvertent shock *might* detonate it. I know, handloaders—particularly handloading tool manufacturers—don't like to talk about this; it's bad press, and they see it as an admission that handloading is dangerous. Hogwash. This dusting is dangerous *only* if one ignores it. If the priming system is inspected frequently and wiped out with a solvent-soaked swab every few thousand rounds, there'll be no cause to worry about a primer-dust explosion. The dust is visible, generally being green, and therefore easily seen against the smoothly-polished steel or brass surfaces of primer feeds. Simply accept the fact that extensive use will inevitably produce *some* accumulation of priming-compound dust. Once you do that, you've got the problem whipped, because you are aware of the need for watching for it and cleaning it out. Believe me, I'm not theorizing—I have been consulted in more than one such explosion, and experienced a very disconcerting "blow" in the first power-operated automatic loader I obtained nearly 20 years ago. It *can* happen, and it eventually *will* happen if you don't take the proper preventive actions.

Above and beyond the manually-operated progressive tools that will turn out several hundred rounds per hour, we have power-operated loading *machines* with capacities up to 5,000 or more rounds per hour. These machines, of which the Camdex is the only current one I know much about, cost upward of several thousand dollars; and, when properly set up and operated, they will turn out factory-quality ammunition from either fired or new cases. Such machines are far less flexible than any of the manually-operated loaders, and in my experience, require a great deal of operator skill and attention to operate continuously and efficiently. Machines like this are nice to dream about, but they are in no way suited to the needs of the *shooter*. Only large, custom-loading shops or small, loading factories can justify their cost or employ them efficiently. For these reasons, I don't think there is much point in pursuing either their construction or operating characteristics. If you are seriously interested in one, you need only contact the manufacturers, and they'll be most happy to supply you with all the information you want, perhaps even to the extent of arranging a demonstration. After all, not that many people are seriously interested in buying a loading machine that costs more than a new Chevrolet.

I think we have covered most aspects of high-volume handloading. As you know, there are literally dozens of ways that you can speed up the output of whatever equipment you already have; and, further, there are bigger, better, faster, and more expensive pieces of equipment readily available if you are dissatisfied with the speed at which you currently load ammunition.

CHAPTER 12

Handloading for Maximum Accuracy

WHILE THE AVERAGE well-assembled handload generally shoots pretty much where the gun is aimed, the ne plus ultra of pin-point accuracy more often than not eludes the inveterate hull-stuffer. The word "accuracy" is rather imprecise, and it usually means whatever the speaker wants it to under a given set of circumstances. I once watched a fellow out in Oklahoma unlimber an old .38-40 Colt SAA six-gun at an out-of-town dump and select a rusty, old, 5-gallon, grease bucket about 20-odd yards away as his target. Earring that big hammer back with a clickety-click-clack, he assumed a manly pose with off hand on hip, let drive with his first shot, then a second and a third, holstered his gun with a flourish, then walked up and proudly pointed to three holes scattered over the surface of the bucket, grinned broadly and said, "Pretty damned accurate ain't she." Then, I once watched a friend of mine stand spraddle-legged on a muddy pond dam down in Tennessee, very carefully two-hand his 9mm Browning auto, and let drive at a 1-gallon, bleach bottle sitting 125 yards away at an angle across the pond on a sandy bank. His first shot splashed low into the water, he took a bit more up front sight, and each of the next three rounds were fired only a couple seconds apart causing the jug to jump a bit as the bullets passed through it. He lowered the gun a bit, grinned (with just a bit of self-satisfaction) and remarked to no one in particular, "That's a close-shooting load." Then, for a more scientific look at accuracy, I have stood in a ballistics laboratory while the operator very carefully loaded and fired (by remote control) 10 consecutive rounds from an accuracy-test barrel clamped in a machine rest that was bolted immovably to a concrete pier sunk through the floor into the earth below the frost line—and when one shot opened the group up from 1¼ inches to 1½ inches, he muttered with disappointment, "This lot isn't very accurate." Each of those people had his own idea of accuracy with a particular gun under a particular set of circumstances. The casual, frontier-style shooter was satisfied to hit a 5-gallon can within spitting distance; the outdoor handgun buff was pleased when his shots stayed on a gallon jug at 125 yards; but the serious tester wasn't satisfied with a 1.5-inch group at 50 yards. The first shooter wouldn't have done any better, considering his skill and his gun, with the finest ammunition made; the second surely un-

"Accuracy" to one person might mean an entirely different thing to another. The *casual* plinker might be satisfied with, say, a 3-to 5-inch group at 25 yards, however, the gentleman pictured here might not be satisfied unless his group ran *under* 2 inches at the same range.

derstood that shooter, gun and ammunition all had to be finely accurate to pop a bleach bottle consistently at that distance; and, the lab technician knew that when nine rounds went into a circular, homogenous group and the tenth went out ¼-inch that the ammunition wasn't really up up to snuff.

There are two kinds of accuracy: The *mechanical* accuracy of the ammunition alone, undiluted by the effects of shooter and gun; and then there is *practical* accuracy which combines ammunition error, gun error, and shooter error, not to mention other external factors such as wind and weather. The loader of the ammunition can't help the gun shoot any more accurately when those same cartridges *must* be suitable for a wide variety of handgun makes and models. All the factory can do is build in as much mechanical accuracy as possible, leaving the final results at the mercy of the gun and the shooter. The handloader faces pretty much the same problem except that in some instances he may be able to match his loads to a particular gun and a particular set of circumstances, but even so, he can't overcome inherent gun inaccuracies. The handloader can't do much about the shooter, either.

The factors which affect ammunition accuracy have been well identified by valid, scientific testing. The eminent experimenter and author of the classic work *The Bullet's Flight From Muzzle To Target,* Dr. Mann, isolated a number of factors and ascertained that of all the components, the bullet was by far the most important to accuracy. Even though the good doctor made these determinations back around the turn of the century, they're still valid today!

Some of these accuracy factors you can't control at all, others you can control slightly and,

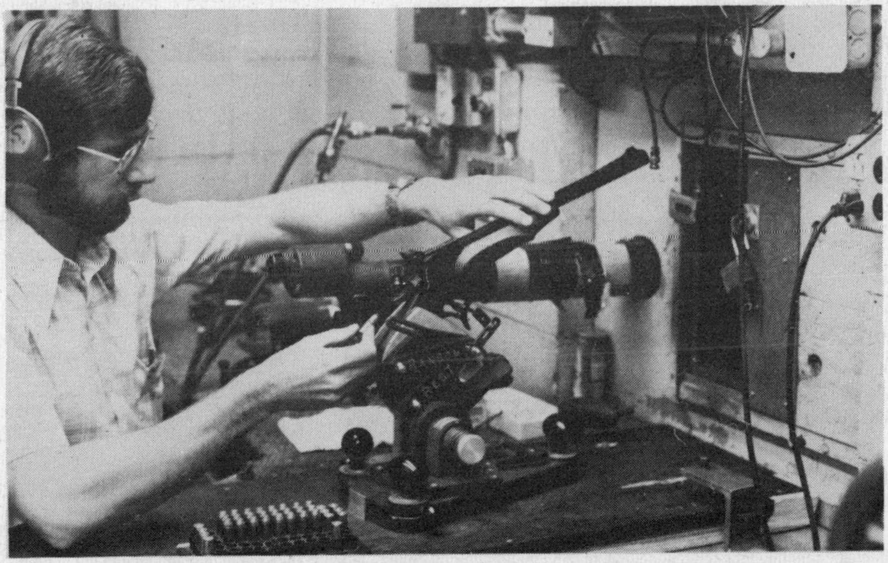

Ballistics labs are always interested in the "question of accuracy." The ballistician here is using a Ransom rest, in conjunction with a T/C Super-14. Under these controlled conditions accuracy is the name of the game.

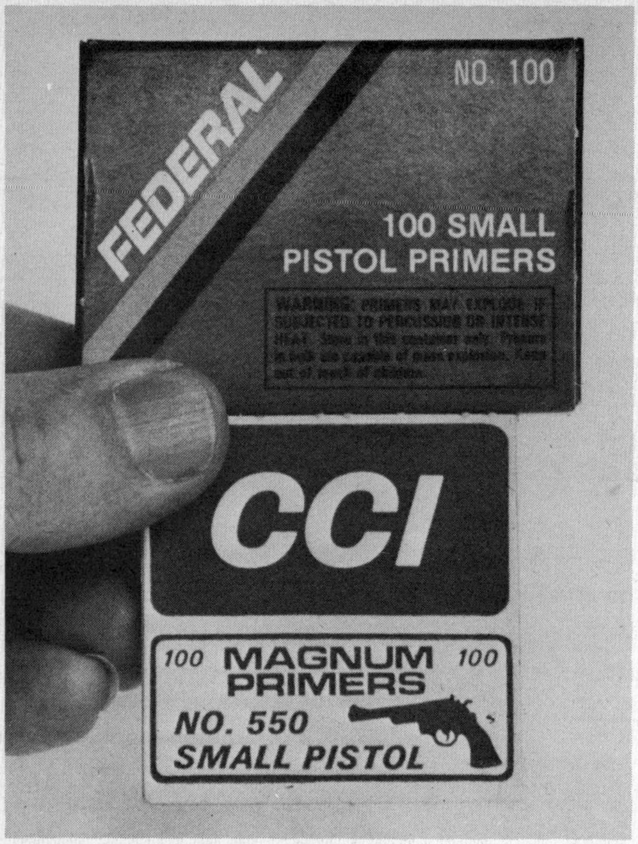

Primers, be they standard or magnum, happen to be the one component over which the reloader has no control. Fortunately, the manufacturers build superb quality and consistency into these items.

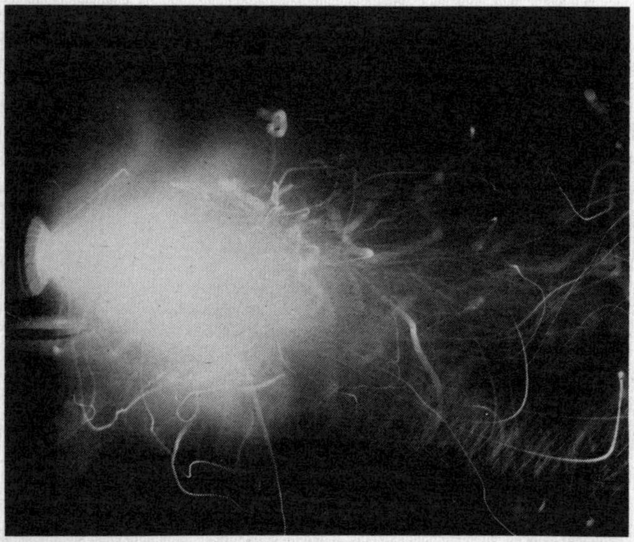

This particular photo nicely depicts the "flashing" of a primer. It's this sort of "controlled explosion" that gets your bullet down the barrel.

fortunately, the most critical factor—the bullet—is almost entirely within your control if you're willing to work hard enough at it. Let's begin with those factors over which you have the least control and get them out of the way so we can concentrate on serious business.

PRIMERS

The primer is the one component which you can not change whatsoever. All of the major primer makes are equally good and equally uniform; and, with the single exception of maximum magnum-type loads with the slowest-burning powders, the "standard" primer will produce superb accuracy. With those exceptional loads, the magnum primer is required to produce adequate and uniform ignition and combustion of the propellant charge. Though you must work with the standard factory primer, the manner in which you seat it in the case can affect accuracy. Ideally, the primer should be pressed into its pocket until the legs of the anvil make firm contact with the bottom of the pocket. In this position, the anvil provides the most uniform resistance possible to the firing-pin blow, thus producing uniform crushing and ignition of the priming pellet. If every primer in a given batch of loads *is seated in this fashion with equal pressure,* you'll have the most uniform primer function possible. If the primer is not seated to the bottom of the pocket, it will yield somewhat, moving deeper in the pocket, under the firing-pin blow, and less of that blow's energy will be available to crush the priming compound. On the other hand, if the primer is seated with too much force, the priming pellet will be placed under considerable pressure, perhaps even cracked or broken to bits, as it is crushed between the crest of the anvil and the cup. Under those conditions, the priming compound is ignited erratically by the firing-pin blow. Erratic primer function means erratic ignition of the propellant charge, and that brings about erratic ignition of the propellant charge, and that brings about erratic propellant combustion, and that produces wider-than-normal velocity variations, Those variations show up on the target by increased vertical spread of the shot group, commonly referred to as "vertical stringing." Dimensional variations in primers, primer pockets, case rims, shell holders and primer-seating devices make it impossible to obtain that ideal primer/case relationship

Consistency of primer seating is the key to consistent cartridge ignition. Be sure to seat all primers in the bottom of the primer pocket; however, don't apply so much force that the primer pellet is crushed.

Powder charges, too, must be uniform, otherwise accuracy will suffer. Weighing every charge may not be the best idea; however, it's a good idea to weigh every 5th or 6th cartridge to insure uniformity and safety.

by use of any mechanical stop. You'll have to do it by carefully "feeling" the primer into the pocket, noting the increase in resistance as the anvil contacts the bottom of the pocket. With a little practice, you'll not only be able to seat them to the bottom of the pocket, but with uniform pressure. And that, friends, is what will produce maximum accuracy insofar as primer functioning affects it.

PROPELLANT

The propellant comes next; and, the only options you have here is to select the one best suited to the type of load you're using, and to further use the correct and most uniform amount of it in every round you load. Chapter 7 on primer and propellant selection covers these details; and, once you've selected the right propellant in the right amount, all that remains is to insure reasonable uniformity of charge weight. There is no need for weighing *every* individual charge; in fact, if you try to do so with the smaller charges, the attendant tedium and frustration may produce charges that vary more than you'd obtain from a properly-manipulated measure. Remember, the keys to maximum charge uniformity from any propellant measure are solid mounting, a uniform head of propellant or alternatively a baffle in the reservoir, and absolute uniformity in speed and force of handle operation.

CARTRIDGE CASE

Next comes the case. First of all, the case must be in good condition, without excessive bulges or dents, and there must not be any cracks—not even the most minute—in the mouth. Tests have shown quite clearly that even if all other factors are exactly as they should be, a case with even the slightest crack in its mouth will throw its bullet out of the group. Minor variations in case volume that will be encountered between manufacturing lots (which you'll not be able to identify anyway) have no measurable effect on accuracy. Furthermore, the small variations in volume that sometimes exist between different makes of brass—with the possible exception of certain military-contract .38 Special cases whose volume is significantly less than that of the commercial variety—won't have much effect. The large volume variation to be found between modern, solid-head cases and old, pre-war, balloon-

Above—While actual case length has little effect upon a *revolver's* ability to function, it does play an important role in the functioning of an auto—be sure they are trimmed to a proper overall length.

Left—When checking your cases over, be sure to separate the damaged cases (right) from the good ones (left). It has been proven that a case with the slightest crack at its mouth will throw a slug out of the main group.

head cases will affect accuracy to the extent only that mixing the two types will substantially enlarge groups. Sorting cases by manufacture and headstamp has not been proven to increase accuracy, but it requires little effort, and it certainly can't do any harm.

Cartridge cases vary a good deal more in length than one might expect. In fact, one reference cites lengths ranging from 1.120 inches to 1.168 inches for the .38 Special and .884-inch to .906-inch for the .45 ACP. I have personally encountered variations very nearly that great, and have seen at least half that much variation in the *same box* of factory-loaded ammunition. In revolvers, even fairly large length variations affect accuracy only inasmuch as they affect the degree of crimp. Since with a given die adjustment a longer case will be given a heavier crimp, reasonable length uniformity is desired. Since case length has nothing to do with functioning in revolvers, the cases may be trimmed to uniform length and thus insure maximum crimp uniformity. This applies equally to rimmed revolver cases employed in certain target autoloaders. Rimless and semi-rimmed cases intended for autoloaders are another matter. These cases usually headspace upon their mouths against a shoulder at the front of the chamber; and, it is this relationship that supports the entire cartridge against the firing-pin blow. Thus, if a case is over-short, ignition will suffer. Consequently, we can't trim autoloader cases to achieve uniform length without interfering with proper primer ignition. Therefore, the effect of case length on accuracy can be minimized by selecting the longest cases from any given batch, and reserving them for loads from which you must have maximum accuracy. Several gun modifications have been developed over the years to minimize the effect of case-length variations on accuracy, but we are concerned here only with what you can do in loading the ammunition, so that's another story for another time. However, if you really want to look into it in detail, may I suggest a copy of my book *Pistolsmithing,* published by the Stackpole Company and available most places gun books are sold.

CRIMPING

The manner and degree in which the case mouth is crimped upon the bullet may produce a measurable effect on accuracy. Two types of crimp exist, "roll" and "taper." The former consists of turning the case mouth smoothly in a curve or radius into a groove in the surface of the bullet. This is accomplished by forcing the case mouth against a curved shoulder in a crimping die. The taper crimp consists simply of giving a shallow taper to the forward portion of the case by forcing it into a tapered die cavity. Applied in the proper degree, both types of crimp appear to produce equally accurate ammunition. However, in my experience, the *average* roll-crimp die is more likely to produce an eccentric crimp than is the average taper-crimp die. The latter is self-centering, without any need for close fit upon the case, while the former may turn in a bit more case on one side than the other unless it fits the resized case very closely. That degree of fit isn't practical in a production die made to be sold at a reasonable price. Up to this point these remarks apply to light target loads, and to typical, non-magnum, full-charge loads. In magnum and high-performance loads in revolvers, a heavy, tight, roll crimp is essential to prevent the bullet from shifting in the case under the influence of recoil. Consequently, you'll have to use a roll crimp on loads in that class.

As already mentioned, autoloading pistol cartridges headspace upon the squarish case mouth and any significant degree of roll crimp removes that surface and allows the case to go a bit deeper into the chamber with the effect that ignition may be impaired. Some handloaders use a roll crimp anyway and depend upon bullet contact in the chamber throat to support the cartridge against the firing-pin blow. This can get a bit sticky, though, and thus a taper crimp is generally chosen for such cartridges.

Light, target loads in revolvers require relatively little crimp. When a roll crimp is employed, running the case into the die until the first .020-inch of the case mouth is turned over is about right. Applying excessive crimp beyond this point will often reduce the load's accuracy. When a taper crimp is used, measure case diameter over the seated bullet just far enough back from the mouth so that the residual flare doesn't give a false reading, then run the case into the die until .015- to .020-inch reduction of that diameter is obtained. Again, with light loads, excessive crimp will often reduce accuracy. In autoloader target loads, apply taper crimp to reduce mouth diameter .015-inch.

Another factor rears its head when striving for accurate autoloader loads. The bullet may be held tightly enough for good accuracy; but, if feeding impact forces the bullet deeper into the case, then loading density will be altered, and that particular cartridge will display a greater vertical deviation (due to velocity deviation) than one with the bullet held solidly. With some batches of brass, it may be necessary to use an undersize expander plug and/or resizing die to make the case hold the bullet a bit tighter. If even that doesn't work, then expanding the case only to the depth of the normal bullet seat will usually cure the problem. By expanding the case in this fashion, you will create a slight shoulder (internally) against which the heel of the bullet can rest securely. If even *this* approach fails, I would recommend putting a cannelure on the outside of the case at the point where the bullet comes to rest when seated (see Chapter 10 for details on applying a cannelure). In any event, as every cartridge is fired, its bullet must be seated to the same depth or excessive vertical stringing may result.

Left—Crimping for auto cartridges is usually called a "taper crimp" as shown. Note that the mouth of the cartridge case is not turned into the bullet as is done on revolver cartridges.

Right—On the left you can see a .38 Special that's been *heavily* roll crimped (a good idea for heavy loads). On the right is the same cartridge with a *light* roll crimp (more suitable for light loads).

Above—If a bullet is going to give you top accuracy, it must be perfectly cast. The slug shown here comes from an RCBS mold and usually weighs a nominal 124 grains. As you can see the weight of the slug is about 120.7 grains; the slug has a void—it *will* affect accuracy and should be put back in the pot.

Right—Leading can contribute to poor accuracy. Be sure to use a quality bullet lubricant and you'll get the results you're after.

BULLETS

And now we come to the most important factor of all. The four factors we've already discussed probably account for only 10 percent of the accuracy a given load will produce, and the *bullet* accounts for the rest; it is all-important.

To begin with, the bullet must be perfectly cast; no internal voids, a perfectly flat and square base, with every sharp corner and edge filled out cleanly. Internal voids cannot be seen, and the only way to detect them is to weigh the bullets. Generally speaking, those which vary more than 2 grains either way from the average contain voids and are not suitable for really accurate loads. Exact bullet weight isn't important—a 140-grain projectile may shoot just as accurately as a 150-grain bullet—but *uniformity* of weight is. If the extreme weight spread is held to less than 2 grains, fine accuracy will be obtained; theoretically, holding this to ½ grain is better, but machine-rest tests have not shown any predictable difference.

As-cast bullet diameter is more important than most people realize. Bullets up to .0015-inch larger than sizing-die diameter produce very fine accuracy. When as-cast diameter exceeds that amount, accuracy begins to deteriorate—bullets .002- to .003-inch larger than the sizing die can increase group size 1 inch or more. If the bullets start larger than that, then genuine target accuracy is beyond reach. The reason for this is fairly simple if you understand the problem. The lead ribs forming the bearing surface between the lubricating grooves are relatively soft, fragile, and easily deformed. They will withstand a moderate reduction (up to .0015-inch) in being forced into a sizing die, but beyond that point, they are likely to swage down unevenly, under the much greater pressure required, and thus the bullet is made eccentric. In short, the center of mass moves away from the mechanical centerline, throwing the bullet out of balance, causing it to travel in a bit of a spiral as it heads downrange; and the more it's out of balance, the wider the spiral will be. It is also quite important to accuracy that the bullet be accurately aligned with the sizing die as it is pressed in. The slightest amount of tipping will make the bullet considerably eccentric, and accuracy suffers badly. Always use the more-modern type sizing die that features a long taper joining its start and finish diameters, and make certain that the top punch is accurately aligned with the die cavity and that it closely conforms to the nose of the bullet you're sizing.

Modern Alox-type lubricants produce accuracy substantially superior to any of the older formulas. The old home-brew formula of beeswax, petroleum jelly, and graphite just won't cut it any more. Generally speaking, maximum accuracy will be obtained with the minimum lubrication that prevents bore leading. Most revolver target bullets contain two or three lube grooves, and if filling only one with lube will prevent lead-

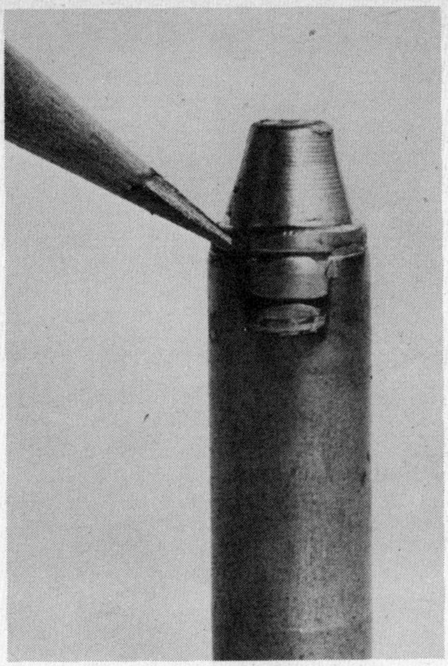

Left—When flaring a case, be sure it's flared sufficiently to allow smooth entry of the slug prior to seating. If you don't, you'll end up with a shaved bullet as shown. (If this should happen, discard the round and attempt to salvage the components.)

Above, left and right—If you choose not to cast your own slugs, you can purchase commercial swaged bullets from the major manufacturers. On the left is a Hornady .38 caliber hollow-base slug; on the right is a Speer .38 caliber SWC—both are excellent performers.

ing, don't use any more. In fact, if one prevents leading consistently, filling the other grooves will probably reduce accuracy somewhat. With autoloader bullets, there is seldom any choice, since they usually contain only one lubricant groove. Lubricant on the base of the bullet may not necessarily affect accuracy of itself, but under some conditions, it might contaminate the propellant and cause a reduction in velocity which will raise that old bugaboo "vertical stringing." Adjust the lubricator/sizer so that no grease is left on the bullet base, and if you can't avoid it entirely, wipe completed bullets across a cloth lightly moistened with lighter fluid or solvent to get rid of the grease.

Bullets should be protected from damage between casting and sizing, and even more so between sizing and loading. Slight nicks and mars on the cast surface will usually be ironed out okay during sizing, but a damaged base will not, and it will certainly reduce accuracy.

The hardness of the bullet alloy can have an effect upon accuracy. While many revolvers will shoot quite well with fairly soft bullets, autoloaders prefer them fairly hard, not only for accuracy, but for functional reliability. Extensive NRA tests some years back established that *18 parts lead, one part tin,* and *one part antimony* make an ideal, all-purpose alloy which produces best accuracy in both types of guns. *Fresh* linotype metal is just slightly harder than this, and shoots equally well. Use of an alloy too soft can cause feeding problems in autos and the amount of lubricant required to avoid leading may cause it to shoot a little bit less accurately than the harder bullet.

Bullets must be seated straight in order to be delivered into the rifling in proper alignment. If a bullet is tipped a bit as it enters the rifling, it is thrown out of balance, and will take off from the muzzle in that accuracy-reducing, wide, spiral we've already mentioned. Straight seating requires that the bullet: (1) be started straight before entering the die; (2) that the die fit the case closely; (3) that the seating stem fit the bullet nose closely; and (4) that the stem be centered in the die. The elasticity of the case will compensate for minor misalignment, pulling the bullet back into proper position as it is seated. In uncannelured cases, an amount of misalignment that will reduce accuracy can usually be seen or felt in the form of a very slight bulge or ridge over the bullet base. Any cartridge showing this symptom should be culled for practice or plinking. If unusually little or great resistance is felt as the bullet is seated, cull that particular cartridge for practice or plinking; wide variations in bullet pull produced by tight or loose case mouths will also cause vertical stringing.

Providing the case mouth was properly flared in preparation, the bullet will not be damaged in seating. If there's insufficient flare (and this will occur often with an abnormally short case), the sharp inner edge of the mouth may gouge one side

Once you select your components and develop an accurate load, you'll get good accuracy. This 25-yard, 5-round group, was fired in the S&W Model 39 shown using 4.6 grains of Unique and the RCBS 124-grain cast slug.

of the bullet or scrape lead up ahead of it. In severe cases, a segment of the mouth may dig deeply into the bullet, perhaps even crumpling the mouth a bit and piling lead up over it. Any of these defects relegate the load to plinking, because the bullet will have been made eccentric and we've already mentioned how that plays hob with accuracy.

If the foregoing seems to apply only to cast, lead bullets, it's simply because this is by far the type most often used by accuracy conscious handgunners. They are cheaper and more available than other types, and if made and prepared properly, they shoot as accurately as any other type. And, as mentioned elsewhere, they put a bit less wear and tear on the gun. If pure target loads are your goal, and you don't care to go to all the trouble of learning to cast perfect bullets and prepare them for loading with equal perfection, then excellent factory-swaged bullets are available at reasonable cost. Providing these commercial bullets are carefully inspected for defects produced by handling, packaging, and shipping, they will do very well by you if all the other factors are given the attention mentioned. Such swaged bullets are normally made of a softer alloy than you'd use for casting, and this makes in-process damage more likely. Look out for it, and relegate to plinking any load in which the bullet gets messed up even slightly during seating. Actually, swaged bullets that haven't picked up handling damage are more uniform in density and weight than the best you can cast; and, not having been subjected to the indignity of being made oversize and then squeezed down, their concentricity is assured.

The accuracy of even the heaviest, lead-bullet, hunting and defense loads—magnums included—can usually be substantially improved by paying particular attention to all the details we've dwelt upon here. To go a step farther, exactly the same may be said of high-performance loads with jacketed, expanding bullets, allowing, of course, for the special techniques applicable to such loads.

Though few handloaders will admit it, lack of attention to at least some of the factors we've discussed in the preceding pages is the reason that the *average* handloader's product is substantially inferior in accuracy to run-of-the-mill factory ammunition. We all like to think that every cartridge we load, no matter how carelessly, is at least as good and maybe better than what comes out of East Alton or Lone Oak. Unfortunately, we're only kidding ourselves; and, if we don't pay attention to all the details that make a bullet or load accurate, we'll keep on shooting loads that really aren't quite up to snuff. At least some of the points we've brought up apply to virtually every load you might even think of assembling, so don't get the idea that something a bit offbeat doesn't require the same attention.

CHAPTER 13

High Performance Loads

BY VIRTUE OF the use to which they are put, regardless of caliber, hunting and defense loads are intended to develop maximum power within the limitations of a particular gun/cartridge combination, and further to transfer the maximum possible percentage of the bullet's energy to the target when it is struck. This means that we must not only strive for a combination of bullet weight and velocity that will generate maximum energy, but for a bullet of such a type and design that it will give up as much of its energy as possible in the target. Ideally, this would mean that regardless of the size of the target, the bullet would come to a stop inside it, and not exit. In that fashion, *all* of the bullet's kinetic energy would be transferred to the target. Unfortunately, wide variations in targets and in the point and angle of bullet impact on the target make it impossible for this ideal result to be achieved more than just occasionally.

The fact remains, though, that whether you are shooting a man to prevent him from harming you or a whitetail deer for meat, the bullet must first stop the target. In the case of the man in a defensive situation, he must immediately prevent his attacker from any further offensive action, and that requires that the attacker be totally incapacitated, at least temporarily. According to some social mores, it would be ideal if the bullet could accomplish this without killing him, but generally a projectile that has a high capability for incapacitation also rates quite high on the lethality scale. In the case of big-game, immediate incapacitation is also our first goal, mainly to prevent a wounded animal from escaping, dying uselessly or experiencing undue suffering. By traditional hunting ethics, the ideal bullet is one that produces instant death, thus serving those three purposes. As a practical matter, there really isn't any difference in the ultimate effect of hunting and defense bullets. The avowed purpose of the former, agreed upon entirely by all, is purely and simply to kill; while others insist that the purpose of the latter is only to "stop" or "incapacitate," but not to kill. Unfortunately it simply isn't possible to shoot someone hard enough to stop him without taking a pretty good chance of killing him.

In years gone by, factory ammunition was generally improved upon for hunting and defense use by handloading with a heavy, cast, lead bullet of semi-wadcutter form driven at the highest, practical velocity—generally around 1100 fps in

Whether you hunt big game (left) or defend the populace as a law enforcement officer (below) you must first *stop the target*.

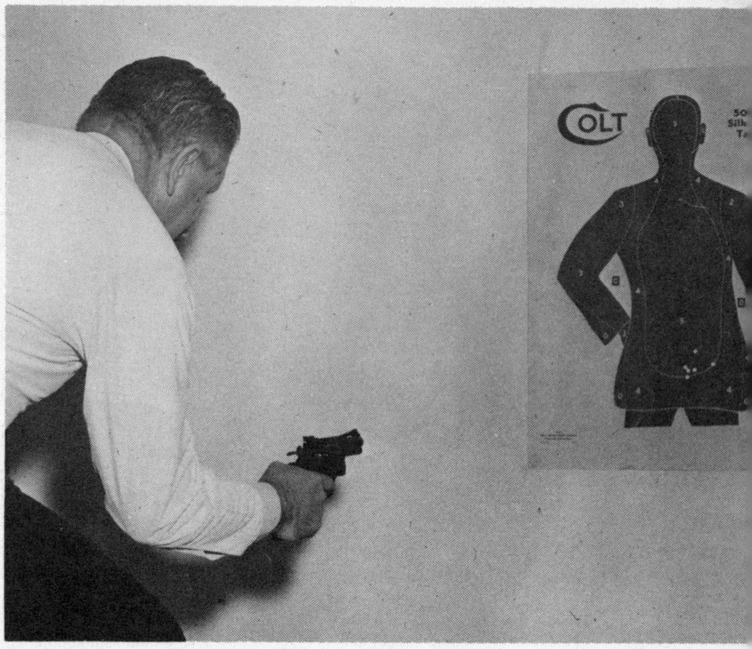

non-magnum revolver calibers. Such loads were generally termed "Keith type," having been heavily publicized and mainly developed in the 1920s and thereabouts by Elmer Keith of Salmon, Idaho. Such loads did offer a definite improvement over typical factory ammunition. They often more than doubled bullet energy, increased tissue destruction and wound-cavity size by virtue of the SWC shapes, had greater disruptive effect, and further produced deeper penetration in tissue. Up until the 1960s, little was done by handloading to improve the performance of autoloading pistol cartridges in this area simply because cartridge length and case volume limitations prevented taking the heavy-bullet/higher-velocity approach. A few handloaders did dabble in this area, mainly with the .45 ACP, employing target-type, semi-wadcutter, lead bullets of lighter-than-standard weight at the highest velocities that could be obtained safely. Such efforts weren't terribly popular, in spite of their effectiveness, because they went counter to the traditional heavy-bullet approach.

Prior to the 1960s, that was all that had been done on a relatively large scale, but certain individuals like the late Jim Harvey (operator of Lakeville Arms) had developed short-jacketed, pure-lead-core bullets of light weight along with loads driving them at abnormally high velocities in revolver calibers. Such bullets and loads were found to produce massive wound cavities and limited penetration, thus offering a higher degree of lethality than had yet been achieved. Bullets of this general type were offered by various custom shops to the handloader, but were never available in sufficient quantity to have any particular impact on handloading in general. Actually, as impressive as their terminal ballistics were, they still possessed a number of disadvantages, particularly in the areas of bore leading, jacket separation, and the like.

In the early 1960s, economical equipment was introduced by C-H Die Company for swaging half-jacketed bullets. For a period of time, such bullets were driven at very high velocities for tremendous terminal ballistics, but in time, their disadvantages too (mainly leading) became apparent, and their use declined. Also in the early 1960s, Super Vel ammunition and bullets appeared on the scene, designed specifically for maximum velocity and lethality on animal targets. About this time, the term "high-

Below — Good penetration coupled with good expansion is the key to successful hunting (or defense work) with any handgun.

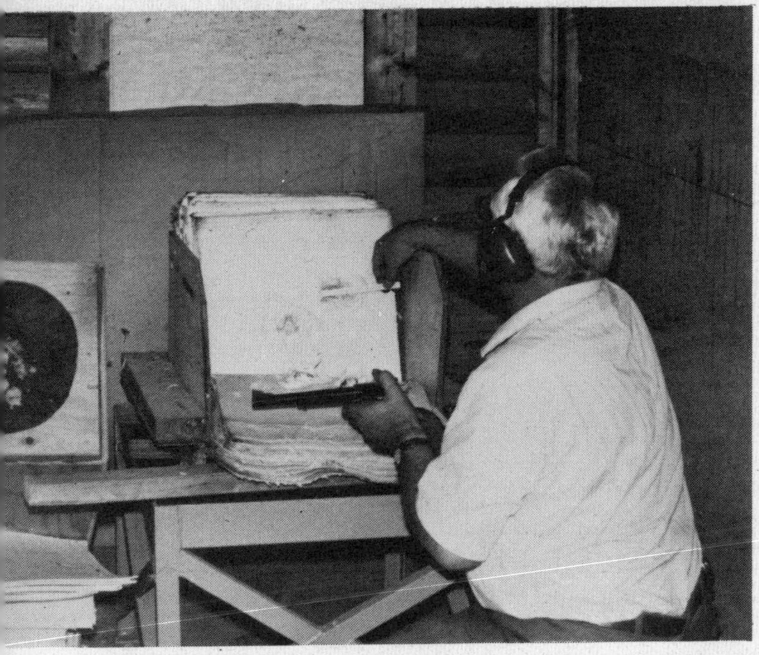

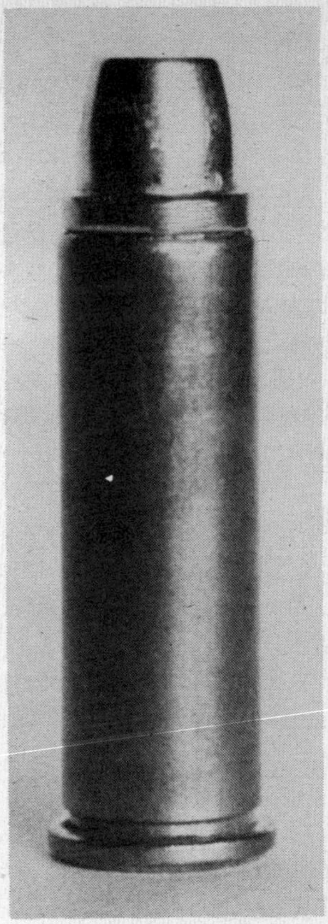

Right — Years ago, Elmer Keith developed what is now referred to as a "Keith-Type" bullet. This slug, as shown, is cast hard and is usually quite heavy, representing the uppermost limits of bullet weights normally utilized in a given caliber/cartridge combination. This sort of slug is a good game- or man-stopper.

performance" was coined to describe handgun ammunition loaded with lightweight, jacketed, expanding bullets (*genuinely* capable of significant expansion) at higher than normal velocities. I really can't say who first used the term in reference to this ammunition, but believe that I was the first one to use it in print, and today it is commonly applied to this type load. Following the advent of Super Vel ammunition and its wide public and governmental acceptance, all major manufacturers developed comparable loads, and today they are quite common. This is the type load we refer to in the title of this chapter.

BULLETS FOR HIGH-PERFORMANCE LOADS

First of all, you can't have "high-performance" loads unless they do actually deliver the performance we've just described; just laying on the name means nothing—and, they can't do that unless you use the proper combination of bullet and propellant. Essentially a load qualifies in the high-performance category if it uses a lighter-than-standard, jacketed, expanding bullet at higher-than-normal velocities; or, if it uses a standard-weight, jacketed, expanding bullet at higher-than-standard velocities. And, it isn't genuinely of the high-performance type unless the bullet actually does expand with a reasonable degree of consistency under normal conditions of use.

Suitable bullets in .38 Special range from 95- to 110-grain weight; in .357 Magnum, from 110 to 140 grains; in .41 Magnum, in the vicinity of 170 grains; in .44 Special and .44 Magnum, around 180 grains; in .45 Colt, 190-200 grains; and, in the autoloading calibers, in .380 caliber, 80-90 grains; 9mm P., 90-110 grains; .38 ACP/Super, 90-110 grains; and in .45 ACP, 185-200 grains. Such bullets are available from a wide variety of independent makers, both large and small, and to a lesser extent from our major ammunition manufacturers. They are generally available in both hollow-point and soft-point form, and as a practical matter, the latter will usually give slightly greater penetration and slightly less expansion than the former when weights and velocities are approximately equal.

True, high-performance, jacketed bullets differ from the traditional expanding type in that they consist of a very soft or pure lead core combined with a very thin and soft or malleable copper jacket. The softest, practical core and jacket are absolutely essential if reliable expansion is to be obtained at the velocities we can produce at safe pressures in handguns. Generally speaking, reliable expansion of any degree cannot be obtained at *impact* velocities of less than 1000 fps. Thus if the desired performance is to be obtained, these bullets must be driven at muzzle velocities well above 1000 fps. After all, the sectional density value of short, light bullets is low, and the rate at which they shed velocity as they travel downrange is high. Just remember that bullets do not strike the target at muzzle velocities; their velocity must be high enough to insure that the velocity remains above 1000 fps at the maximum range to which the load will normally be employed.

The best high-performance bullets are those which combine the previous characteristics with a relatively long bearing surface and a major diameter that is just a wee bit less than standard. For example, .357-inch is considered standard diameter for .38 Special and .357 Magnum, but some of the best H-P bullets measure only .356- or .3555-inch. Being slightly undersize permits these bullets to accelerate out of the cartridge case a bit more rapidly, thus delaying the pressure peak a bit; yet, at the same time, allowing their *soft* construction to upset very quickly and thus seal the bore fully to prevent gas-cutting and to produce maximum accuracy. The relatively long bearing surface (for the bullet's weight) aids in obtaining a heavy, consistent bullet pull* which is essential for proper ignition and combustion of the rather large charges and types of propellent such loads require. In addition, the long bearing surface tends to align the bullet better in the bore and thus improves accuracy.

PROPELLANTS FOR HIGH-PERFORMANCE BULLETS

Propellant selection for high-performance, light-bullet loads is quite important, We've discussed this to some extent already in Chapter 7 but a bit more attention is in order. W-W 231 ball propellant is probably the best choice for the .380 ACP, and with the lightest bullets, is a good choice in 9mm Parabellum. For the .38 Super, .38 Special and .357 Magnum, Unique is as good a propellant as can be had; and, it performs nearly as well in 9mm Parabellum, and extremely well in the .45 ACP. Unique is also an excellent choice for the long .45 Colt cartridge. In the .41 and .44 Magnums, Hodgdon's H110 produces superb performance with the lighter bullets, and W-W 296 does nearly as well. There are, of course, other propellants that can be used to produce nearly the same performance levels, but these are the ones which have proven to be the most efficient in our own considerable testing.

Little choice is available in primers, with the standard type (of the proper diameter, of course) being the best in all calibers except in the big

In the early 1960's Lee Juras pioneered the now famous Super Vel line of ammunition. Super Vel ammo features very high velocities and good bullet expansion — it was, and is, some of the best "high-performance" ammo made.

Hercules 2400 and Hodgdon's H110 have long been favorites with handgunners who shoot the magnums; however, the selection of propellant should be balanced against the weight and type of projectile to be loaded.

*Bullet pull is the force required to extract the bullet from the loaded case, measured as the static force in pounds necessary to pull the bullet.

magnums with large charges and slower propellants. Ignition will be improved in those if the switch is made to magnum-type *handgun* primers. *Don't* try to use rifle primers.

CASES FOR HIGH-PERFORMANCE BULLETS

Quite effective high-performance loads can be assembled in good-condition fired cases. However, extensive tests conducted by Lee Jurras several years ago clearly established that the ability of a given case to grip the bullet adequately and consistently decreases somewhat with every firing and reloading. Inasmuch as a heavy bullet pull is necessary for most efficient ignition and combustion of large charges of fast-burning powder behind light bullets, a case becomes less desirable for this use after several firings. A load developed to produce maximum velocity within allowable pressure limits in *new* cases will therefore produce less velocity and energy in cases that have been fired and reloaded several times.

(Above) From left to right: .357 Magnum; .38 Special; .38 Super; 9mm Parabellum and the .380 ACP. (Below) From left to right: .44 Magnum; .44 Special; .45 L.C., and the .45 ACP. All of these cartridges utilize bullets whose weight and construction vary in accordance with the cartridge's intended use.

The degree of this loss will vary according to the load, but generally, the heavier the load, the more rapid the deterioration of the case in this respect. Consequently, whether your loads are intended for hunting or defense, there is considerable merit in purchasing factory-new cases, firing the first shot from them for your most serious shooting, then relegating them to less potent loads, and finally reserving them for ordinary service, target, or plinking use. Since the number of shots one might fire in serious hunting or defense situations will be rather few over a given period of time, the added cost of buying new cases really isn't enough to worry about. After all, it isn't as if you were going to throw the cases away after a couple firings; they'll be used for lesser purposes until they actually become defective.

There is another factor to consider in the use of new cases for high-performance loads in autoloading pistols. That portion of the case which is unsupported by virtue of the feed ramp entry into the chamber will almost always bulge slightly under the pressure of a heavy load. This *can* weaken the case, and if it is subsequently used for a high-performance load, the likelihood of a case failure becomes somewhat greater, though not necessarily highly probable. Actually, we're dealing in abstracts here, because some autoloader cases will remain entirely serviceable in this respect through a great many loadings. However, the probability of failure is definitely greater than in a revolver, and if your life or a costly hunt is hanging on the absolute reliability of that case, then I for one would much prefer that it be new, not used and worn. If my personal practices are of interest, let it be known that the H-P loads I will carry for serious social intercourse or for any hunting more than the most casual are assembled in factory-new cases.

Regardless of whether new or fired cases are used, bullets must be assembled into them quite tightly to produce the necessary bullet pull. With at least some high-performance bullets being undersize, this can present a problem. Fired cases resized and expanded on ordinary dies may not grip the bullet tightly or even at all. With some lots of brass, a standard resizing die may not reduce the case mouth sufficiently, or a standard expander plug may open it up too much. After seating but before crimping, the bullet should be held tight enough by the case that it can neither be rotated nor pushed deeper in the case by hand. If it can be moved in this fashion, then a smaller-diameter expander plug and/or resizing die is needed.

Above — Every firing and reloading of a case tends to decrease its suitability for use in high-performance loads. This cartridge case has been loaded several times and heavily crimped during each loading — not the best choice for a H-P load. Stick with new cases.

Below — When loading up high-performance ammo for autoloader use, stick with new brass. The case on the left is new, the one on the right old. Note the bulging on the unsupported portion of the case on the right — while this amount of bulging is common, several loadings will weaken the case making it unsuitable for H-P loads.

CRIMPING OF HIGH-PERFORMANCE LOADS

High-performance bullets intended for use in revolver cartridges contain a cannelure into which the mouth of the case must be crimped. This crimp should be as heavy as the cannelure will accept without wrinkling the case. The mouth should be turned in far enough that it seats tightly in the bottom of the cannelure. The combination of crimp and tight assembly provides the desired bullet pull, and the more uniform this pull value is, the more consistent velocities—and therefore accuracy—will be. Consequently, significant variations in case length, which will affect the degree of crimp, should be avoided. If a given batch of cases intended for revolver high-performance loads varies more than .005-inch in length, you'll be well advised to trim.

Autoloader cartridges—especially of the high-performance type—are subject to heavy impact during feeding. If bullet/case assembly is not tight enough, this can cause the bullet to be pushed deeper into the case—with the inevitable result that failures to feed will occur and that cartridges with such receded bullets will not shoot to the same point of impact as those seated to the correct depth. Even ammunition factories encounter problems in this area, and often find it necessary to employ not only tight assembly and tight crimp, but a glue-like mouth sealer between case and bullet—in extreme cases, even a cannelure in the case supporting the base of the bullet. New cases are seldom supplied with a cannelure, simply because the position of it will vary according to the bullet loaded. A cannelure position chosen by the manufacturer probably wouldn't satisfy more than a minute percentage of handloaders, and the rest would beef about it. To be absolutely certain that you won't encounter the problem of receded bullets, roll a shallow cannelure into the case (as described in Chapter 10) about 1/32-inch above the position of the base of the seated bullet. Located thusly, the shoulder formed inside the case by the cannelure will be dislocated very slightly by the bullet, and will, therefore, provide maximum support. Don't make the cannelure too high on the case, for if the

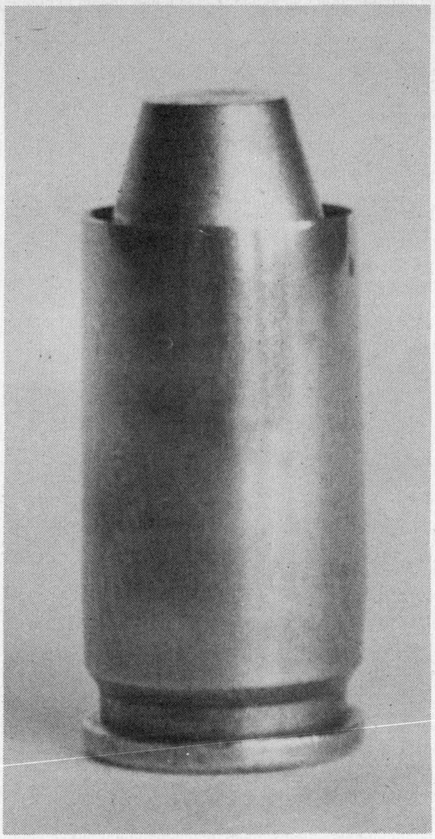

In high-performance loads for autos, the handloader should be sure that his case/bullet assembly is tight enough to avoid having the bullet pushed deeper into the case (left) through impact on the feed ramp. To avoid this problem simply roll a cannelure (right) into the case about 1/32-inch above the base of the seated bullet.

bullet dislocates it too much, a bulged ring will be produced on the outside of the cases, and this can interfere with chambering to cause even greater problems. After the bullet is seated, a moderate, taper crimp should be applied, reducing case-mouth diameter approximately .015-inch. A greater degree of crimp might hold the bullet a bit more tightly, but since our aim is absolute functional reliability, we must not reduce the headspacing shoulder of the case mouth below minimum limits.

INSPECTION OF HIGH-PERFORMANCE LOADS

High-performance loads—for either revolver or auto—which will be used for serious social purposes or important hunting should *always* be inspected with a good deal more than the usual care, to include actual function-firing of a modest sample of the batch, and manually cycling through the gun all of the cartridges that you'll be carrying afield. Only a combination of these latter two tests will guarantee (as far as anything can be guaranteed) that the cartridges will feed with 100 percent reliability, and only the firing test will guarantee that the bullets are held tightly enough in the case to avoid being dislodged by feeding or recoil impacts. The ultimate in load testing is velocity measurement on a good chronograph, and if you don't own one, doubtless you can locate another handloader who does and who will be able to perform the tests for you. The ultimate in accuracy testing is a series of 10-shot groups fired in the gun for which the loads are intended from one of the better machine rests such as the Ransom or Broadway. Few handloaders possess such equipment, but if you can't locate someone locally who does, and you are really serious about determining the accuracy of your gun/load, there are a number of shops around the country who will conduct such tests for a nominal fee; you just have to send the gun and ammunition to them.

Now, all this doesn't mean that every time you load 100 rounds you have to burn up 50 of them in testing. When you first select a high-performance load, particularly if it is for defense, and refine it for the ultimate in power, accuracy, and reliability, all of these tests should be conducted. However, once you have ascertained that the load does perform to your requirements, you can use it for years without requiring more than occasion-

After you've made up a batch of high performance loads, inspect them closely, and, fire-function a sample number of rounds to insure desired accuracy, feeding (in the case of autoloaders) and reliability.

al additional testing, just to be sure. However, if at any time you change the load or the manner in which it is assembled, the tests should be rerun. This should be done if, for example, you switch from new to fired cases; from one brand of cases to another; change makes, types, or weights of bullets; etc.

Choosing, refining, and assembling high-performance, jacketed bullet loads can be even more demanding than making up the most accurate loads. This is true simply because high-performance loads must not only be superbly accurate, but must meet other requirements that aren't necessarily placed upon target ammunition.

MAKING UP HEAVY LEAD BULLET LOADS

There are, of course, many among you who probably still prefer the older approach of a heavy, lead bullet at maximum velocity for hunting and defense; so, we can hardly avoid devoting some space and attention to that type. When using heavy, lead bullets, virtually everything we've said about the primer, propellant, and case applies except for the selection of *proper* propellants. The fast-burning propellants best suited for light bullets are naturally not ideal for heavy, lead bullets in revolver cartridges, though they are the best choice for light, lead bullets in the basic autoloader cartridges.

There really isn't an ideal propellant currently available for heavy, lead bullets in the .38 Special. The Keith/Lyman, 173-grain, SWC bullet is probably the best choice for this cartridge; and, Unique propellant burns just a bit too fast while H110 is just a bit too slow. Unique is the best of all the conventional pistol propellants, producing velocities a bit over 900 fps without exceeding safe pressures. Traditionally, Hercules 2400 has been used with this bullet in charge weights that completely fill the case, and it will produce slightly greater velocities without excess pressures. However, it burns very inefficiently under these conditions, and velocity spread is more than we'd like. Since H110 burns a bit faster than 2400, it should be more ideally suited, but at this point in time, we do not have valid laboratory tests to verify this. In .357 Magnum caliber, with its larger case capacity, the same bullet is ideal, and H110 propellant begins to come into its own. A charge of 13.0 grains of H110 will drive this bullet at over 1300 fps while pressures remain at least 5,000 CUP below the 40,000 CUP working level of .357 factory loads. This performance is superior to any other propellant we've tested in the .357 Magnum with lead bullets of this weight. The heaviest lead bullet available for the .41 Magnum is Lyman 410426, weighing 240 grains, which gives it a relatively high sectional density and superb penetration in heavy game. The case capacity of the .41 Magnum and this bullet weight create conditions ideal for the use of Hercules 2400 propellant, and it will drive the 240-grain bullet at well over 1400 fps from the longest revolver barrels (8⅜ inches) at approximately the same pressure level (40,000 CUP) as the full-charge, factory load. This load (20.0 grains of 2400 and the 240-grain bullet) is the finest, long-range, revolver load of all the scores I have tried. It produces about 100 fps less (under the same conditions) as the factory load does with a 210-grain bullet; but, the additional 30 grains of bullet weight gives it virtually the same energy as the factory load, and a trajectory that is actually flatter over the longer ranges.

In both .44 Special and .44 Magnum, the most common heavy-bullet choice is Keith/Lyman #429421 weighing nominally 245-250 grains. In the case of the .44 Special we have a case capacity a bit too small to take advantage of 2400 with this much bullet weight; H110 will drive the bullet at approximately 1200 fps from a 6½-inch barrel at pressures not exceeding 14,000 CUP. This is beautiful performance, slightly better than can be obtained from 2400 at pressures of nearly 10,000 CUP greater. The larger .44 Magnum case, with the same bullet, will handle a wider variety of propellant, with H110 producing very nearly 1600 fps at pressures well below those of factory loads; Hercules 2400 produces about 150 fps less and Hodgdon's H4227 will produce about 100 fps less than factory loads at pressures about 5,000 CUP below those of factory loads. W-W 296 is also a suitable propellant for this type of load, approaching fairly close to H110's efficiency, but we don't have extensive test data available on it at this time.

Because these bullets are heavy for their caliber, keeping them immovable in the case under recoil forces becomes a problem. Cases must be resized and expanded so that they grip the bullet very tightly; a mouth diameter as much as .005-inch less than bullet diameter may be required. In addition, a *very heavy* roll crimp is needed, and the bullet must contain a fairly deep crimping groove to accommodate it. Since we require the ultimate in crimp force, cases should be trimmed to uniform length so an extra-long case doesn't become crumpled in the crimping process.

Because all loads of this type contain quite large charges of relatively slow-burning propellant—often a ball propellant at that—magnum-type primers are generally indicated. Of the lot, standard primers might give adequate ignition in the .357 Magnum, while magnum primers are indicated for all of the larger calibers.

It's important that you not turn up your nose at the heavy, lead-bullet loads of the past simply because of all the favorable publicity given modern high-performance, jacketed-bullet loads. The old lead-bullet loads kill more by deep penetration to vital organs and by fracturing and punch-

If you like wheel guns chambered for .357 Magnum, the Colt Python is a good choice. The author found 13-grains of H110 and the Keith/Lyman 173-grain SWC work nicely in the .357 — you'll get over 1300 fps and about 35,000 CUP (5,000 CUP below the working levels of factory loads).

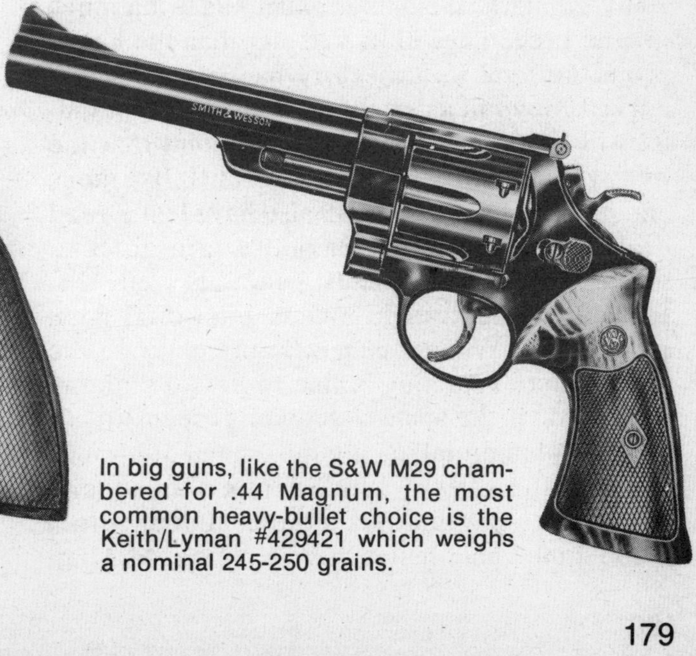

In big guns, like the S&W M29 chambered for .44 Magnum, the most common heavy-bullet choice is the Keith/Lyman #429421 which weighs a nominal 245-250 grains.

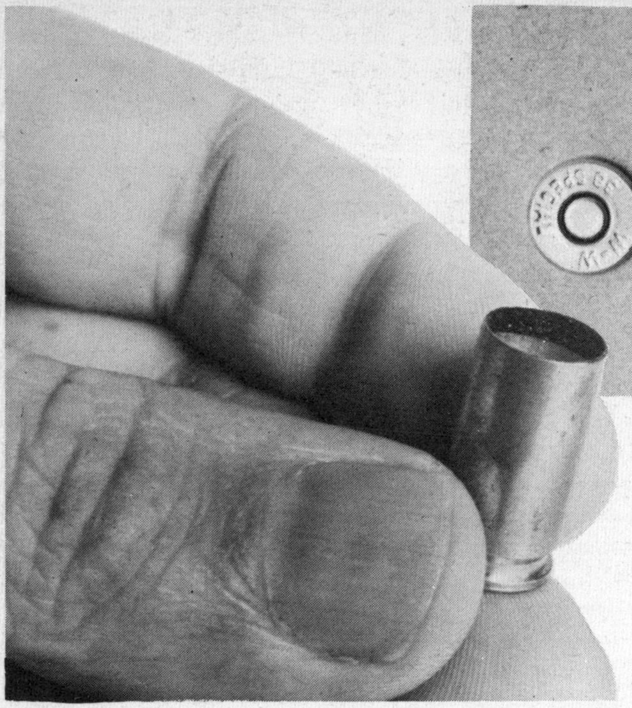

Above — You can easily waterproof the primer area of a cartridge by spraying the heads (after loading) with a thin coat of spray lacquer. Punching holes in cardboard and running the rounds in to their heads provides a good "holder" when doing the spraying — it's less messy.

Left — Much of the factory ammo made today uses an asphalt-base mouth sealer to insure the bullet and case are sealed tightly. This sealant can be seen in the form of the dark ring just inside the mouth of the case.

ing through heavy bones; whereas the jacketed bullets kill by massive expansion and tissue destruction accompanied generally by less penetration. The heavy-bullet loads do have an advantage for some types of shots on the larger animals where light bullets won't penetrate deeply enough. They also have the advantage that in some instances their trajectory is flatter over longer ranges. Quite obviously, a 110-grain bullet in .357 caliber will lose velocity much more rapidly than one weighing 173 grains. At 50 or even 100 yards, the trajectories may not be different enough to interfere with hitting, but at the 200-yard mark, the lighter bullet has lost a much greater precentage of its velocity than the heavy projectile, and its trajectory has become more curved. Beyond a certain distance, the light bullet will have slowed down to the point that the heavy bullet possesses substantially more energy. The light bullet will also have lost a good deal of its expansion capability so, in order to produce a clean kill it must penetrate well. Unfortunately, lightweight bullets don't offer good penetration. While I *generally* prefer to go the light-bullet/expanding-bullet route for hunting loads, I do make some exceptions; especially if I were to plan on taking a large animal at a range greater than 100 yards. Under these circumstances I would feel more confident with a heavy-bullet/lead-bullet load in .41 or .44 Magnum than with anything else.

Heavy, lead-bullets at highest practical velocities do perform quite well for defensive use against humans. Unfortunately, these projectiles possess the singular disadvantage that they will almost invariably penetrate the target completely; and, when they exit they retain a high percentage of their velocity and energy. This means simply that they possess a high potential for injuring innocent bystanders and property either by striking directly or by ricochet. Also, in cases of over-penetration or misses, the lead bullet is much more prone to ricochet than the soft-construction, jacketed bullet at high velocity.

WATERPROOFING/SEALING H-P LOADS

When assembling high-performance loads for hunting or defensive use, we must assume that at some point they may be exposed to extremes of weather and temperature and in particular to rain, snow, sleet, etc. In fact, considering the conditions in which some hunting and defensive gun work takes place, one might well find his handgun and entire available supply of ammunition temporarily immersed in water and yet be required to function perfectly immediately thereafter. That brings up the question of waterproofing ammunition.

Most factory-loaded handgun ammunition is labeled as "oil-proof" or in some such manner as

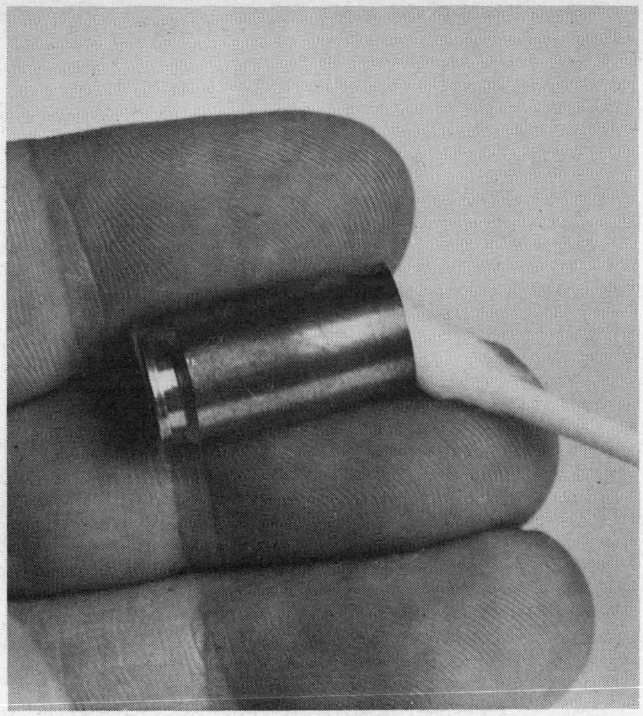

Above — Waterproofing ammo is a good move if you're going in, on or near water. If you should get your ammo wet, it's fairly easy to dry off; however, if it ends up submerged in 50 feet or more of water, chances are the best waterproofing techniques in the world will not help.

Left — After you mix your own asphalt-base sealer, it can be easily applied with a cotton swab or Q-Tip as shown.

to imply that its performance will not suffer from considerable wetting. A U.S. military specification for waterproofing small arms ammunition does exist, and many law enforcement agencies require that it be met for service ammunition. Factories meet this specification by using an asphalt-base mouth sealer applied inside the case mouth before loading, and by applying lacquer at the junction of case head and primer. The lacquer runs into the case/primer joint, hardens there, and seals out moisture. The mouth sealer is formulated so that while it becomes dry to the touch with exposure to air, it never becomes completely hard. Thus, when the bullet is seated, it seals any gaps between bullet and case and prevents the entry of moisture.

Sealing the primers of your handloads is really quite simple. Any good grade of thin clear lacquer can be applied with a pencil-point artist's brush after priming, but the case must be free of grease or oil. So long as the lacquer is suitably thin, simply touching the joint with the brush will deposit a drop which will run around the primer perimeter and be sucked into the joint by capillary attraction. If the lacquer is a bit thick, this will not occur, and it must be brushed into the entire joint. Alternatively, primed cases or loaded ammunition may be placed head-up in a loading block or perforated rack, after which the heads are sprayed with clear lacquer from an aerosol can or other spraying so that no lacquer is applied to the bodies. This is accomplished most easily by drilling holes in a sheet of cardboard or plastic just large enough to barely accept the case. This allows rimmed cases to be supported by their rims, so that only the head can be reached by the spray. For rimless cases the same perforated sheet is used, but some sort of support beneath is required to hold the cartridges or cases at the proper level. Sealing the bullet/case joint doesn't present much of a problem when lubricated lead bullets are employed. A very tight case/bullet fit combined with the malleable and stable nature of lead and the lubricant-filled grooves will generally prevent the entry of moisture. However, if you aren't willing to trust this method, seat the bullet first without crimping, then carefully brush fairly thick lacquer into the space between bullet and case, and immediately crimp the case mouth before the lacquer hardens. After crimping, immediately wipe off any squeezed-out lacquer with a solvent-wet cloth or swab to prevent chambering difficulties that might occur if a blob hardens there; it's also a good idea to swab out the crimping die with solvent to prevent any lacquer buildup which might cause erratic crimping.

Waterproofing a jacketed bullet requires a bit more care, inasmuch as even when case/bullet assembly is quite tight, minor irregularities in the

When going after big game with a handgun you should consider the Keith-Type, hard cast, SWC slug. It insures good penetration, through heavy bone, on into the vital organs.

two surfaces might leave open tiny passages through which moisture could pass by capillary attraction. The best method is the use of asphalt-base mouth sealer employed by the factories. Unfortunately, this material is not generally available to handloaders; however, a reasonably acceptable substitute can be made by dissolving ordinary asphalt in a suitable solvent. Its consistency should be such that a thin uniform coat can be applied to the inside of the case mouth and the case set mouth-up without the sealer running down toward the head of the case. The sealer is best applied with a cotton swab similar to a Q-Tip and should be applied to the primed case, then allowed to become dry to the touch (but not completely hard) before charging with powder. The bullet is then immediately seated and crimped—a perfect waterproof seal results. When employing this method, it must be kept in mind that the mouth sealer increases bullet pull substantially and this causes pressure and velocity to be increased. With moderate loads this won't cause *excessive* pressures to be generated, but if you're dealing with a real ring-tailed roarer load whose pressures are already against the roof, adding mouth sealer could very well produce excessive pressures. If you don't want to go to the trouble of using mouth sealer, then seat and crimp bullets separately, applying lacquer between bullet and case mouth before crimping as outlined above for lead bullets.

Incidentally, *clear* lacquer isn't mandatory—colors seal just as well and you might use them for indentification. Say, red for hot loads, black for "normal," and green for the light target type.

Once cartridges have been waterproofed as described, they will normally withstand any wetting they might receive. The seals aren't likely to be penetrated except in circumstances where the ammunition is submerged for a long period of time under many feet of water where water pressure would become great enough to break through the seals. If your handloads wind up in 50 feet of water on the bottom of a lake, I wouldn't advise trying to use them, no matter how well they were waterproofed in the beginning.

There is one common cause of seal destruction that is all too often overlooked. It not only destroys the effectiveness of the seal, but usually results in direct contamination of the primer and/or propellant as well. I might add that this same problem exists to the same degree with factory loaded ammunition. Most modern spray-can lubricants and preservatives employ a very volatile solvent and/or vehicle which will *dissolve* the lacquer seal around the primer. Some will destroy it in a matter of hours, some may require days or weeks. In the past, a good many pistoleros—police officers in particular—have made a practice of spraying their guns heavily with this type of lubricant/preservative to the extent that quantities of the fluid remain in the chambers, under the extractor, and on the recoil shield after the surplus was wiped off the outside. When cartridges were placed in the gun, the heads were wetted, as well as the case/bullet joint, and the solvent attacked and dissolved the seal. Some shooters went even farther, to the extent of spraying *loaded* guns, with the result that the lubricant and solvent were applied directly to the primer/case joint. A considerable rash of misfires—encountered by police officers after the use of spray lubricants became widespread—caused investigation by several agencies. It was determined that the problem was the spray lubricant—more correctly, it was careless and improper *use* of the spray lubricant/preservative. The material is a great boon to field maintenance and preservation of a handgun, but it must be applied sparingly, and it

Commercial or home-brewed waterproofing seals are designed to keep out moisture, not solvents. *NEVER* spray a cylinderful of ammo with any high-grade solvent. If you do, you can be sure your cartridge(s) won't fire!

must be applied in such a manner that it *cannot* wet the head of the case or the case/bullet joint.

Incidentally, it should be kept in mind that with each loading of a particular case the probability increases that moisture can enter. With each firing/decapping/repriming, the primer pocket becomes rougher and more irregular and in some instances larger, thus opening up passages; and, with each firing/resizing/expanding/seating/crimping, the same occurs to the case mouth; and, in addition, it loses some of its ability to grip the bullet tightly and to retain a proper crimp. A case which was perfectly waterproofed for the first couple firings will not necessarily be so afterward and will require sealing to prevent the entry of moisture.

Insofar as the type of high-performance load you select, "you pays your money and you takes you choice." The modern, light, jacketed bullet approach probably has the most adherents today, and it does offer superb performance within certain limitations; yet, the heavy lead bullet offers advantages of its own, and it may very well suit your needs best. The important thing to remember is that the ultimate choice must be made based upon the inherent ballistic characteristics of the two different types of loads as they apply the particular conditions of target, range, and surroundings under which *your* shooting will generally be expected to take place.

In the end, any load, high-performance or not, is only as good as you make it by your selection of components and manner of assembly. If you keep in mind all of the factors we have brought out in foregoing pages, rest assured that the loads you shoot will be superior to most.

CHAPTER 14

Shot and Multi-Ball Loads

TRADITIONALLY, from the very beginning, as far back as the earliest concept of firearms, the one-hand gun has fired a single ball or bullet with each shot. All the same, human nature being what it is, doubtless even some of those earliest pistols were occasionally charged with multiple projectiles or shot of some sort as a matter of convenience, necessity, or in hopes of enhancing the probability of hitting the target. During the early days of the black powder cartridge, manufacturers offered assorted shot-loaded cartridges for rifles in calibers such as .44-40, and .38-40. They were intended primarily to permit the taking of small-game for the pot at short range under circumstances where it was either inadvisable to fire ball ammunition or marksmanship wasn't quite up to the task of hitting a grouse or cottontail with a single bullet. Since revolvers were also chambered for these same cartridges, it was only logical that this shot ammunition be used in them. In fact, it was probably more practical for use in six-guns than in rifles. We have no way of knowing now how extensive their use was, but they continued to be cataloged by the major manufacturers for many decades. And as time went on, similar loads were offered in other calibers of purely revolver persuasion.

The earliest handgun shot cartridges were loaded with black powder, of course, and plain, spherical, lead shot of the smaller sizes. Apparently shot as large as today's number 9 was sometimes used, but so were smaller sizes such as 12 and "dust." Even reduced charges of black powder occupied a great deal of the case capacity, leaving little room for shot between an over-powder wad and the mouth of the case. Some loads were assembled in this fashion, containing only the amount of shot that would fill the case to the roll crimp. Early in the game, though, bullet-shaped containers of wood or pressed paper were employed. At this point in time, it's perhaps a bit difficult to say whether their primary purpose was to allow a greater quantity of shot or to insure better feeding in repeating rifles. Some of the older, long-gun designs wouldn't handle a case without a bullet, and in them, a shot load was useless to the average shooter unless it would feed from the magazine. A sturdy, wood or other bullet-shaped container assembled tightly to the case solved this problem, and, at the same time, its hollow interior allowed a good deal more shot to be used. No such feeding problems existed in the revolver, but surely the extra shot was appreciated by those pistoleros who chose for whatever reason to use shot-loaded cartridges.

In the latter part of the 19th century, men set great store in demonstrations of their marksmanship prowess with a six-gun. Shooting exhibitions were quite popular, and probably the most impressive stunt of all consisted of breaking glass or pitch globes "on the wing" so to speak as they were tossed in the air by an assistant. While we all know this can be done with bulleted ammunition if a fellow wishes to practice long enough and hard enough, it's a helluva lot easier with a charge of shot; and while it's doubtful this ever fooled the spectators much, there's no evidence that they complained. Of course, when such demonstrations were carried on indoors in auditoriums, opera houses, music halls, and even huge tents, ball ammunition was out of the question. Cartridges loaded with dust shot could be fired with more or less impunity under those conditions. With these light powder charges, the worst damage to be expected was a slight dimpling and chipping of paint and perhaps occasionally a light, harmless pattering of tiny, lead particles on the audience. Today spectators would sue, but then it was just considered part of the game.

Today, commercial shot loads are no longer offered by the major manufacturers, though Speer produces them as "Snake Loads" in some revolver calibers of .38 Special and above. Of course, we shouldn't overlook the .22 rimfire shot loads, commonly referred to as "rat shot" or "bird shot" which are still standard production items and are widely sold. We don't reload them, so there's no point in discussing them further.

SHOT LOADS

Thompson Center "Hot Shot"

By far the most potent and effective handgun shot loads available commercially are the "Hot Shot" .357 Magnum and .44 Magnum cartridges sold by Thompson/Center for use in special, choke-equipped, shot barrels in the "Contender" single-shot pistol. These cartridges contain an abnormally large quantity of shot enclosed in a bullet-like, plastic capsule which extends so far beyond the case mouth that they may not be used in revolvers. The loaded .357 Magnum Hot Shot

Above — For those who want shot shell performance, Thompson/Center offers their "Hot Shots" in both .357 and .44 Magnum calibers. They are, however, designed for use *only* in T/C handguns as the overall length of these shells makes them unsuitable for revolvers.

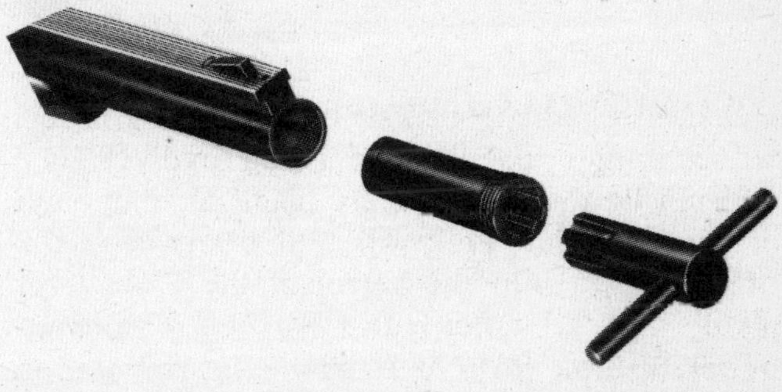

To make their Hot Shot cartridges perform to the maximum, Thompson/Center offers a barrel complete with special, recessed, screw-in choke (above) that assures good patterning. The older version of this gun (below) is no longer available — it utilized a special choke that *extended* from the muzzle as shown.

Here we have (above) the T/C Contender with standard octagon barrel and fully adjustable sights; to the right is the round barreled version (with simple fold-down rear sight) used in conjunction with Hot Shot cartridges.

measures a monstrous 2.65 inches in length, far, far too long for use in any existing revolver. The T/C plastic shot capsules are made small enough in diameter that they will seat easily atop the rifling, down into the barrel of the Contender pistol. The T/C shot barrel is rifled in conventional fashion (to avoid federal law/regulation problems) and is equipped with a screwed-on choke device containing straight lands and grooves of exceptional dimensions that serve to insure breakup of the plastic shot capsule as it leaves the muzzle.

The lands stand much higher than usual in this choke device, increasing in height from rear to front; and, as the capsule passes through, they serve to cut or fracture it lengthways so that the plastic fragments separate from the shot charge. Those fragments decelerate quite rapidly because of their light weight, and leave the shot free to travel on unimpeded. To one who has experienced nothing more than the old-style, factory shot loads in a revolver, performance of the Hot Shot is truly amazing. In .44 caliber, it rivals the performance of the old standby 2½-inch, .410-gauge, shotgun load, and I have seen expert marksmen shoot very respectable Skeet scores with it. Both the .44 and .357 Hot Shot loads are very good for taking small-game on the ground and upland game birds with the special Thompson/Center barrel and choke.

The Thompson/Center Hot Shot represents the epitome of handgun shot loads. Unfortunately, it can be used only in the Contender pistol with full effect, and though there are other single-shot pistols and rifles in .357 and .44 Magnum chambering that will permit their use, they lack the special ribbed choke needed for full performance. "Hot Shot" handloads can be assembled in .357 or .44 Magnum cases (and .44 Special cases), providing one obtains the special shot capsules from Thompson/Center. Loading procedures are conventional except that an ordinary seating die won't handle the unusually long capsule. I prefer to seat the capsule to the proper depth by hand, then crimp the case on the capsule with a conventional crimp die or a seat/crimp die from which the bullet-seating stem has been removed to provide full-length clearance for the capsule. Some makes of seating dies use a small-diameter bullet-seating stem threaded directly into the die, and this type cannot be used unless made especially for the Hot Shot load. Of

When loading the long Hot Shot capsules it's best to remove the stem from the seating die. This allows the capsule/case combination to be crimped without interference or damage to the capsule itself.

course, such a die can be reamed or ground out to provide full-length clearance for the capsule, but that spoils it for seating ordinary bullets. Other dies are made with a bullet-diameter seating stem assembled to the die by a large-diameter thread. Removing the stem from this type die provides adequate clearance for the capsule to be properly crimped. Cases should not be crimped too heavily on these capsules. Since they are fired only single-shot, only enough crimp to secure the capsule for handling is needed. Excessive crimp will only damage the capsule and perhaps impair its performance.

Remco and Speer "Shotcaps"

Two other types of shot capsules are available. Oldest is the Remco, a clear, plastic cylinder filled at the factory with shot and sealed at both ends. It is available in .38/.357, .44 and .45 calibers. The .45 capsule is offered in two lengths (and holding different amounts of shot) so that loads may be made up for both revolvers and autoloaders. A shot capsule with limited shot capacity is supplied for the .45 ACP because of cartridge length limitations. A much longer capsule, containing more shot, is supplied for .45 revolver loads where more space is available.

These Remco "Shotcaps" are loaded just like bullets, and may be seated and crimped in ordinary loading dies with the seating stem simply adjusted to produce the proper cartridge length. One problem does arise with shotcaps: The capsule will be deformed if the case mouth is too tight upon it. The inside diameter of the case mouth must be adjusted so that the capsule can be seated easily; and, of course, this may require a special-diameter, expander plug. When too much seating pressure is applied, the front of the capsule may crumple and break open. Excessive crimp can also damage the capsule. These two factors limit the powder charges that can be employed with the Remco Shotcaps. Due to the light crimp and loose fit on the capsule, the case doesn't have much grip with which to resist recoil forces. Since we can't do anything about this, recoil must be kept below the level which will cause the capsules to move forward out of the cases, and that can be done only by keeping propellant charges relatively light. Loading data is supplied in each box of shotcaps, but in my experience, even these moderate charges are sometimes too much. Of course, this doesn't apply to loading the .45 ACP since the shotcap cartridge will not lie in the magazine, and even if it did, the cartridge could not feed because of its wadcutter shape. Further, the round does not normally develop sufficient recoil to drive the slide fully to the rear.

Based on several years experience, dating

Remco "Shot Caps" are designed to be used primarily in revolvers and can be loaded like a normal bullet. They're available in .38 Special, .357 Magnum, .44 Magnum, .45 Colt and .45 ACP.

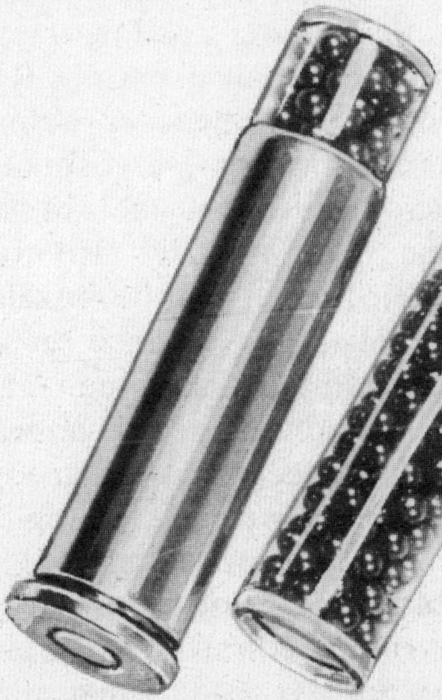

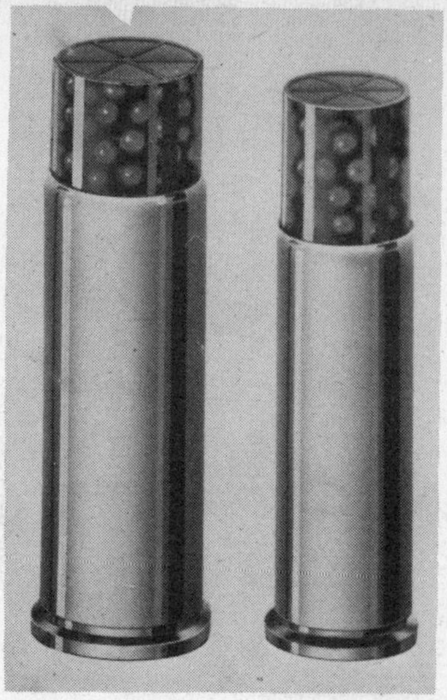

As you can see (left), Speer's Shot capsules are scored to insure breakup of the plastic cap as it exits the muzzle. Right: two loaded cases — the .44 Magnum (left) and the .357 (right) loaded with a .38 Special Shot capsule.

back to about 1967, I have found that Remco Shotcaps generally perform well, but have about a 20 percent failure rate. By that I mean that the plastic capsule is intended to break up as it leaves muzzle, and when it does so, short-range patterns are quite adequate for small-game and unpleasant beasties like copperheads and rattlers in camp. Occasionally, though, the capsule does not break up immediately at the muzzle (if at all) and even out to 10 to 15 feet may strike (or miss) the target as a single projectile, with the shot still enclosed. This isn't a tremendous problem, because when it does occur, you can simply fire a second shot quickly, and I have *never* had two of the capsules malfunction consecutively.

Speer also offers shot capsules for the man who wants to load his own. They can be purchased in either .38 Special or .44 Magnum and come packaged 50 to the box.

Speer also offers plastic shot capsules for reloading as well as their factory-loaded shot cartridges. As a practical matter, loading the Speer capsules is no different than using the Remco Shotcaps. However, the Speer capsules are made of a different material, are somewhat tougher, and therefore give less trouble in seating and crimping. The Speer capsules also contain serrations to promote break up at the muzzle and appear to have a better performance record in this area. Speer shot capsules have been made in more than one form, and the one that appeals to me the most contains a deep crimping groove so that a very heavy roll crimp may be applied to resist recoil when fired in revolvers.

Placing the shot charge inside a plastic capsule serves several purposes. First of all, it greatly simplifies loading. As you read farther along in this chapter, loading with loose shot and wads involves a great many more operations. If the capsule serves no other purpose than to simplify assembling loads, it would still be worth its cost. Shot loads are almost invariably fired in rifled barrels (for reasons which will become apparent farther along) and the plastic capsule protects the shot and avoids the considerable mutilation of the individual pellets that the rifling would cause otherwise. When naked lead shot is driven down a rifled barrel, it is subject to deformation from "scrubbing" the bore surface, just as in a smoothbored shotgun barrel. In addition, the rifling tends to rotate the charge of shot, and the raised edges of the lands mutilate most of the outer layer of pellets badly. Mutilated pellets are likely to take off in most any direction when they leave the muzzle, spoiling the pattern; and,

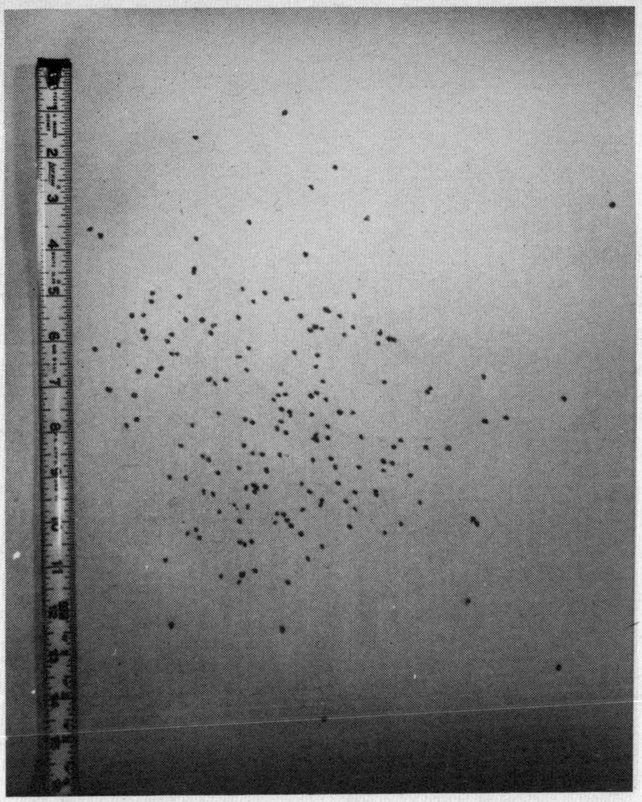

This pattern was fired with a Speer .44 Magnum shot capsule at a distance of 15 feet. The density is quite good.

the rotation tends to enlarge the pattern and give it a hollow center. The rotation effect is still present with plastic capsules, but pellet mutilation is virtually eliminated. By extending beyond the case mouth, the capsule allows a substantially greater amount of shot to be employed in the load, and since the base of the capsule seals powder gases in the bore, it eliminates the need for over-powder wads. There are, of course, problems in the use of plastic shot capsules, which we've already touched upon. In a conventional rifled barrel, capsule break up is not totally reliable. Successful break up has been achieved only with the special ribbed choke employed by Thompson/Center. And, as we've pointed out, retaining the capsule in the case (for revolver use) can become a problem with loads developing significant recoil.

It should be readily apparent that we could expect better patterns especially from unprotected shot charges if a smoothbored barrel were used. The ardent pistolero with a use for shot loads would certainly not object to this smoothbored barrel on one of his guns, particularly in these days with the Dan Wesson revolver design available to permit rapid barrel interchange. Were smoothbored six-gun barrels *allowed,* rest assured I'd have a new Dan Wesson .44 Magnum accompanied by a spare smooth tube which I could swap for the rifled barrel in a matter of moments when I wanted to use shot loads. If smoothbored barrels were legal, then Thompson/Center wouldn't have to employ that complex and costly ribbed choke to obtain maximum performance with its Hot Shot loads.

The law—or, more properly, *regulations*—is the problem. Back in the early 1930s, anti-gun and anti-crime forces combined to cause enactment of the "National Firearms Act" and "Federal Firearms Act," both intended to reduce criminal use of firearms. Since then, certain firearms have been either prohibited for private possession or have been so heavily taxed and controlled that they can't, as a practical matter, be owned by the average individual. Among them were machine guns and sawed-off shotguns. At the time, and for many years thereafter, this had no effect upon those few individuals who chose to smoothbore a handgun barrel to obtain optimum performance from shot cartridges. Bud Dalrymple, to mention one such, wrote of smoothboring a single-action, .45 caliber Colt and using it to shoot jackrabbits and the like from horseback. In the 1950s, the late Jim Harvey (Lakeville Arms) was offering custom, shot conversions of S&W revolvers. Working with either the .44 or .45 caliber N-frame guns, Harvey smoothbored and choked the barrel, thereby providing optimum shot-charge performance. This barrel could be used with the standard cylinder and standard-length cases loaded with shot; however, the ne plus ultra conversion utilized a cylinder with its chambers bored straight through to accommodate a cylindrical, brass case the same length as the cylinder. This extra-long case would accommodate a good deal more shot, as much as ½-ounce, the same amount employed in the standard 2½-inch, .410 gauge, shot shell. The combination of the special case and choked, smoothbored barrel made Harvey's S&W conversions the epitome of shot-shooting revolvers.

Then the trouble arose. The Alcohol, Tobacco, and Firearms Division of the U.S. Treasury ruled that a smoothbored shot revolver—in fact, any handgun with a smoothbored barrel—would thereafter be classified as a sawed-off

One of Harvey's conversions of N-Frame Smiths called for the use of extra-long cases, and, amazingly, it contained up to a ½-ounce of shot — the same amount used in the 2½-inch .410 shell shown.

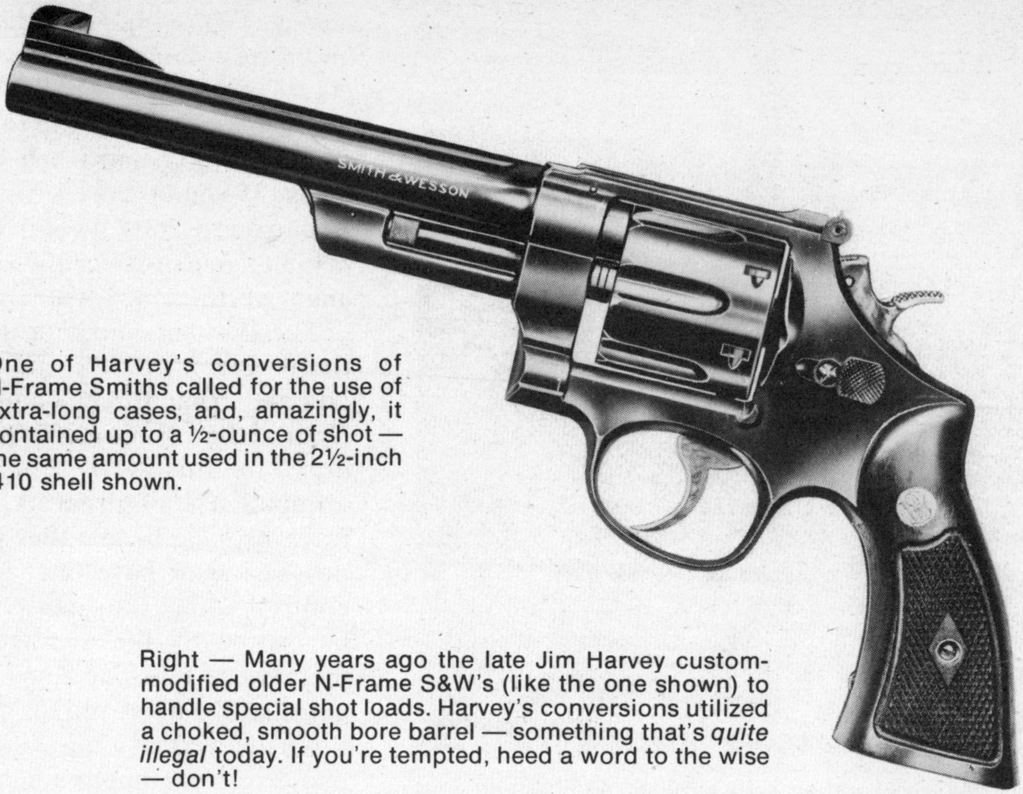

Right — Many years ago the late Jim Harvey custom-modified older N-Frame S&W's (like the one shown) to handle special shot loads. Harvey's conversions utilized a choked, smooth bore barrel — something that's *quite illegal* today. If you're tempted, heed a word to the wise — don't!

shotgun and subject to the same taxing and regulations. While technically this did not prevent Harvey from continuing with the conversions, its practical effect was to eliminate any further production of such guns simply because shooters would not pay the special tax or submit themselves to the onerous restrictions and surveillance that possession of such a gun brought about. That ruling by the ATF remains in effect today, and is justified by the BATF (successor to the ATF) on the basis that a handgun with an unrifled barrel constitutes a shotgun, and further that it constitutes a sawed-off shotgun inasmuch as the barrel is of less than 18 inches in length. In the absence of evidence to the contrary it must be presumed that the original ruling was made on the basis that a shot-shooting handgun would be used for the same criminal purposes as a genuine sawed-off shotgun. And, the ruling was handed down in spite of the fact that to this date I have been unable to uncover a *single reported instance* of a smoothbored shot revolver being used in the commission of a crime.

That ruling remains in effect, so it is simple fact that you can't own a smoothbored shot revolver (or autoloader for that matter) except by complying with all the taxes and restrictions laid upon sawed-off shotguns. However, recent discussions with BATF indicates that the presence of *some* rifling in the barrel would keep the gun out of the sawed-off shotgun classification. We don't know just what constitutes ''some'' rifling, but it might be possible to get by with only an inch or so at the breech or muzzle without having the gun classified as a sawed-off shotgun.

In any event, Harvey's approach was/is the one that produces the most effective shot loads in big-bore revolvers. We'll deal here only with preparation of cases and loading cartridges for use in such guns, and if you are really interested in the gun modifications necessary to obtain the very best performance possible from a rifled barrel, we refer you to the book *Pistolsmithing*.

HANDLOADING SHOT LOADS FOR REVOLVERS

Assuming you have a .44 Special or .44 Magnum revolver, and do not wish to alter it, you can make up cylinder-length cases which will accommodate nearly ½-ounce of shot. Begin with new or like-new .30-.40 Krag rifle cases, and cut them off to a length of 1.62 inches. Bell the mouths taking care not to split or crumple the case. If your particular batch of cases is prone to

In some instances, certain rifle cases (left) can be trimmed back to specific lengths for use in high-performance shot loads in handguns. Once a case has been prepared, prime it with a standard pistol primer (of proper size) and assemble powder, wads and shot as indicated in the text.

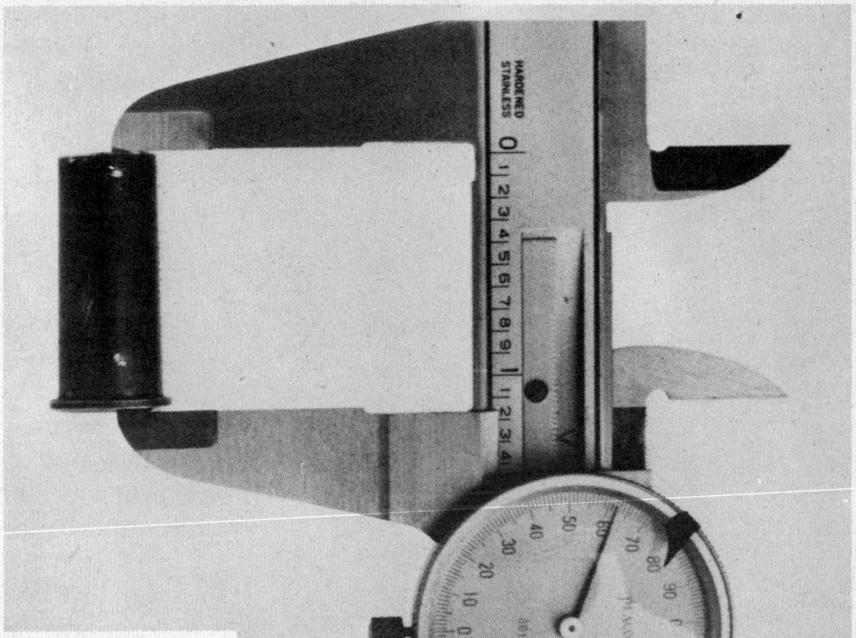

split, anneal the mouths about halfway back toward the rim. The most practical way to anneal is to take the decapped case, dip it in light oil or finely-powdered graphite, mouth first almost up to the head, then plunge the front half into molten bullet-casting lead for about 5 seconds, withdraw it, give it a quick flip of the wrist to throw off any clinging lead, and drop it instantly into cold water. This will soften the front half of the case sufficiently to eliminate splitting. The mouth flaring should be sufficient that it becomes a close but not tight fit in the chamber throat. Almost any tapered plug or punch will serve for the job, but a .41 Magnum expander plug and die is ideal for the purpose. Don't attempt to expand the front of the case full depth.

Make certain the cases will enter the chambers freely. Fired cases may not do so unless resized. Also, those .30-40 rims may require thickness or diameter reduction in some guns; certainly so in the small-cylinder Charter Arms .44 Bulldog.

Now, prime the case with a standard, large pistol primer and drop in a charge of 6.5 to 7.0 grains of Hercules Unique propellant. Follow this with a wad about ⅛-inch thick seated tightly on the powder charge. It's a bit too much trouble to force a conventional wad through the small-

In some cases, you won't be able to apply a crimp to a shot cartridge using a conventional seating die. You can, however, make your own special crimping tool from short lengths of dowel and pipe as described by the author. These two items cost less than a buck and are available at most hardware stores.

diameter section of the case, so I normally use a tightly-rolled ball of tissue rammed hard against the powder with a dowel. Pour No.9 or smaller shot into the case to a level about 3/32-inch below the mouth. The exact amount of shot isn't important, just enough to fill the case to this level. Remember, this is just a case-forming load, and we aren't concerned with its ballistic performance. The simplest way I've found to keep the shot in place until firing is to press a cake of cheap, fairly-soft soap over the mouth, and twist the case to free it, pulling out a soap plug that seals the mouth. That completes the case-forming load.

Fire these loads in your .44 six-gun, chambering only one at a time. If you try to load more than one, recoil will dislodge the soap plug and scatter your shot charge. Assuming the chambers are properly dimensioned and finished and are undamaged, the fired cases should extract easily, and will be properly fire-formed. What you should have is a two-diameter case, the smaller front portion filling the chamber throat. The case mouth may be a bit out of square from flaring and fire-forming, and if so, true it up with a file or case trimmer. I usually finish up fire-formed cases so that the mouth falls at least 1/32-inch but no more than 1/16-inch inside the mouth of the chamber throat. If the brass in the cases isn't too soft and if the chambers don't vary too much in your gun, the cases should be suitable for use without resizing. I find I can usually get several loadings out of a case before sizing becomes necessary, but this will depend upon both the case and the chamber, not to mention the potency of the load you are using.

If and when resizing becomes necessary, you may go the expensive route and special-order a resizing die from RCBS, accompanying your order with three fired cases. However, we assume that you'll have a .44 resizing die available, and it will serve admirably for sizing the major portion of the case. If after this, the neck is still too big for free chambering, use a .41 Magnum sizing die, running the case in only to the small shoulder. It may be necessary to push the shoulder back just a tiny bit for easy chambering, but don't go any farther than necessary because you'll just be reducing shot capacity.

Whether fire-formed or resized, loading full-power shot loads is the same. You'll need hard, card wads of nominally .44-inch and .42-inch diameter for over-powder and over-shot positions respectively. If you can't buy this size locally, make up a simple wadcutter punch from steel tubing, lay about three sheets of stiff cardboard over a block of lead or an end-grain chunk of hardwood, and tap the punch with a piece of dowel and you're in business. While additional wads aren't really necessary, a composition wad ¼-inch thick on top of the over-powder wad will cushion the shot against the forces of acceleration and reduce shot deformation.

Use a charge of 6.5 grains of Unique and seat the larger-diameter card wad firmly on top of it with a flat-ended dowel. The wad will have to go through the necked portion edgeways, and this sometimes makes it difficult to seat it without flipping part of the powder charge on top. We need *all* of the powder *beneath* the wad, so keep an old crochet hook or something similar handy to pull out any wads which scoop up powder.

Once the powder and wads are properly in place, fill the case with No. 9 shot to within 3/32-inch of the mouth. Level the shot and seat the smaller of the card wads firmly in the mouth. Secure the wad with a fairly stiff roll crimp and the job is finished.

The only trouble with that last sentence is that applying the crimp requires a die or tool that isn't available unless you've spent quite a few shekels to have one specially made by RCBS. As is usually the case in problems of this sort, there *is* a simpler and cheaper way. Unless you plan on loading lots of shot cartridges, a simple hand-type crimping die will do quite nicely. This can be made from nothing more than a piece of dowel and a short length of ⅜-inch water pipe picked up at the local hardware store. Cut or have cut in one end of the pipe a deep 45-degree bevel. This can be done beautifully on a lathe, or even on a tool grinder, but you can also do it with a countersink wielded by hand. As a last resort, you can even do it with a round or half-round needle file, but it'll take lots of patience and perseverance. In any event, polish this bevel as smooth as possible with successively-finer grits of abrasive cloth. It needs to be smooth enough that the brass case mouth will pass easily over it without snagging. Now, carve out a piece of dowel that is a snug fit inside the pipe and just a bit longer. Position the dowel in the pipe so that when the conical recess (formed by the bevel) rests solidly on the case mouth, the flat end of the dowel reaches just barely inside the mouth. Hold it in that position, and run a few drops of cyanoacrylate adhesive in between dowel and pipe to secure the two together temporarily. The C-A adhesive will set up in only a few seconds, permitting this rudimentary "crimping die" to be handled and finished. Drill a couple ⅛-inch holes through pipe and dowel together, and drive in tight-fitting pins. File the pins off flush and trim off any excess dowel protruding from the flat end of the pipe.

To use this tool, set the charged case in the shell holder of your loading press and run the ram to the top of its stroke without any die in position. This will hold the cartridge solidly without any possibility of pressure or impact being applied to the primer. Make certain the over-shot wad is level and solidly against the shot, then set the crimping tool over the case mouth. Make certain the tool is in line with the case, then give it a couple sharp taps with a soft mallet. Do this gingerly at first, examining the result, until you can determine exactly how much force is required to turn a neat crimp without damaging the case. You may find that you'll have to adjust the shot level a bit or perhaps modify the amount of dowel protruding inside the crimp tool to obtain the best job. An alternative to this home-made hand crimper is available by shortening a .41 Magnum seat-crimp die so the necked portion of your shot case will enter it deep enough to be crimped. If you're fortunate enough to have access to someone who can do internal hole grinding, leave the die full length and simply open up the lower portion to a diameter of about .456-inch, (or until the rear of the case enters freely) deep enough that the case mouth can reach the crimp shoulder. Either way, a wadcutter seating punch can then be adjusted to hold the over-shot wad in place while the crimp is being formed. Just make sure that the case mouth isn't jammed tight between the stem and the crimp shoulder.

If you're willing to spend time and money to modify the cylinder, you can obtain slightly greater shot capacity and achieve simpler loading procedures (and if you do a lot of shooting, a spare cylinder and crane modified just for shot could be a worthwhile investment). To do this, then, the chamber throats should be reamed out to chamber diameter, so that straight, cylindrical, cylinder-length cases can be accommodated. Making cases is done exactly as before, except you'll have to flare the mouth a bit more, and final loading of fire-formed cases will require only one size wad. More important, though, is that you'll be able to crimp the cylindrical case in a standard .44 seat-crimp die. All that's required is to back it out of the press to accommodate the additional length (just as when loading .44 Magnum cases in .44 Special dies) and installing and adjusting a wadcutter seating punch to control the over-shot wad. If the gun in which you wish to use maximum capacity, cylinder-length cases happens to be .45 Colt caliber, the procedures remain exactly the same except that you will use .444 Marlin brass as the basic case, duly shortened, flared, and fire-formed. In modern double-action revolvers, rim diameter shouldn't give you any trouble; however, an upper-limit .444 might have a rim large enough in diameter to cause some interference with the closer chamber spacing found in the Single-Action Colt. In that event, rim diameter might have to be reduced very slightly, a job most conveniently done on a lathe, but with such a small amount to be removed, you can do it by hand with a file quickly enough.

At one time M1917, .45 ACP, Colt and S&W revolvers were quite plentiful and low in cost. I recall buying them in fine condition for less than $25 during the early 1960s. Unfortunately, their value has escalated tremendously since then, but they still make fine guns for use with shot with the proper treatment. The best route to go with these guns is just as described for the .44 Special, with either unaltered or reamed-out chamber throats. Cases are formed in exactly the same manner, but are made from .30-06 or .308 Winchester brass. Beyond that, the only difference is that the rimless cases must be used with S&W, three-shot clips, just as the cartridge for which the guns were originally intended. However, there is one problem that you might encounter in the gun. The sharp shoulder at the front of the chamber, intended to headspace the cartridges when clips are not used, may be burred or so sharp as to dig into the case a bit and cause hard extraction. This doesn't occur in the other calibers mentioned simply because the transition from chamber to throat is tapered. Simply light chamfering the edges of that shoulder with a scraper or abrasive cloth glued around a dowel will cure the problem.

Other calibers can be given the same treatment, but some problems will be encountered. The .44-40 works well enough with unaltered chambers, even though slightly bottle-necked, but it does not lend itself to reaming out the chamber to a straight hole because the wad/shot column will then be .45 caliber and will encounter problems in entering the .44 caliber barrel. The same may be said of the .38-40, and due to its more severe bottle-neck, it becomes difficult to load — trying to force the shot charge out through that bottle neck raises pressures substantially, and I don't really recommend it very strongly. The .41 Magnum can be given exactly the same treatment as the .44 Special or Magnum, but we don't have any rifle case that is ideally suited for reworking into cylinder-length cases. The .30-30 Winchester series will work well enough, and the fact that it is about .015-inch undersize at the base doesn't cause any safety problem, it just looks sloppy after fire-forming. The .30-30 rim will be a good bit larger than that of the .41 Magnum; and, in revolvers with countersunk chambers, it will have to be turned down to a nominal diameter of .488-inch. Actually, I don't feel that any cartridge under .44 caliber will contain enough shot to be worth this extra effort. However, the same procedure can be used to make up cylinder-length cases in the smaller calibers if the amount of shot they contain will suit your needs.

There is another point that perhaps we should bring out before leaving the subject of shot loads for revolvers. For many years Remington-Peters has supplied a case known as "Five-In-One-Blank." It is not normally catalogued or available through the usual outlets, being made expressly for the entertainment industry. It is responsible for the smooth, rapid functioning of single-action Colts and Winchester lever-action rifles you see in films and on television. The case has the basic external shape and dimensions of the .38-40 cartridge, but is made a full 1.510 inches long, approximately the same length as loaded .38-40 and .44-40 cartridges. The purpose of the extra length is to permit smooth functioning through lever action rifles which will not ordinarily feed a bulletless case reliably. The added case length substitutes for the bullet, so when primed and charged with special blank powder designed to produce the amount of flash and smoke the cameraman wants, this case feeds merrily through .38-40 and .44-40 rifles. Naturally, then, blanks loaded in the same case will chamber in .38-40 and .44-40 revolvers; and, because the base diameter and length of the .45 Colt case is the same as those two calibers, it will function in guns of that caliber as well. So while the name of this case is "Five-In-One," it really fits only three chambers—.38-40, .44-40 and .45 Colt. The "Five" in its name comes from the fact that it will function in two different caliber rifles and in three different caliber revolvers, even though only three different caliber chambers are actually involved.

Though not a stock item, this case can be special-ordered through Remington dealers. Though it isn't quite full cylinder length, it can be fire-formed as previously described for use in .45 Colt, .44-40, and .38-40 revolvers and will then hold substantially more shot than the standard cases. Especially in the first two calibers, it offers an alternative to the laborious task of making long cases from rifle brass.

If a revolver cylinder is bored straight through to accommodate cylindrical shot cases, it might seem that it is ruined for use with ball ammunition. It is, as far as conventional factory or handloads are concerned, but not at all for specially adapted hand loads. Such loads consist of a wadcutter-type lead bullet seated entirely (or very nearly so) within the case mouth over a suitable powder charge and crimped moderately in place. If a wadcutter bullet of the desired weight and diameter isn't available, simply invert almost any form of bullet so that its flat base is forward and it can be seated flush with the case mouth. Alternatively, any ordinary bullet can be seated deep in the case mouth so that it does not protrude to interfere with cylinder rotation. The only problem with this is that there isn't any good way to crimp such a deep-seated bullet solidly so that it won't be moved forward by recoil to tie up the gun at an inopportune moment. However, deep-seated bullets can be held moderately well by a three- or four-point stab crimp applied with a small screwdriver and plastic mallet. In any event, a powder charge suitable for the same weight bullet loaded conventionally in that caliber will be correct for loads of this sort. *Just don't try to magnumize them.*

When such long-case ball loads are fired, functioning is normal except that during its travel to the barrel, the bullet is guided by the front portion of the case, rather than the chamber throat. It's conceivable that accuracy might not be quite as

If you should decide to use your favorite 1917 Colt or S&W for shot cartridges, and use the modified .30-06 or .308 brass as discussed, you'll have to use the half-moon clips as shown — fortunately, they are in fairly good supply and available at gun shops or through surplus dealers.

One alternative to modifying rifle cases for use in shot shell loads for handguns is the 5-in-1 blank as sold by Remington. The case is a strong one and has an overall length of 1.510 inches — the approximate overall length of regular .44-40 and .38-40 cartridges. This case can be used to make up excellent shot loads for the cartridges mentioned, and, the .45 Colt. Because of its length, it will hold substantially more shot than conventional cases of the calibers mentioned.

good as a conventional load in an unaltered cylinder, but it will certainly be adequate for close-up use in an emergency.

On the other hand, if you attempt to fire conventional loads in such an altered cylinder, the bullet will upset badly in the oversize chamber throat; it will then be oversized when it reaches the barrel forcing cone. The result will be not only excessive pressures and perhaps eventual damage to the barrel, but will cause small bullet fragments to be thrown off to the sides, endangering bystanders. Still, in an emergency, like if your life was threatened, the bullet would get out the barrel with adequate velocity and energy to do the job, and most probably without damaging a good gun.

HANDLOADING SHOT LOADS FOR AUTOS

Shot loads have never been *commercially* available for autoloading pistol cartridges. In fact, only the .45 ACP has the capability of containing enough shot to serve any worthwhile purpose. During WWII, special .45 ACP shot loads were made and included in aircrew survival kits. The first type utilized a red, laminated, waxed-paper shot container of bullet shape. Unfortunately, this container was rather fragile and did not stand up to handling stresses, particularly in the tropics. It was replaced by a long-case type (utilizing a full-length case 1.28 inches long) with the forward portion necked down, leaving a shallow shoulder to serve the headspacing function of the mouth in the conventional case. Shot filled the case from the over-powder wad to the mouth and was secured by a varnished, card wad held in position with a roll crimp. This load was never sold commercially, but quantities were apparently "liberated," and for a few years after WWII might have been encountered from time to time. This load was (is) effective, but its length and shape prevent it from entering the magazine, and it must be single-loaded by hand; further, it will not always drive the slide far enough rearward to extract and eject. (If you must have shot shells for the .45 auto and won't settle for the Remco shotcaps, then this military-type case would be by far the best to use. Diligent searching and query might turn up a modest quantity of loaded cartridges. They are available occasionally from dealers in collectors cartridges, but the price is prohibitive for shooting.)

Shot loads for autos have never been commercially available; however, during WWII, the military had some made up in .45 ACP for air crew survival kits. The round shown here is the laminated, waxed-paper type discussed by the author. The second type (not shown) did away with the paper shot cup and used a longer case with the forward portion necked down. Both are in the collector's realm today; however, if you can find cases of the "second-type," they can be reloaded.

You *can* make a similar case from .308 Winchester or .30-06 brass. Because of its thicker walls, it won't hold quite as much shot as the military survival cartridge, but it will work. I prefer to do this by first shortening .308 cases and then fire-forming them in an M1917 .45 ACP revolver as already described. This is substantially simpler and easier than attempting to fire-form in the autoloading pistol. Then, trim the cases to a length of 1.40 inches; and, if you want maximum reloading life, ream the neck to a wall thickness of .010- to .012-inch, and then anneal back about ⅛-inch past the shoulder. The reaming operation not only increases case life, but adds somewhat to shot capacity because the .308/.30-06 case walls are much thicker than needed at this point.

Once the case has been formed in this fashion, it is loaded just like the cases described earlier for the unaltered .45 M1917 revolver. When resizing for subsequent loadings, the standard .45 ACP die will handle the body, and running the necked portion into a .44 Special/Magnum resizing die will reduce it correctly. Take care when doing this to insure that the headspacing shoulder is not pushed too far back.

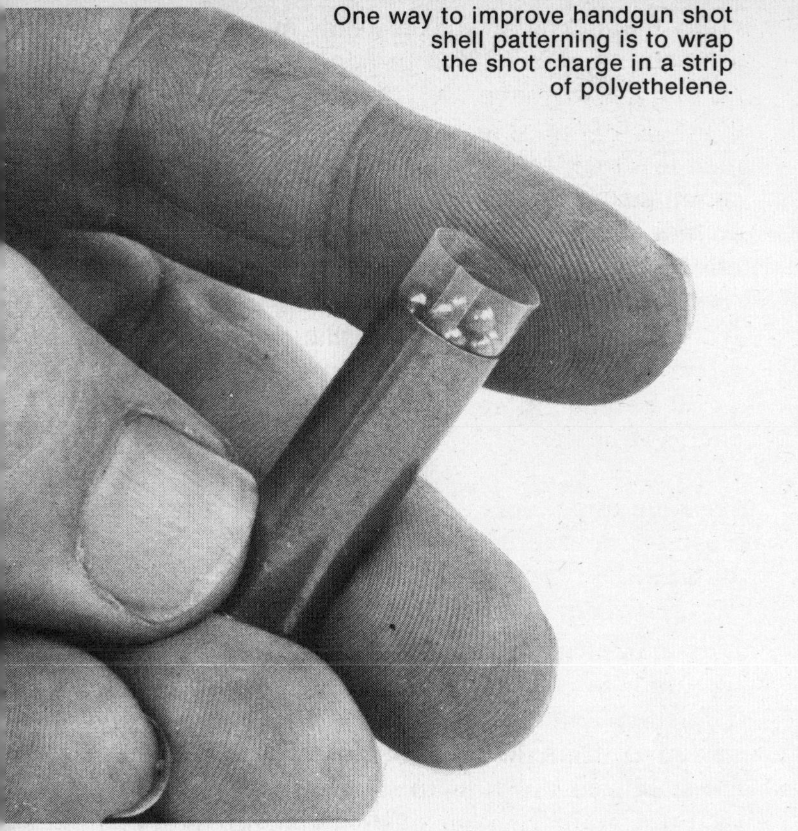

One way to improve handgun shot shell patterning is to wrap the shot charge in a strip of polyethelene.

Once you've formed cases in this fashion, it *is* possible to rework them a bit more and produce a load which will feed through the Colt magazine, even though it will not generate sufficient recoil to produce reliable semiautomatic functioning. The process consists of first further shortening the case to a length of 1.30 inches, priming it and charging it with propellant, and seating the over-powder wad. At this point, I prefer to go ahead and give the bullet portion of the case most of its reshaping so as to avoid spillage of shot that can result if it's done all in one operation later. Again, take a piece of ¼-inch pipe and make up a "coning" tool, as was done for crimping, described earlier in this chapter. Simply cut or have cut a 60-degree included angle chamfer in one end of the length of pipe and polish the surface of the cut as smooth as possible. Lubricate the cone surface uniformly but sparingly with resizing lube or STP and then tap it down over the case mouth until the hole remaining at the center will just barely admit a .25 caliber gas check. Then, pour shot in until it reaches a level between 3/32- and ⅛-inch below the mouth. This level will probably have to be adjusted as you go along. Insert the .25 caliber gas check open side down, make certain it is below the mouth, and level, then place the coning tool over the case again and give it one sharp but light tap. This should close the hole just enough to keep the gas check from escaping; reposition the check if necessary to keep it level, and give the tool a couple more taps to close the hole enough to hold the gas check securely in place. If the shot charge is exactly right, this slight additional coning of the case will have shrunk the interior so that it crowds right up against the gas check, forming a tight assembly from which you might get just the slightest shot rattle if you shake it. The coning will have also locked the gas check securely in place where it serves the function of an over-shot wad.

The resulting cartridge has the silhouette of a conventional case assembled with a truncated-cone bullet and is of the correct length to move through the magazine freely. It doesn't hold as much shot as the long, straight-case load, but it will feed from magazine to chamber when the slide is cycled manually—for whatever value that might have in view of the fact that there isn't enough recoil to cycle the action.

IMPROVING SHOT PATTERNS

There is one other thing you can do to improve patterns that we shouldn't overlook. When loads are assembled in straight cases (or those only very slightly necked to fit into the chamber throat), the shot pellets are totally without protection from the bore surface and the rifling lands. As we pointed out several pages back, this produces a high percentage of badly deformed pellets with the inevitable result that patterns are not as good as we might want. A reasonable amount of protection can be given the pellets by enclosing them in a polyethelene wrap as is done in the Winchester MKV shotshell load. This requires only that you cut rectangular pieces of polyethelene film of such a length and width that they will extend from the over-powder wad to just below the crimp and also encircle the shot charge. I've seen the material from dry-cleaning bags recommended for this purpose, but in my experience, it is simply too thin; the pellets rub holes right through it. I prefer to use the thicker material sold by hardware stores to tack over windows and doors as a storm-window substitute. However, don't get the heaviest grade of this material, unless you're willing to sacrifice a bit more shot capacity. The thicker the plastic, the less shot deformation you'll have. It's worth trying several thicknesses to arrive at the combination that suits you best. The rectangle is simply rolled up, placed in the case, and then positioned with a dowel so that it unrolls against the case walls. The shot is then added and the remaining operations completed. The result is the best revolver shot load you can possibly come up with.

Shot loads for your handguns are not only fun, they satisfy a specific need for some forms of pest or small-game shooting. If you do your part, they'll bag small upland game on the wing and are great for potting venomous snakes around camp and the like. Nor should they be overlooked for defensive use; they can be lethal at table-top range in your home without the likelihood of penetrating walls (as ball ammo would) to injure some innocent, unseen party.

MULTI-SHOT LOADS

Cartridges containing more than one full-caliber bullet are almost as old as the modern, metallic cartridge, with numerous references being made to them over 100 years ago. The object, of course, of launching more than one projectile from a single cartridge is to increase the probability of obtaining hits—particularly first shot hits—on animal targets. Such "multi-ball" loads are intended to compensate for the shooter's inability to obtain first-shot hits with a single projectile. It is anticipated that even a good marksman may not, under the stress conditions of a fight, be able to obtain a lethal hit with his first shot.

Interest has also been increased in these so-called "multi-ball" loads through the theory that the total effect of multiple wounds upon an aggressor opponent is far greater than the sum of their individual effects. The theory reasons generally that if a bullet strikes a man in a non-lethal area and transfers 200 foot-pounds of energy to his body, it will have "X" amount of incapacitating effect. It reasons further that if, instead, the man is struck by four, smaller or lighter bullets simultaneously, each of which transfers 50 foot-pounds of energy to the target, the incapacitating effect will be much greater, even though the same amount of energy has been transferred. Adherents to this theory feel that the combined effect of the four smaller bullets may be as much as 50 percent more than the same amount of energy transferred by a single bullet. I don't know of any valid evidence that will prove or disprove this theory, but it has been one of the reasons for interest in multiple-projectile ammunition.

Unfortunately, when more than one projectile is assembled in a case, the spread of those projectiles upon firing cannot be accurately predicted in handguns, and no single one of them can be expected to strike the point of aim. The projectiles are stabilized by rifling to some extent, and thus their deviation from line of sight may not be very great at normal handgunning ranges, but the fact remains that the capability of pin-point accuracy possessed by single-projectile loads is lost. Further, because of the obvious limitations, each projectile in a multi-ball load must be quite light in order that the total projectile weight be within reasonable limits. This has the very decided disadvantage in that if an opponent is struck by only one or two such projectiles, their limited energy and penetration will have relatively little incapacitation effect. Proponents of the multi-ball load that I have talked to seem to be laboring under the assumption that at least most of the projectiles from a given cartridge will strike the target; but my thinking is simply that if conditions

Multi-projectile loads may consist of two or three — depending on case length — round balls. The ball on the left measures a true .360-inch while the ball on the right has been sized down to .3570-inch.

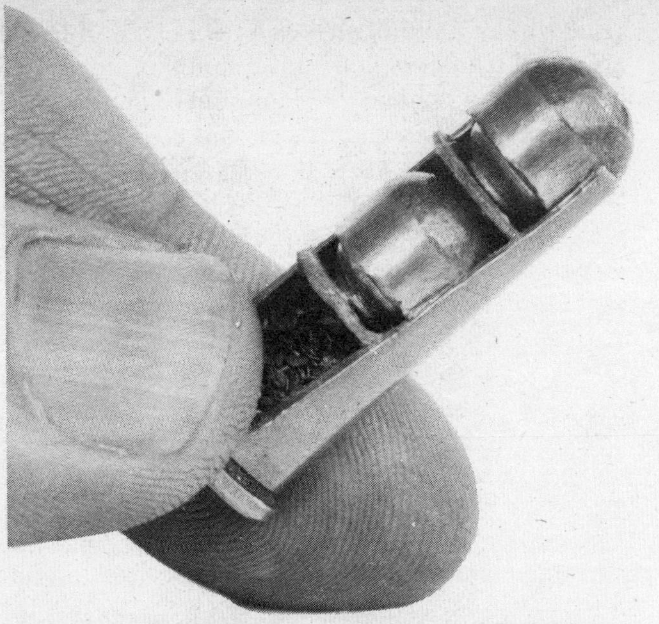

Here's an internal look at the construction of a 2-projectile load for a .38 Special. Both projectiles are 000 buckshot (.360-inch) sized to .3570-inch. The wad over the powder may be a bit superfluous; however, the wad between the projectiles will prevent slug deformation during firing.

are such that this can be accomplished, that it would be equally easy to place the more powerful and accurate projectile of a single-ball load in the target with greater effect.

The simplest form of multiple-projectile handloads consists simply of two or more round balls seated on top of the powder charge. Depending upon the ratio of case length to diameter, two or three balls may be accommodated. One ball is seated directly down upon the powder, followed by another if space is available, with the uppermost ball seated one-half its depth in the case mouth and crimped there securely. Upon firing, the rearmost ball will accelerate first, causing it to be flattened somewhat between the propellant gases and the second ball. The second ball in turn will be flattened between the first and the third and the rear of the third ball will also be flattened. Thus, when the projectiles leave the muzzle they will resemble short, stubby bullets more than round balls. In the case of unprotected lead balls, there is some probability that the projectiles will stick tightly together and not necessarily separate and diverge in flight. This can be corrected by placing a thin card or plastic wad over the first and second balls so that there is no lead-to-lead contact. A layer of insulation or protective material between the individual bullets insures their separation in flight. Plain, unprotected, lead bullets will sometimes literally weld themselves together under the pressure of initial acceleration. While this problem can be partially solved by placing lubricant between the bullets, an insulating wad does a better job.

Bullets of conventional shape may also be used, but we run into a problem of availability. Molds simply aren't available to cast bullets of light enough weight and short enough length. Once we depart from round balls, length and weight become critical. For the sake of argument, let's say that 200 grains represents the upper limits of *total* bullet weight that can be accommodated in the .38 Special or .357 Magnum and still be given effective velocity without exceeding safe pressures. That means that the total weight of all the projectiles loaded cannot exceed 200 grains. If we can settle for only two bullets, that allows us 100 grains each; if we use three bullets, they can only weigh 66 grains each. A few molds are available in the 90-100 grain range, but none for casting bullets as light as 66 grains. Bullet shape also becomes important because we have a limited length into which the several bullets must be crowded. If they're of typical round-nose form, that space won't be long enough. The shortest bullet for its weight is of pure wadcutter form, and it will be necessary to employ this type of bullet/shape if the maximum number of bullets is to be crowded into the case.

If round-nose or similar lead bullets are used, we'll find that upon firing the noses of all but the front bullet will be badly deformed, squashed to almost wadcutter shape by the bullet ahead. Again, as with round balls, the individual bullets must be separated from one another by an insulating wad if they are not to jam together and to insure separation as they leave the muzzle. Thin, hard cardboard works fairly well, but I prefer .010-inch thick styrene or polyethelene sheet.

The most successful multi-ball loads I've used employed swaged, gas-check or short-jacketed bullets. Some dies for making half-jacketed or short-jacketed bullets can be adjusted to produce projectiles of very light weight, as little as 60 grains in .38/.357 caliber. If you're serious about multi-ball loads, I suggest this approach. One such set of dies obtained many years ago from C-H were made particularly to produce bullets for multi-ball loads. The base punch was shaped to produce a shallow concavity in the bullet base and, likewise, the nose punch produced an identical convex profile. Consequently, the completed bullets nested very nicely together, nose-to-tail. Gas-check bullets swaged in this die set separated reasonably well at the muzzle, but did even better when a second gas check was swaged over the nose of each bullet. We then had copper-to-copper contact between bullets, and there was none of the "welding" or fusing of bullets that occurs frequently with the unjacketed type.

There is another form of multiple-bullet load which usually delivers better single-bullet accuracy and still employs a second projectile to increase hit probability. This type load consists of a single, round ball seated down inside the case, then a conventional, jacketed bullet of relatively light weight for the caliber is seated and crimped in the normal fashion. When this load is fired, the front bullet seems to give reasonably good single-bullet accuracy, while the extra round ball (no longer quite spherical after its passage through the bore) emerging from the muzzle behind it, is most likely to strike somewhere within a few inches of the first-bullet point of impact. This type load is the simplest and cheapest of all to assemble, requiring no special tools or special bullets. You need only select a front bullet which is light enough in weight to compensate at least partially for the weight of the lead ball inside the case. In .38 or .357 caliber, for example, a conventional bullet of 90- to 100-grain weight compensates for the 75 or so grains of the round ball without *significantly* exceeding the weight of the typical 158-grain bullet used in these calibers. If you want to go the limit and accept 200 grains as maximum projectile weight, then subtracting the weight of the round ball leaves us 125 grains for the front bullet, and there are several excellent designs in this range, both jacketed and lead types.

After having toyed with numerous multi-ball loads (including several for rifles) over the past several decades, I have come to the conclusion that while they are most interesting and seem at first to offer advantages, they really aren't very practical for defense use, and are virtually useless for any other purpose. *If* the dispersion of the several projectiles could be controlled and predicted as we are able to do with shot loads in smoothbored barrels, then the multi-ball load would have certain utility. Unfortunately, there is one helluva lot of difference between attempting to control the dispersion of two, three or four, full-caliber bullets from a rifled barrel than in controlling several hundred lead spheres fired through a smooth barrel fitted with appropriate choke. The eagerness with which many people have jumped into multi-ball handgun loads is usually based upon the mistaken belief that the individual projectiles can be controlled as in a shotgun, and this simply isn't true. Even if you do come up with a load which shoots close enough to point of aim and which does not spread too widely, it's still necessary to place roughly 75 percent of those projectiles on the target to equal the effect of a conventional, single-bullet load; and, as I said before, if conditions allow you to do that, then they'll allow you to obtain a solid hit with ordinary ammunition. That being the case, there is little justification for the complexity, the cost, and other disadvantages of the multi-ball load.

The 2-ball load on the left is compared to a regular, single-slug load on the right. Accuracy with multi-projectile loads may be sufficient for your purposes, but they won't equal the accuracy of 5 or 6 well-placed regular slugs.

As a practical matter, it is my opinion that multi-ball loads can be justified only in the case where the principal goal is to fill a given area with a maximum number of projectiles in a minimum

The 9mm cartridge on the right has much less overall case length than the .38 Special on the left. Unfortunately, most autoloading cartridges are not top candidates for multi-projectile loads.

length of time. Those conditions exist only in military combat, where heavy covering fire is being laid down or where a large volume of fire is directed at unseen targets, depending upon a literal rain of bullets to insure that the maximum number of targets is hit. It's interesting to note that in the early 1960s the U.S. Army adopted a two-bullet load in 7.62mm caliber for employment in Indo-China for just this reason. When used in a conventional machine gun or the Gatling-type Mini-Gun, this cartridge permitted doubling the number of projectiles placed in a given area within a given period of time. This permitted, for example, the Mini-Gun, firing at the rate of 100 rounds per second (6,000 rounds per minute) to saturate a piece of jungle of bush with 2,000 projectiles from only a 10-second burst of fire. It is also worthy of note that the 7.62mm, two-bullet load placed its front bullet on point of aim consistently, while the rear bullet gave "random dispersion within known limits." The quoted term means simply that while the direction of the rear bullet's displacement from the front bullet could not be predicted, the amount of that dispersion *could*. This was achieved by making the base of the rear bullet at a particular angle, that angle being adjusted to produce the degree of dispersion desired. Theoretically, you could do this in a handgun, but I don't feel it would be practical.

Another factor which limits the utility of multi-ball loads is that they cannot be assembled in short cases. Most autoloading cartridges use cases which simply don't have the space to accommodate any additional projectiles, and still allow any room at all for propellant. Revolver cartridges are another matter, though even there space becomes a limiting factor below .38/.357 caliber. Assembling multi-ball loads doesn't present any significant problems, nor does it require any costly special tools except where special jacketed bullets are to be swaged. Even then it is usually possible to simply obtain new base and nose punches for existing dies and thus keep the cost down.

HANDLOADING MULTI-BALL LOADS

As we've already said, the simplest form of multi-ball loads consists of two or three round lead balls, one seated half its diameter into the case mouth and crimped there, with the other(s) beneath it inside the case. The balls themselves are no problem, either: in .38/.357 caliber, use 000 buckshot pellets (which are swaged, and thus more uniform than cast balls) or obtain a mold .430-inch diameter respectively; in .45 caliber, swaged balls are available in a suitable size, or, cast them .452- or .454-inch respectively for .45 ACP and .45 Colt.

Cases are prepared by sizing, priming, and expanding in the usual fashion; mouths are flared just enough to allow forcing through a ball without shaving off any lead. Add the propellant charge, and you're ready to get on with the different parts of the operation.

An over-powder wad isn't really necessary, but I prefer to use one. Use a home-made or store-bought wadcutter punch to cut out thin, stiff, cardboard wads large enough to fit tightly inside the case. About .365-inch .38/.357, .415-inch .41, .435- to .440-inch in .44, and .460-inch in .45 caliber. Force the wad down tightly on the propellant with a flat-ended length of dowel as large in diameter as will slip easily into the case. A wood or plastic knob about 1½ inches or so in diameter fitted to the end of the dowel will ease the wear and tear on your hand; in fact, it's al-

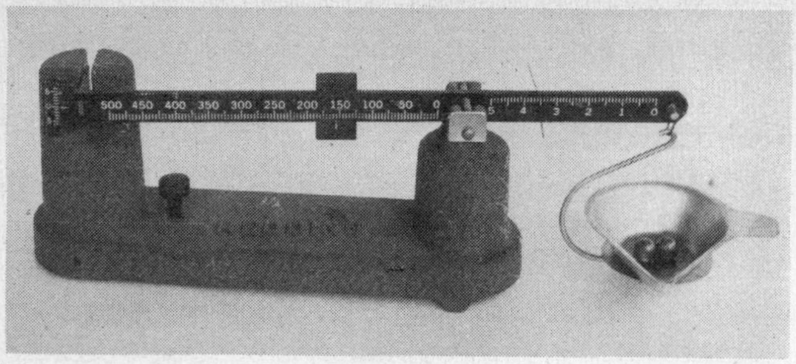

When making up multi-ball loads, check the combined weight of the projectiles to be used — in this case the total weight of the two sized 000 buck shot is 158.5 grains and they will be used in a .38 Special. You can probably select a given type of powder (and charge) for this amount of weight from the .38 Special section of any of the loading manuals. Select the powder (slower-burning powders are best) and a *moderate* load for the weight of the projectiles to be fired, then reduce the powder charge by 10 percent for safety.

most a necessity if you'll be loading many multi-ball cartridges.

If you are using .360-inch diameter 000 buck in .38/.357 cases, a problem will arise immediately in that balls seated inside the case may bulge its wall slightly so that they may not chamber freely. This effect will be aggravated if case walls are thick and sharply tapered as in some military-contract .38 Special cases. Solve this by sizing the pellets down a bit, either by pushing them through a conventional bullet-sizing die without lubricant, or, lacking that, by simply forcing them through the smallest-diameter chamber throat of the revolver in which they're to be fired. This sizing converts the spherical lead ball to a very short, round-nose/round-base bullet, but no matter. On the other hand, if you're using cast lead balls of the diameter mentioned earlier, they'll not normally bulge the case, and no sizing will be required. In .45 calibers, swaged balls will not normally require any sizing, though it might become necessary if you are afflicted with tight chambers and/or thicker-than-usual case walls.

Measure the space left in the case between the over-powder wad and mouth. From this determine whether you'll load three balls or two. The front ball will occupy an amount of case length equal to half its diameter; what is left may or may not accommodate two balls inside. If two balls are to be placed inside the case, start the first by hand, then seat it down snug against the wad with the dowel used for seating the latter. Smear a small lump of bullet lubricant inside the case mouth, then start the second ball and seat it solidly against the first; the lubricant will flow to fill the spaces between the two balls. Take care not to use too much lubricant or it will hold the balls apart and possibly not leave enough space at the case mouth to accommodate the front ball. Place another dab of lubricant in the case mouth, then start the top or front ball by hand. To finish, adjust an ordinary bullet-seating die to seat the top ball one-half its diameter in the case mouth and crimp the case snugly upon it. Use a roll crimp and, by experimentation if necessary, make sure there is enough crimp to keep all the balls solidly in place under recoil forces. Don't overcrimp or the front ball will be squashed out of shape; though when we consider the use to which such loads will be put, any resultant loss in accuracy probably won't be important.

If for preference or forced by case length to use only two balls, seat the first only slightly below the case mouth, then add the lubricant, followed by seating and crimping the front ball. This allows the rear ball to be in solid contact with the front ball when the propellant starts to burn, thus avoiding the effect of a bore obstruction which would occur if there were significant space between them.

Cast or swaged bullets of other shapes need somewhat different treatment. Naked lead bullets of ordinary configuration and shape should be sized and lubricated normally to prevent leading and to avoid the separate application of lubricant described for round balls. Again measure the amount of space available in the case for bullets and determine how many you'll use *inside;* usually it will be only one unless they (the bullets) are abnormally short and light to permit more. Using the dowel as before, seat the first bullet just inside the case mouth, then place a thin card wad over it; place the second bullet in the mouth, then seat it similarly, simultaneously forcing the first bullet deeper; repeat if there is to be a third bullet inside. To finish the load, place the front bullet in the case mouth then adjust a bullet-seating die to seat and crimp it in the usual fashion. The front bullet may be seated very shallowly, so long as the completed cartridge doesn't come out too long for the gun; it may also be of different shape and weight than those inside the case. In fact it might well be desirable to have the front bullet of round-nose or semi-wadcutter

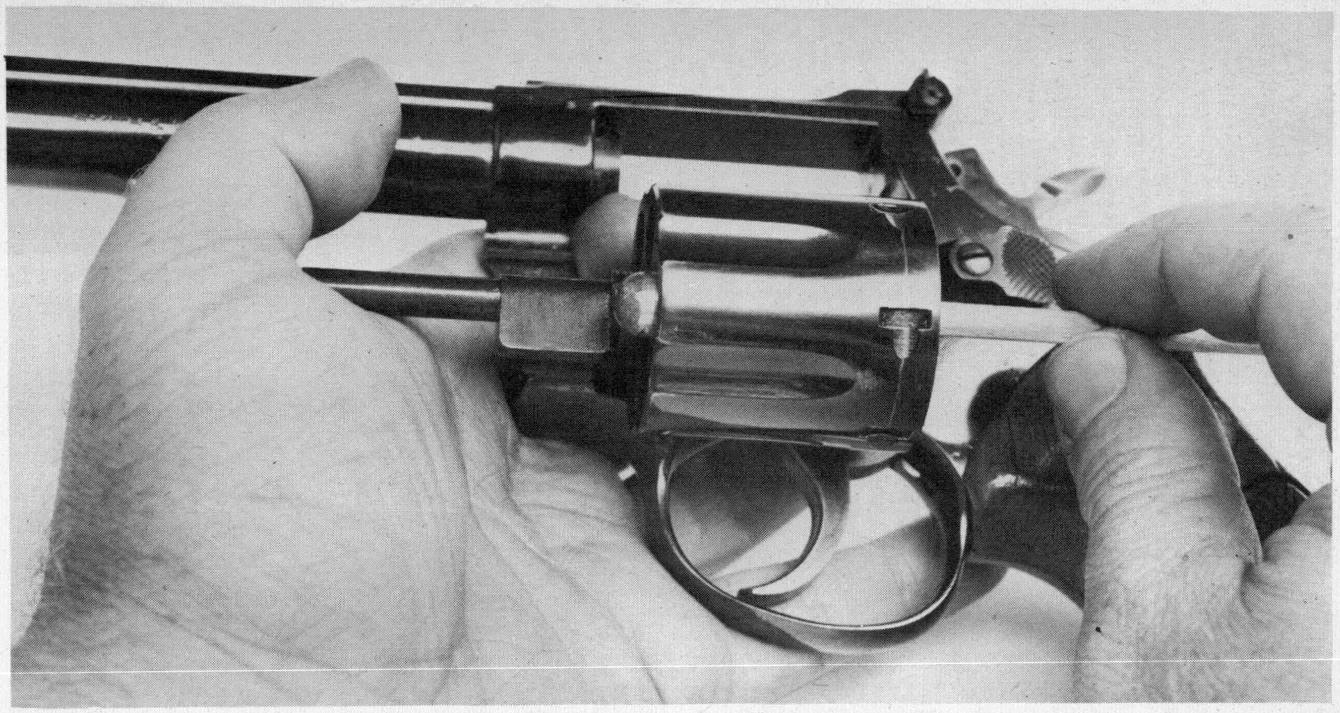

If you live in the "outback" and don't have access to a sizing die, 000 buckshot (for .38 Special or .357 loads) may be sized by forcing them through the smallest chamber in your gun. When doing this, stabilize the crane and cylinder as best you can and tap the ball through the chamber gently. Better yet, remove the cylinder from the gun.

shape while all the others are of pure wadcutter profile. After the *first* cartridge is completed, make certain that you haven't made a mistake which allows the propellant to be compressed by the rearmost bullets. The only way I know to do this is to cut away one side of the case near the rear; easily done in a moment or two with a medium-size, half-round, fine-cut file.

When jacketed bullets are used, we can dispense with the between-bullets card wad, and with the lubrication, but all the other operations remain the same. However, if you're swaging half-jacket or gas-check bullets especially for the job, the large amount of exposed lead may allow bullets to fuse together or to lead the bore. To avoid this, use both the card wad and a very small amount of lubricant between bullets. Alternatively, swage a copper gas check over the nose of each bullet that will be loaded inside the case. Copper-to-copper contact will prevent fusing and leading, and no lubricant or wads will be needed.

There are variations of the loads we've mentioned; you might like to try one or two round balls beneath a conventional lead or jacketed bullet; or, perhaps, several very short wadcutters (almost disc-like) inside the case and a round ball out front.

I know of no valid chamber-pressure tests conducted with multi-ball loads. Consequently, we are working pretty much in the dark when selecting propellants and charge weights. The rule of thumb mentioned earlier in these pages, has proven safe thus far; work from the total projectile weight, selecting from a published data a propellant and charge weight which produces *moderate* performance with a single bullet of the same or greater weight. A greater safety factor will be available also if the slower-burning propellants are chosen. Then, *to be even more safe* I would recommend reducing the charge weight at least 10 percent to start, then work back up slowly in very small increments while watching closely for any evidence of excessive pressures.

As you've no doubt noticed, I am not terribly fond of multi-ball loads, even though they appear on the surface to possess some advantages for defensive use. Try them if you will, but do not expect them to substitute for precise marksmanship and good gun handling. Always remember that such loads are not capable of precise accuracy, no matter how good the gun or the shooter.

CHAPTER 15

Loads and Barrel Lengths

THERE IS A WIDE variation in the velocity that a given load will produce in different barrel lengths. Even factory loads—in fact, especially factory loads—which are designed specifically to produce the best possible all-around ballistics in all standard barrel lengths, show a very wide spread in velocity.

This is essentially a revolver problem, for the very simple reason that center fire, autoloading pistols do not vary a great deal in barrel length within given calibers. For example, among the current crop of .380 pistols, barrel lengths range from 2 inches to 3⅞ inches—no more; 9mm pistols cover a spread from 4 inches to 6 inches, and in .45 caliber, the spread is even less, from 4¼ inches to 5 inches. Of course, longer and shorter barrels are found in customized guns and some limited-production models, but it's the major makes and models with which we must be concerned.

Revolvers are another matter, with factory .38 Specials ranging from barrel lengths a wee bit under 2 inches up to the 8⅜ inches of the longer S&W K-frame variation. There is also the simple fact that the specified barrel length of an autoloader represents the *true* length from breech face to muzzle, while the true or "effective" length of a revolver barrel is the specified length *plus* the length of the cylinder. This is quite often overlooked in comparing performance of the two types.

It is an inescapable fact that none of the propellants available to handloaders (or, for that matter, to the factories) are suitable for producing approximately the same velocities or even performing efficiently in this wide range of barrel lengths. The factories have chosen—and this all happened really before most of us were born—to load for standard performance in the longest, *standard* barrel length. Magnum calibers are, therefore, loaded to produce standard velocity at established working pressures in specified 8⅜-inch barrels—but wait a minute—that means in an *effective* barrel length of about 10 inches. With a few exceptions, non-magnum calibers are loaded to produce standard velocities and pressures in specified barrel lengths of 6 inches, but again, that really means in an effective barrel length of about 8⁹⁄₁₆ inches. One exception is the .45 Colt whose velocity is normally listed in a 7½-inch barrel, this having been established over a century ago when the cartridge was intro-

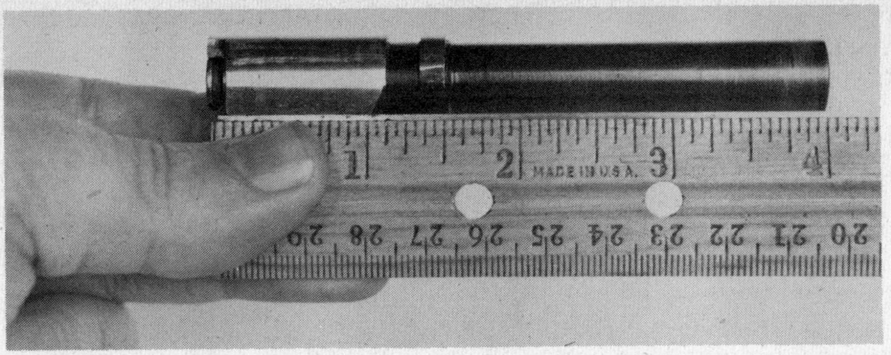

When measuring effective barrel length, remember there are certain differences to take into consideration. On the left, a Smith & Wesson Model 39 autoloader barrel measures out at about 4 inches. Below: An older S&W chambered for .32-20. While the stated length of the barrel *itself* is 6 inches, you have to consider the almost 1⅝ inches of cylinder length that gives the gun an effective barrel length of almost 7⅝ inches.

duced in a gun of that barrel length, the venerable Single-Action Army Colt.

Because the factories publish velocities in these barrel lengths, the majority of the handloading data that has circulated for all these decades is based on the same lengths. This means very simply that if you pick a load from a manual based upon its velocity, in anticipation of using it in a 2- or 2½-inch barreled six-gun, you aren't going to get anywhere near the published velocity—with the single exception of those instances where the data is based upon a shorter-than-standard barrel and is also specified.

All this isn't terribly important for ordinary shooting. So, most .38 Special loads show published velocities obtained in 6-inch barrels, and if we shoot them in a 4-inch tube, it really can't make that much difference if the velocity is a bit less when we're shooting at tin cans, paper, small-game, aerial targets, or whatever. As long as the load barks and kicks about the same as the factory load, and as long as it shoots where it's aimed, we really don't worry about it. But, if you're shopping for a load for personal defense or serious social intercourse, there is little doubt that it will be for use in a barrel somewhere between 2 and 4 inches long. On the other hand, if you'll be hunting or shooting at long range, you'll be concerned with the longest practical barrel. In either instance, maximum velocity and energy is needed.

Ideally, we should have a fast-burning powder for barrel lengths of 2 to 4 inches, one burning not quite so fast for 4- to 6-inch barrels, and an even slower-burning propellant for barrels longer than 6 inches. That sounds simple enough, and if it were no more complex than it seems at first glance, we'd probably have those propellants suited to each barrel-length range.

Unfortunately, it is by no means that simple. To obtain optimum results (maximum velocity with minimum pressure), the variations in case capacity, bullet weight, and other factors would make nearly a score of different propellants necessary, rather than just three. In a 2-inch .38 Special, for example, a different propellant burning rate is needed to produce maximum velocity with a 95-grain bullet than for one in the 135- to 140-grain range, and yet another for the heavier bullets in 155- to 175-grain range. Obviously, the 95-grain bullet will not offer nearly as much resistance as will the heavier bullets to the buildup of gases as the propellant burns. A propellant ideal for the lightest bullet in a 2-inch barrel would have its burning rate balanced with the resistance offered by the bullet so that gas generation will increase in volume more rapidly than the moving bullet opens up space for it, only up to the point the bullet exits the muzzle. This will produce maximum propellant efficiency and projectile velocity. However, if the same propellant is used with a heavier bullet, the greater resistance the latter offers will have the effect of increasing the burning rate and thus the peak pressure. Excessive pressures can result. If that same propellant is used with the same intended bullet weight in a substantially longer barrel, performance will start out essentially the same and continue so through the first 2 inches of barrel. But after the initial 2 inches, the gas generation rate will have fallen off to the point that space in the bore opens up behind the moving bullet more rapidly than gas can be generated to fill it and maintain useful working pressure on the bullet base. The effect is that in extreme instances, the bullet actually begins to slow down while still in the bore. At the point in time and space where working pressure applied to the base of the bullet drops below the level of the resistance existing to the bullet's passage, velocity begins to drop. We have de-

Top to Bottom: S&W Model 29, 8-inch barrel; S&W K-38, 6-inch barrel; S&W Model 10, 2-inch barrel. To get optimum results from each of these barrel lengths requires that the handloader take into consideration case capacity, bullet weight and other factors *before* selecting the "right" propellant.

For loads in short barrel revolvers (from left to right) Bullseye. W-W 231, Herco and Unique are perhaps the most useful. Again, the right selection of powder is dependent upon the weight of the slug and case capacity of the cartridge to be loaded.

monstrated this many times with factory ammunition in extensive velocity tests where loads showed a significantly higher velocity in 6-inch barrels than in 7½-inch and 8⅜-inch tubes. The effect is less and more difficult to measure between 2- and 4-inch barrels, but it exists nevertheless. The fact remains that *if* there were a propellant *ideal* for a particular bullet weight in a 2-inch barrel, it would produce less velocity in a 4-inch barrel, and even less in those of greater lengths. The reverse is also true. For if we use a propellant ideally suited for a particular bullet in a 4-inch or 6-inch barrel, it will produce less velocity in the shorter lengths.

However, at least to a degree, this discussion is of mainly academic interest. We simply do not have propellants with a sufficiently wide variety of burning rates to meet all the conditions of bullet weight, type, and barrel lengths that are encountered. This is by no means derogatory to the propellant manufacturers. To simply have such a variety of propellants would require more different propellants than could be marketed profitably. Such an effort would be comparable to the petroleum industry attempting to produce a different gasoline carefully matched to each different type and class of automobile engine on the market. It simply isn't economically practical.

As a result, we are limited to the propellants available, and while there is a large variety, their suitability for different bullet weights, types, and barrel lengths is often more a result of chance than intent. For example, the propellant that performs best (Hercules 2400) and is most widely used in long-barreled, magnum revolvers (above .357) was originally designed specifically for small-capacity rifle cartridge use. We must also consider the fact that there have been no new true *pistol* propellants placed on the market for many years except the Winchester-Western ball numbers. All of the others have been in existence since long before the current jacketed-bullet/high-velocity handloading trends developed.

The ammunition loading plants are by no means as restricted in propellant choice as the handloader. Propellants sold for handloading are carefully tested and blended to produce a specific performance level so that, for example, the Unique you bought last week will perform the same as what you were buying and using in 1957 or earlier. Loading plants don't even use that same Unique in most instances; they take Unique direct from the propellant manufacturer and, when preparing to use a particular new lot number, they conduct pressure and velocity tests to ascertain the charge weight that will produce the pressure and velocity required by the cartridge/load specifications. As a practical matter, the loading plant couldn't care less what the propellant charge weighs as long as it produces the desired velocity at pressures not exceeding established limits. Propellant lots as delivered to loading plants vary in burning characteristics, so for a specific load, a lot used during February may do the job with 4.5 grains; and, the next lot purchased in, say, June may do the same job with 4.2 or 4.7 grains. In addition, for a loading-plant customer prepared to buy an entire lot, the propellant manufacturer can increase or decrease the burning rate of any particular propellant to meet specific needs. Therefore, if a loading plant wishes to introduce a new load that requires a propellant burning a bit faster than Unique but not so fast as Bullseye or 231, a special lot of Unique can be manufactured and blended to meet that need. It is in this manner that ammunition manufacturers can often achieve performance levels which simply cannot be duplicated by the handloader. Unfortunately, such special lots, and even special propellants, simply cannot be used *safely* in developing loads unless both pressure and velocity measuring equipment is available, along with sufficient expertise to operate them correctly and consistently. The result of all this is simply that we are restricted to commerically-available "cannister grade" propellants and published loading data.

As we've already intimated, there is a wide choice of propellants suitable for all practical bullet weights in the middle range of barrel lengths, approximately 4 inches to 6 inches; and, propellants that will efficiently produce the entire range of performance from very light bullets at extremely high velocities up to quite heavy bullets at moderate velocities of around 1200 fps. Not nearly so many propellants are suitable for best performance in the long magnum barrels, but there are at least three which allow sufficient production of very high velocities with the entire range of practical bullet weights and types; and to my way of thinking, that's plenty. Numerous loading manuals and a wide variety of other data sources contain thousands of tested loads that perform well in these longer barrel lengths.

In regard to the shorter barrels (2-inch to 4-inch) we are not nearly so well supplied with efficient propellants. In fact, in my considered opinion, there is not a single propellant available which is ideally suited to full-charge and high-performance loads in short barrels. Bullseye, W-W 231, Herco, and Unique are the most useful. Bullseye burns fastest of the lot and is the least useful, performing best only with the very lightest bullets and in the smallest calibers. Next fastest is 231, very nearly as limited as Bullseye. Herco comes next, being between 231 and Unique in burning rate, and in this scribe's opinion is probably the best all-around choice for light bullets in 2- to 2½-inch barrels. Unique also performs well, though it is a bit slow-burning for 2-inch barrels except in the larger calibers. Unique is also excellent (perhaps even a bit better) in barrels up to 6 inches in length.

Even though loading manuals do not generally indicate velocities for barrels in the 2- to 4-inch range, by keeping the foregoing in mind, you can select from them loads which will be as efficient and powerful as it is possible to obtain in such shorty guns. With the very lightest bullets (90-100 grains) in .38 Special, this means loads utilizing 231 or Herco; in the 4-inch tubes, Herco or Unique loads. In the .357, Herco or Unique loads for the shortest barrels and Unique for those over 2½ inches. You aren't likely to run into any real short barrels in the bigger bores, except the 3-inch Charter Arms .44 Special Bulldog, and in it, you'll find the Unique loads the most productive. If you prefer the heavier bullets—150-170 grains—in the .38 and .357, Unique loads will serve the shorter barrels best. Virtually all of the loading data published on .41 and .44 Magnum loads is designed for maximum performance in 7½- to 8⅜-inch barrels; so, if you want performance out of the 4-inch tube (nobody in his right mind would shoot anything shorter in those calibers), look for loads using Herco, Unique, or HS5. Though these loads will show less velocity in the longer barrels, they'll normally do fully as well in the 4-inch tube, and do it more efficiently and with less report and flash. For 4-inch barrels (nothing shorter was standard) in calibers like .45 Colt, .45 A-R, .44-40, etc., again look for Herco and Unique loads. They'll burn cleaner and give less report and flash—not to mention use only half as much propellant—as the long-barrel loads containing 2400, IMR 4227, H110, etc.

Barrel length *is* important to the choice of load. Fortunately, existing loading data takes good care of those wishing to use their handloads in 6-inch and longer barrels. You can't really go wrong in selecting from a good data source there. However, if you want a really powerful short-barrel load, you're going to have to take all the factors we've discussed into consideration and then choose a load carefully by powder type or number.

CHAPTER 16

Ultra-Light Loads

WE'VE TALKED ABOUT especially accurate target loads and discussed high-performance handloads producing the ultimate in power and lethality, plus we have covered the subject of handloading in general in the assembly of perfectly ordinary loads for both autos and six-guns. However, there is another especially useful segment of handloading that we feel requires a bit of special attention. In fact, it becomes even more important in these times when more and more prime informal shooting area is being swallowed up by housing developments, shopping centers, and expressways and is simply being declared off limits to shooters and hunters alike. There are, to be sure, still millions of square miles of sparsely-settled or completely uninhabited country where one may go and shoot to his heart's content with whatever manner of ammunition pleases him. Unfortunately, the majority of our modern pistoleros simply do not have access to those wide open spaces. Sure, there are commercial ranges, club ranges, and even some park departments and police ranges which can occasionally be made available; but, lumped all together, they aren't a drop in the bucket insofar as satisfying the shooting-place needs of literally tens of millions of handgun owners. Many solutions have been put forth to the problem of finding a place to shoot—but seldom is one's home considered one of those solutions. This is a reasonable enough attitude, considering the fact that unless you've spent thousands of dollars for a heavily baffled soundproofed, insulated, and ventilated indoor range, there's no safe way you could shoot ordinary loads inside without calling down the wrath of family, neighbors, and police alike. Even if the neighbors and cops don't come running, triggering off a half-dozen full-charge loads in the basement of the average home will puff dust from every crack, rattle all the vents, shimmy the windows, and likely cause the kids to leave home and mama to file for divorce, even though the noise doesn't bother you in between those earmuffs. Even typical target loads of the .38 Special wadcutter variety become exceedingly vicious in a low-ceilinged basement, and the bark is certainly no soothing whisper in a spacious garage.

However, there's no rule that says you have to use that type of ammunition. What we call ultra-light handloads can allow you to shoot indoors without stirring up a lot of fuss, and some of the

Given the right conditions, indoor target shooting with ultra-light loads can really sharpen the handgunner's eye.

loads are so quiet and well-mannered that you can shoot in the den or living room after dinner and not even overpower the audio from your television set. Ultra-light loads require no modifications to the building, and backstops or bullet traps that are entirely safe (as long as you keep your bullets within their confines) can be as cheap as nothing. As far as loading equipment is concerned, that which you already own and use will do the job perfectly—but, it is entirely possible to assemble excellent loads of this type without owning (and therefore without buying) a single piece of what ordinarily passes for handloading equipment. And whenever I think of that, I'm always reminded of the one point in my life when I owned but a single, good .38 Special revolver (a S&W Military & Police Model), had plenty of time and places to shoot, but not enough money for factory ammunition and not enough for both loading gear and components. During that period, I loaded and shot several thousand ultralight loads with no more tools than a piece of dowel, a small mallet, a charge-cup made from a .22 Short case soldered to a paper clip, and a decapping pin consisting of a 20-penny spike with ¼-inch of its tip filed down small enough to pass through the flash hole. Those sophisticated loading tools, an 8-ounce cannister of Bullseye powder, a few 5-pound bags of 000 buckshot, and a supply of primers saw me through better than half a year of shooting. I don't recall the price of the store-bought components in those days, but even now such loads can be assembled for only slightly over 1¢ each—substantially less than the cost of .22 rimfire ammunition. Just recently I dug a bag of buckshot out of the back room, duplicated the homemade tools of so many years ago, and assembled a few of those loads to shoot here in the office. They were just as much fun this time around as before, and I got thinking maybe I had waited 20 or 30 years too long to try it again.

Naturally, there are a lot of things ultra-light loads won't do: They won't knock a charging grizzly dead in his tracks; they won't shoot a 100-10X at 50-yards slow fire; they won't cycle an autoloading pistol; and, if you want to show off busting rocks across the canyon or tree stumps across the river, well, you'll have to use something else. On the other hand, they'll do a lot besides going "pop" indoors and punching holes in paper. They'll kill a cottonmouth in your favorite fishing hole or a rattler invading your camp just as dead and just as quick as the biggest magnum; they're great for Sunday-afternoon barnyard plinking without spooking the livestock; they'll rid your shade tree of starlings and blackbirds according to your marksmanship; they'll help you learn to bust thrown aerial targets, and they'll bag bunnies out of the buckbrush and bushytails out of river-bottom shag-bark hickories for pot or skillet if you do your part. Some of the tastiest turtle soup I ever ate came to the table as a result of wandering along the banks of a meandering, little, muddy river with those "el cheapo," round-ball loads I used so many years ago.

Before going farther, maybe we ought to *attempt* to define "ultra-light" loads. One definition I've seen in print places any load which does not exceed the muzzle energy of the .22 LR in this class; but I'm not too fond of that one because even the lightest, practical projectiles (round balls) in .44 and .45 caliber can easily exceed that energy level before they're traveling fast enough to deliver acceptable accuracy. I don't think we can place a general limitation on either energy or velocity and class everything beneath it as ultra-light. Instead, I prefer a floating definition that says generally, "any load that drives an extra-light (for the caliber) bullet at the *minimum* velocity necessary to give it acceptable accuracy at a range of at least 25 feet." That's rather imprecise, I know, but it seems to fit.

Within this class, then, we have a variety of load types. Using lead bullets, we have the round-ball load which uses either swaged factory-made buckshot, the larger swaged balls made for muzzle loaders, home-cast, round balls, and those assembled with conventional, lubricated, cast bullets or lighter-than-usual weight for the caliber. Another type—one with which I've never had much success but which has apparently seen some use in Europe—uses an *empty* metal jacket as a bullet. One step below these we find the plastic-bullet loads assembled either in special, plastic cases such as those made by Speer, or in ordinary brass cases.

At the bottom of the scale in both cost and complexity, we have the wax-bullet loads popularized quite a few years back during the quick-draw craze.

There's a good deal of difference in the accuracy and effective range that can be obtained from the different types of loads. Those using conventional lead bullets have usually given me the best accuracy, followed very closely by round balls, with plastic bullets next in line, and the wax loads wagging the tail. Lead bullets may perform well to 25 yards, while wax bullets are good out to only 10-12 feet; the others fall between those lines.

Here are three popular types of .38 Special "Light" loads designed for indoor shooting. Left to right: Sized, 000 buckshot seated over a *light* propellant charge; Speer's primer powered plastic-bullet-and-case load; and, a primer powered wax-slug load.

HANDLOADING ULTRA-LIGHT LOADS

Round-Ball Loads

Let's look at the lead-bullet loads first, beginning with round balls. As a practical matter, there's no real difference between swaged buckshot or balls and the leaden spheres you've cast yourself, assuming you've done the job right. It is necessary with cast balls to align the flat sprue *accurately* in front or behind when seating the ball in the case mouth. Obviously, you can't see the sprue if it's inside the case, so it'll have to be out front, and to obtain satisfactory accuracy, it *must* be centered as precisely as possible. Some people very carefully trim off the sprue and true up the ball with a sharp knife, but to me that is a lot more trouble than simply aligning it accurately. Results appear to be equal, so you pays your money and takes your choice.

Assuming the cases you'll use will chamber freely in the gun at hand, there's no need whatever to resize for round-ball loads; in fact, it can be detrimental. Decapping can be done on conventional tools, or you can set the case on a block of wood through which has been drilled a ¼-inch hole and then, with a small mallet, tap the fired primer out with a 1/16-inch punch or a nail you've filed down to fit through the flash hole. Once decapped, you should make certain that the cases are reasonably clean inside and out. The bare-bones method of repriming consists of setting a primer anvil-up on a smooth, flat, solid surface, placing the case carefully over it so that it is started into the primer pocket, then slipping a 5/16-inch dowel into the case, and tapping it with a small mallet or stick to drive the case down over the primer. When the case head comes to rest solidly on the flat surface, the primer is seated flush. While this method of seating is not what we have recommended for other loads elsewhere in this volume, it has always seemed to work perfectly well with ultra-light loads. There are those who have the idea that this method of seating primers is dangerous. It isn't. Relatively little force is required to drive the case down over the primer, and since the primer is shorter in length than the pocket is deep, there cannot be sufficient resistance to the anvil to cause the primer to

Above — Decapping can be accomplished by making a primer knock-out-pin out of a nail — it's that simple. However, if you've got a few extra bucks to spend you can get a Lee knock-out-pin and base from your gun store. The wooden block is home made out of 2x4's. It's counter-sunk and drilled through to accept the base and allow spent primers to fall into any recepticle you may care to put under the block.

detonate, even if you use the mallet more vigorously than is required. However, if it makes you nervous to seat the primers in this fashion, the very economical and extremely simple Lee hand-type priming tool seats primers very nicely in the usual fashion, and it's small enough to fit in the palm of your hand. In addition to all that, its operation is quite fast if that interests you.

After priming, chamfer the case mouth *lightly;* just a twist or two of a deburring tool under very light pressure to remove the sharp inner edge without producing a significant bevel. If you have to do it with a knife, scrape instead of cut, all we want to do is break the sharp edge, not funnel the case mouth.

Bullseye is the propellant to use; there are others that will work, of course, but it has the right burning rate and ignites with extreme ease, which is important where only a pinch is being used in a relatively cavernous case and the charge may assume any position relative to the primer. Under some conditions of use, the primer flash may have to travel nearly a full inch to reach the propellant, and that means we need quick and easy ignition.

Dipped charges of Bullseye may not be quite as consistent as 231 ball propellant, but its favorable ignition characteristics outweigh this minor

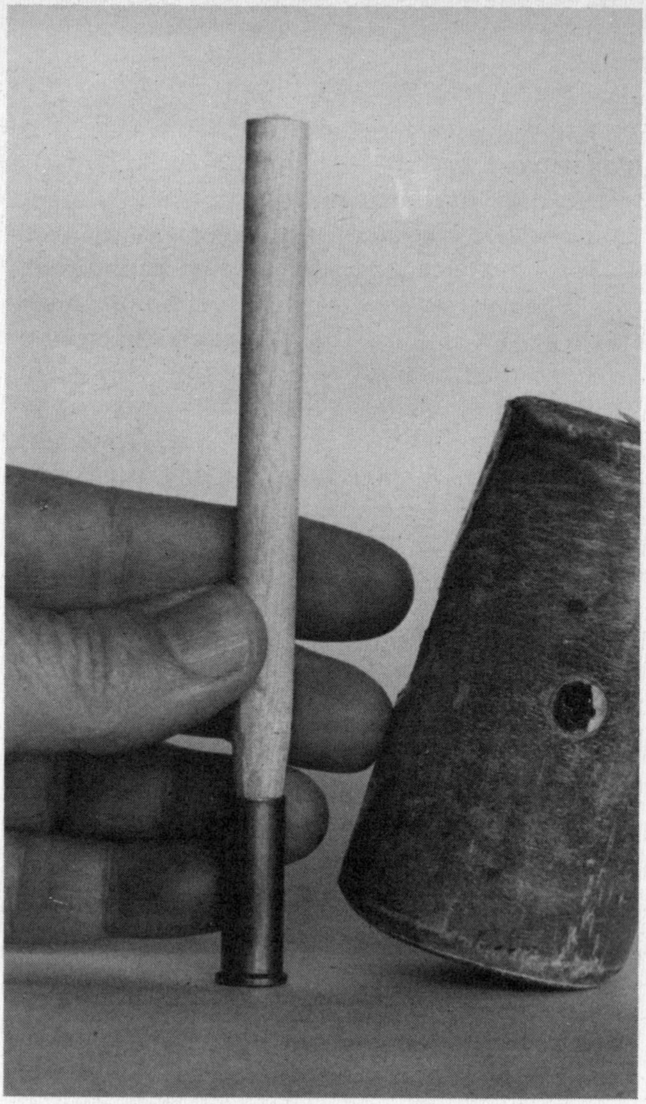

Lacking priming tools, simply find (above) a flat, clean surface, trim down a piece of dowel and you can tap your primer into place. Below, you can see the results.

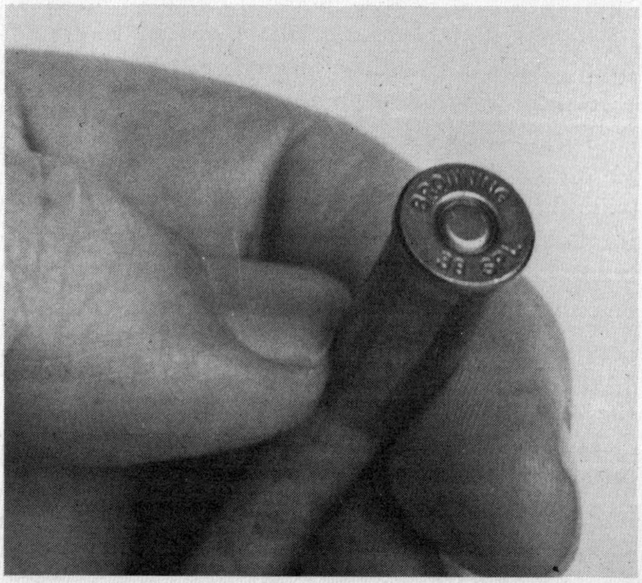

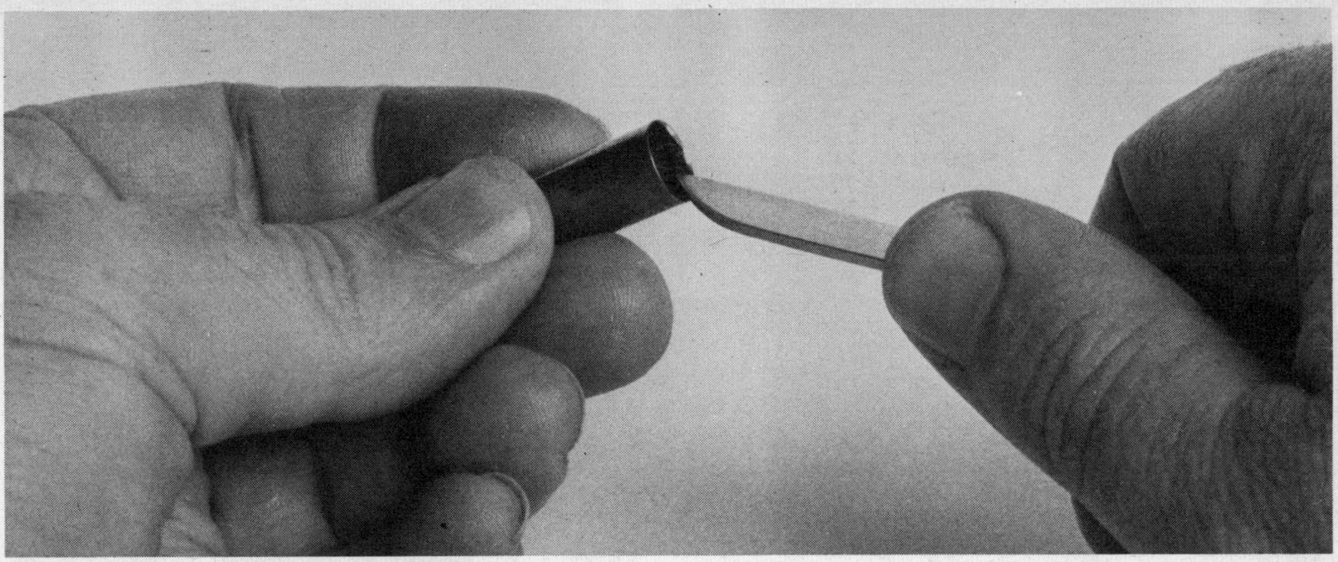

Chamfering the inside of the case mouth can be quickly accomplished with a pocket knife. When doing cases for wax-bullet loads, make the chamfer fairly sharp — the case will cut through the wax like a hot knife through butter.

disadvantage. A *fired* .22 Short case with a wire or paper clip handle soldered or epoxied in place makes an excellent charge cup or powder dipper. The typical Short case measures .425-inch long and holds 2.0 grains of Bullseye; filed to a length of .300-inch it holds 1.4 grains, and further shortened to .250-inch it holds 1.2-grains, but perhaps one of the others will suit you better. In .44 or .45 caliber, 2.1 grains of Bullseye performs very nicely, and for that, use the shortened .25 ACP case as described below.

For any charges of Bullseye greater than what .22 rimfire fired cases will hold, other fired cases shortened and used in the same way will do nicely. The .25 ACP case holds 3.2 grains, and shortened to .415-inch holds 2.1 grains; I've yet to find any need for larger charges than that with round balls in even the largest calibers.

Pour a coffee cup about three-quarters full of Bullseye, then take the charge cup in hand and submerge it upright, straight down into the propellant until it fills and is completely buried; then lift it straight back up out, and with any flat, blade-like instrument strike the surplus off the top of the cup. Dump the cup's contents into the primed and prepared case, tapping it on the case mouth to make certain no propellant hangs up in the bottom of the cup.

The round-ball projectile differs from most others in that it contains no provision to hold lubricant. Even when heavily graphited, as the swaged, commercial balls and buckshot are, some lubricant is still needed to prevent leading and its deleterious effect on accuracy. To solve this little problem, press a lump of bullet lubricant out into a flat sheet no more than 1/16-inch thick. I usually just flatten out a few square inches by hand, but if you wish a bit more neatness, lay some plastic film over the lube and work it into a thin sheet with mama's rolling pin. In any event, press the sheet of lube over the mouth of the charged case, cutting out a disc which clings there and which will provide adequate lubrication for the shot.

When loading .38 Specials, take 000 buckshot, whose nominal diameter is .360-inch, and thumb one into the mouth of the case. Unless the case has been overly expanded by previous firing in an oversize chamber, the ball should not quite enter the case but come to rest with just a bit less than one-half its diameter inside the case. With a stick of wood or a very light mallet, tap the ball gently until just barely over one-half of its diameter is into the case. Normally this will seat it tight enough that it will not become dislodged during normal handling or functioning of the revolver. However, if it seems not tight enough, the edge of a screwdriver bit may be used by hand to impress two or three very light stab crimps around the case mouth for added security. Of course, if you have a crimping die available, a *very light* conventional crimp may be applied.

In .44 or .45 caliber, revolvers, use lead balls of

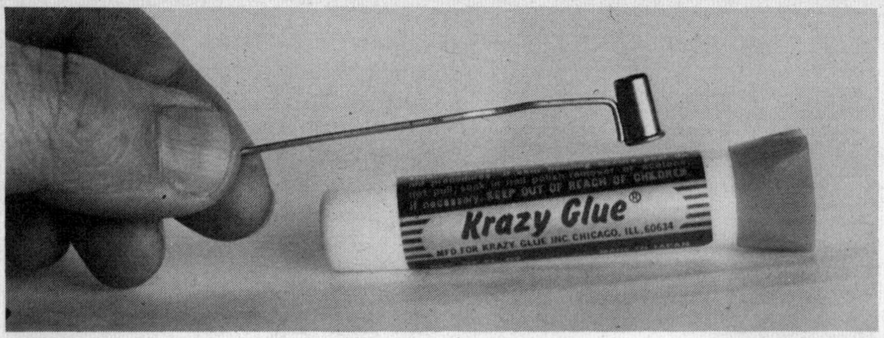

Left — Need a handy charge cup for those ultra-light loads? Just use a bent paper clip with a .22 Short case attached with one of the "miracle" glues commonly available at most hardware stores. A full-length, fired .22 Short case will hold 2.0 grains of Bullseye.

Below — Once you've made your dipper, be sure to lower it *straight-down* into the powder — don't "scoop" it out. Scrape off excess with straight-edge ruler.

nominally .435-inch and .455- to .457-inch respectively; .38 and 9mm autoloader cartridges require the same 000 buck, but the .45 ACP requires a round ball of no more than .454-inch.

You may find that some of your cases have been fired in slightly oversized chambers and are thus too big at the mouth to grip the ball. Slight resizing will handle that problem, but you may choose instead to seat the ball down inside the case. In fact, some handloaders prefer to do this with round balls for the more efficient combustion and use of propellant that results. If the ball falls freely into the case mouth, after powder and lubricant wad are in place, simply seat it deeper with a piece of 5/16-inch dowel; the combination of tapered case walls and less expansion toward the rear will produce a tight seat for the ball some distance inside the case. Naturally, this will vary from case to case, and we need to have all the balls seated to the same depth. Use a mark or a stop on the dowel to obtain uniform seating depth. I prefer to drill a small piece of wood to slip over the dowel, then glue or pin it in place as a stop so that when it contacts the case mouth the ball is seated to the correct depth.

When using deep-seated balls, there is a variation that you might like to try. After the single ball is seated down in the case, apply another lubricant wad, and then seat a second ball down against the first. If it doesn't fit tightly enough to remain in place, stab-crimp the case lightly over the ball with a screwdriver. This double-ball load can be of lot of fun, but generally it won't be as accurate as a single ball.

Lead-Bullet Loads

Conventional bullets for ultra-light loads share one basic characteristic—they are lighter than usual for the caliber. In 9mm/.38 calibers, this means 90-100 grains, about 160-180 grains for the .44s, and about 170-200 grains for the .45s. Wadcutter or semi-wadcutter shapes are best, and while a fairly soft alloy is preferable, even wheel-weight metal isn't too hard. Just because ultra-light loads aren't expected to produce much in the way of top velocity or accuracy is no reason to stint on casting the bullets; the better they are the better they'll shoot, so do the job as if they were intended to win a match.

Depending upon several factors, the bullets may be used either sized normally or loaded as-cast without any sizing whatever. Sizing and lubrication can be done on a lubricator/sizer, of course, but the lack of one need not be any problem. My favorite simple and cheap method of sizing it is to use the old Lyman tong tool and "bullet-sizing chamber." The tool is still made as well as the dies and after all these years, plenty of used tools show up at gun shows, so finding one isn't difficult. Lee also makes a handy little sizing tool employed with a hammer, and it is currently available.

Bullets to be sized can be lubricated by simply smearing the stuff into the grooves with your fingers, but that's a bit messy. Instead, find a flat-bottom tray or pan with sides at least ½-inch high, and set bullets in it, leaving a bit of space between them. Set the tray over low heat on mama's range, and add bullet lubricant to melt

Left—In this instance, the 000 buckshot ball has been sized and seated to about half its length. If you choose to seat an unsized ball, simply thumb it into the case, tap it in with a wood mallet.

Right — Some guns will shoot unsized and lubed pistol bullets as well — or better than — *sized* and lubed slugs; also, the unsized bullets load faster. Try both and check your accuracy results.

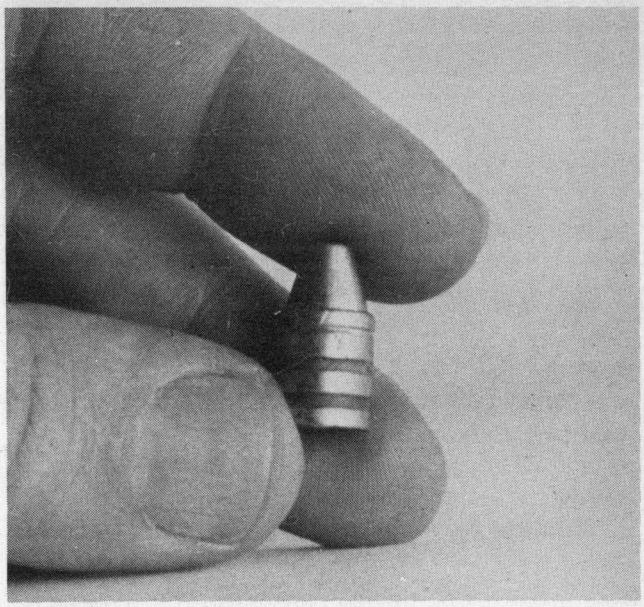

until the level reaches just above the top of the lubricating grooves. When the lube is completely fluid, remove from heat, and let it harden. Careful—that hot lube clings to your skin like glue and can produce nasty burns. A tool called a "cake cutter" is used to remove the bullets from the hardened lubricant. Lyman sells this device in several calibers, but you can make one from a piece of tubing or a fired cartridge case with the head cut off. The inside diameter of the tool should be just a trace larger than the bullet. Press the cake cutter down over one bullet, shearing it free from the lubricant, then another; repeat until they're all done—once the tool is filled, bullets will be automatically ejected from the top. If the cake cutter fits exactly right, the surface of the bullet will be peeled clean, leaving the grooves evenly filled with lubricant. Set a fresh bullet in each hole in the cake, and put the tray back on the stove to heat to prepare a second batch. *Never* go off and leave bullet lubricant over the heat, or you may wind up collecting on your fire insurance.

With the Lyman tool, screw the sizing chamber into the handles, then drop a lubricated bullet *base-first* in the top of the die, drop the push rod on top of it, and close the handles smoothly. This will force the bullet through the die, reducing it to proper diameter, leaving it ready to load as soon as you've wiped the base clean. With the Lee tool, drop the bullet into the die, place the punch on top of it, and tap it through with a hammer.

Using sized bullets, you'll probably have to begin with resized cases. You'll usually be able to get several loadings from them before sizing again becomes necessary. In this instance, the mouth needs to be slightly flared so the bullet can be started easily. An ordinary seating die will do nicely, but if you don't have one, start the bullet as straight as possible, then tap it gently to the proper depth in the case. No crimp is needed, and if you've been careful, there won't be enough flare to interfere with proper chambering.

Some guns will shoot as well—and some even better—with unsized bullets as with those which have been sized. When unsized bullets can be used, loading is simpler and faster, and I like that. Unless cases have been expanded excessively, they won't need resizing for use with unsized bullets. A slight flaring of the case, and bullets lubed either with your fingers or in melted lubricant can be started by hand. Often you'll be able to seat them deep enough by finger pressure, but if not, tap them down to the proper depth. If a case is a bit oversize and the bullet enters too easily, two or three very light stab crimps over the solid portion of the *bullet* (meaning not over a lubricating groove) will hold the bullet. With the lightest loads, sometimes only the stickiness of the lubricant will hold the bullet in place well enough. Nevertheless, I prefer something more substantial, either the stab crimp or a very slight resizing. Again, if you haven't overdone flaring of the case mouth, no crimp is needed.

Generally, a bit more powder than used for

round-ball loads is required to produce acceptable accuracy with cast bullets. They're not only heavier than round balls, but longer, and require a bit more rotational velocity in order to be stabilized—and that means burning a bit more powder behind them. In the 9mm/.38 calibers, propellant charges of 0.75 grains to 1.5 grains of Bullseye are about right; the .44s and .45s require 2.2 to 2.6 grains of the same powder. The correct propellant charge in any case is the minimum amount of Bullseye that will produce acceptable accuracy. To use any more than that is to defeat the basic purpose of ultra-light loads.

Plastic-Bullet Loads

Plastic-bullet loads are quite a different matter. Over the years several different types of plastic bullets have been offered, but at the moment only Speer supplies them, either separately or in conjunction with special plastic cases adapted to them. The Speer bullets are of bore diameter forward, with a reduced-diameter rear section to fit inside a rather thick-walled, plastic case. Bullets are of pure wadcutter shape and made without any grooves. They fit friction-tight into the case, and seating depth is regulated by the shoulder joining the two diameters. Nothing more than

To the right a Speer plastic bullet seated in a brass case. When brass cases are used with the plastic bullet load they should be (below) drilled out to .120-inch for .38/9mm sized cases. (In the photo below, a regular .38 Special case is on the left, with one drilled out to .120-inch on the right.) Once you've collected a batch of cases that have had their flash holes drilled out (bottom right) keep them segregated from other brass — they *never* should be used for full charge loads.

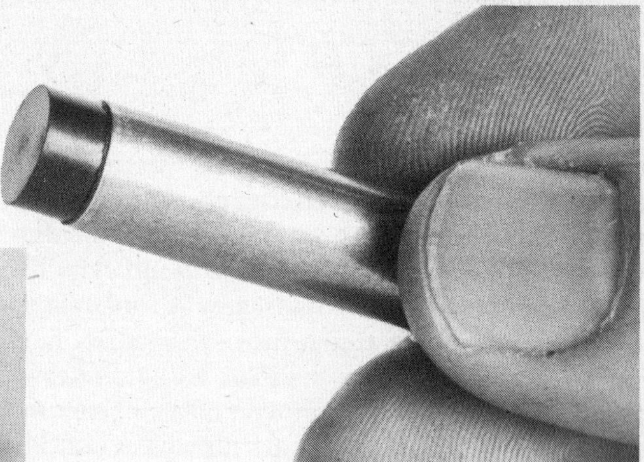

finger pressure is needed to assemble the two components. As we've said, the Speer plastic case has rather thick walls, and its internal capacity is much less than that of the conventional brass case. External rim and head dimensions are the same as for metallic cases, but length is a good bit shorter. The cases are made to be used without a propellant charge, the primer supplying the entire impulse. Toward this end, all calibers accept the large-size pistol primer, and the flash hole is nominally .180-inch in diameter, much larger than the .080-inch maximum in conventional metallic cases. The purpose of the large primer and enlarged flash hole is to obtain adequate impulse without propellant. Toward this same end, magnum primers are recommended. The enlarged flash hole also eliminates primer setback which can tie up the gun.

Loading the Speer cases is simplicity personified. Punch out the fired primer with any convenient instrument, then tap the case down over an anvil-up primer on a flat surface as described earlier, then finger-seat the plastic bullet, and the job is done. Some individuals have employed a very small charge of Bullseye to give the bullet greater velocity. So long as the charge remains small enough, this apparently poses no safety hazard, but I recommend against it. Lacking the plastic cases designed for this bullet, ordinary brass may be used without any particular difficulty. They'll require resizing and a mouth diameter small enough to hold the bullet snugly (over its major diameter); and, if no propellant is to be used—as we recommend—the flash hole in .38/9mm cases should be drilled out to a diameter of .120-inch. The larger calibers require the flash hole to be drilled out to .150-inch. And, as with the plastic cases, use magnum primers.

When flash holes have been drilled oversize, the case *should never again* be used for ordinary lead-bullet loads. The big hole allows too much propellant gas to act against the inside of the primer cup and will drive it rearward out of the case as far as the headspace tolerance of the gun will allow. In autoloaders, this won't cause any functional difficulties unless pressures are sufficiently high to cause leakage or perforation; but in revolvers, the primer jams tightly between the case head and recoil shield, and thus interferes with cylinder rotation. In order to avoid mixing up cases, you might want to mark—red nail polish works nicely—those modified cases.

Using cases with standard flash holes, a very small powder charge can be safely used behind the plastic bullets. However, in my experience, attempting to drive the bullets faster than they were intended generally destroys their accuracy and will also deform the bullets making it impossible to reuse them as can be done when only a primer is used. When loaded as intended, these bullets may be reused many times so long as they are fired into a sufficiently soft backstop that the impact doesn't damage them.

We should point out that Speer produces these plastic cases only in .38, .44, and .45 revolver sizes. If you want to use the bullet in guns which won't accept those cases, you have no alternative but to use the correct brass case.

Wax-Bullet Loads

The last type of ultra-light load we'll discuss uses wax bullets. They represent the absolute maximum in economy, and nothing could be simpler to reload. First of all, the cases must be altered by drilling out the flash holes as already described. As we mentioned above, this oversize flash hole also prevents the primer from backing out of the pocket under the influence of its own gas and impeding revolver cylinder rotation. When a primer is ignited, fairly high gas pressures are developed inside the pocket, and the gas is produced much more rapidly than it can escape through the flash hole. The result is that the pressure acts to drive the primer rearward out

Ordinary paraffin is cheap, reusable, and available at most grocery or hardware stores — It's perfect for the manufacture of ultra-light target loads.

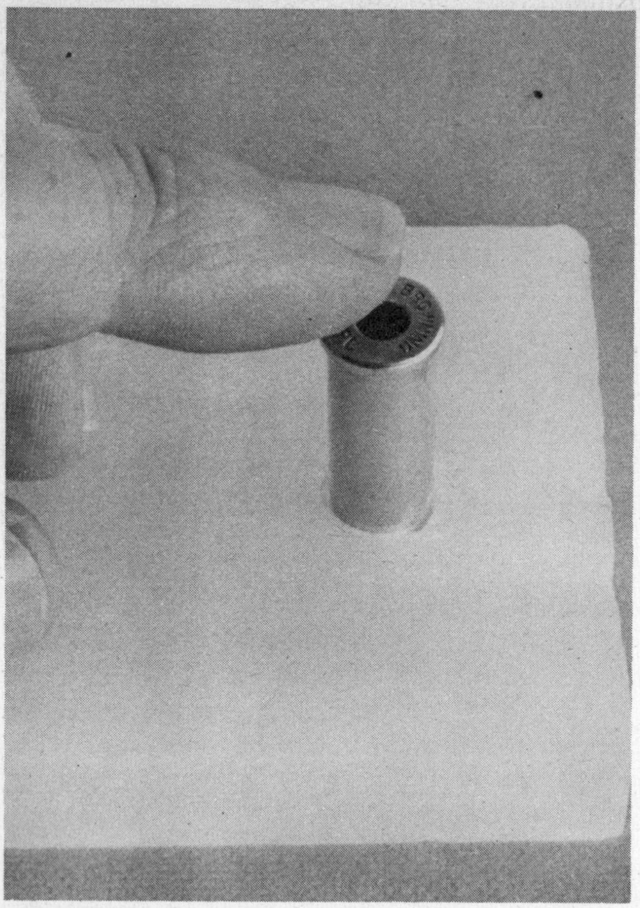

Above — When making your primer-powered wax bullet load, just press a primerless case into a block of wax. (If the case is primed when you do this, the air compressed in the case may push the wax slug back out.)

Below — This group was fired at a distance of 15 feet using wax-bullet loads and the S&W Model 10 shown. As you can see, the accuracy isn't too bad, and, the practice is indeed cheap.

of the pocket against the recoil shield. If there is a substantial propellant charge in the case, this pressure then drives the case back against the recoil shield as well, reseating the primer flush with the head. If that primer isn't reseated in that fashion, it binds on the recoil shield. If you doubt this, prime three or four unaltered *empty* cases, load them in your favorite six-gun, and fire them. First you'll feel the drag of the protruding primers after the first shot, and when you extract the cases, you can see how much the primer has backed out of the pocket.

The "wax bullet" starts out as an ordinary cake of paraffin wax obtained at your supermarket (in the home-canning supplies department) and is brought to final shape by simply pressing the case mouth through the wax cake. Once the case has been pushed through the wax cake, the bullet has not only been formed but seated, and remains in position, flush with the case mouth.

This ridiculously simple method of making and seating a bullet has one drawback—if the case is primed first, sufficient air may be compressed inside the case to push the bullet back out. To eliminate this, it's customary to decap the case first, push it through the wax cake, and then reprime as the final operation. This eliminates our ultra-simple priming method of using a dowel to tap the case down over the primer. Consequently, the little Lee hand-type priming tool comes into its own.

That's really all there is to assembling wax-bullet loads. If they are shot into a backstop which permits recovery of the wax, then it can be reused almost indefinitely. Just melt it, remove debris by skimming or straining, and pour it ½-inch deep into a suitable container to solidify.

SAFETY PRECAUTIONS

That any of the various lead-bullet loads are dangerous should be obvious. Simply slowing a bullet down to perhaps 300-400 fps doesn't make it a harmless toy. Even the lightest round-ball loads are capable of producing fatal injuries if negligence causes someone to be struck. After all, at the beginning of this chapter, we mentioned using them for taking small-game, and if they'll do that, they'll make a mess out of people. Such low-velocity lead projectiles are also terribly prone to ricocheting. Such a bullet going astray (not into your backstop) in the basement may well impact several times against wall, ceiling, floor, etc., before coming to rest—and nobody can predict the path a ricocheting bullet will take.

The wax and plastic bullets are likely to give the impression of being harmless since no propelling charge is used. Don't you believe it—while they won't ordinarily penetrate the human body, they will raise a hellava welt, and may break the skin; I've never heard of one striking an eye, but it is patently obvious that if this did occur, the eye would surely be destroyed. Both wax and plastic bullets are quite capable of breaking windows, busting up china and damaging furniture. While they won't penetrate a heavy drape hung loosely, They will often penetrate upholstery cloth which is less yielding by virture of being supported by padding and springs. Such loads should *not* be treated with any less care and respect than lead-bullet loads. They can cause a lot of trouble *if* you get careless.

NEW ULTRA-LIGHT LOADS

Since the first draft of this book was prepared, a new type of ultra-light load has drifted on to the scene. It employs reusable bullets and was derived from one long-time shooter's dissatisfaction with plastic-bullet and wax-bullet loads for short-range indoor shooting. Carrying the rather esoteric name "Hoffman's Elastomeric Ringed Rocket," this load employs unaltered .38 Special cases (meaning the flash hole isn't enlarged), a very small charge (0.6-grain) of W-W 630 propellant, and an unusual undersize aluminum-alloy bullet fitted with two "elastomeric" plastic O-rings to seal the bore. That description doesn't do justice to the bullet; it is .785-inch long, with a major diameter of .340-inch and three deep circumferential grooves into which fit O-rings to seal the bore. As can be seen, the bullet's major diameter is slightly less than the bore diameters of 9mm/.38/.357 barrels; when in place, the outside diameter of the O-rings is .358- to .360-inch, sufficient that the rings seal the grooves quite effectively. More important, the material from which the rings are made is quite tough and flexible so that they are not cut or permanently deformed by the rifling. Even though the rings begin to look a little chewed-upon after a few dozen firings, they still seal the bore well, and accuracy doesn't seem to be reduced measurably.

Otherwise, the front of the bullet is of plain wadcutter form, but the base is unusual. It is reduced to .250-inch diameter and contains a small, .105-inch, centered cavity into which the propellant charge is placed in a rather unusual manner. It goes like this: The fired case is decapped by a simple lever and pin on the *wood* loading tool;* the mouth of the case is then cleaned with a bore brush, the bullet is started, and is seated clear back against the case head by being forced down over a short length of dowel set in the tool base. This places the front of the bullet inside the case mouth. The case is left over the bullet-

*The bullets, a loading tool and other accessories are all available in kit form directly from the manufacturer.

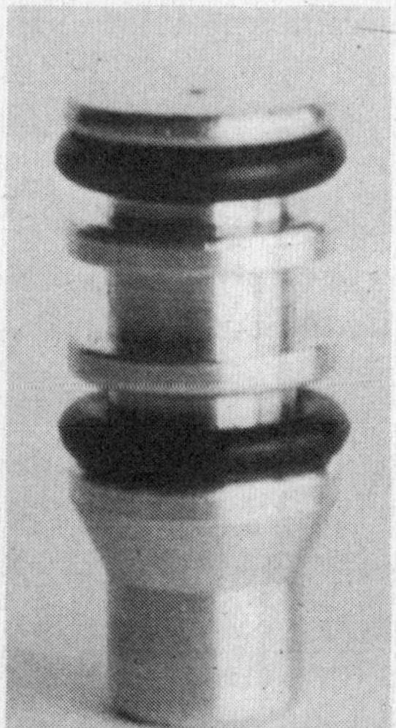

Left — The name: "Hoffman's Elastomeric Ringed Rocket." This is the latest thing in .38 Special ultra-light, indoor, practice projectiles — the inventor claims 1-inch accuracy at 25 feet. (Ed. Note: The *Gun Digest* staff has fired this new practice slug, and, it is *extremely* accurate — perhaps the most accurate practice slug on the market.)

Right — The "Ringed Rocket's" base is hollow and designed to hold a minute (0.6-grain) charge of W-W 630.

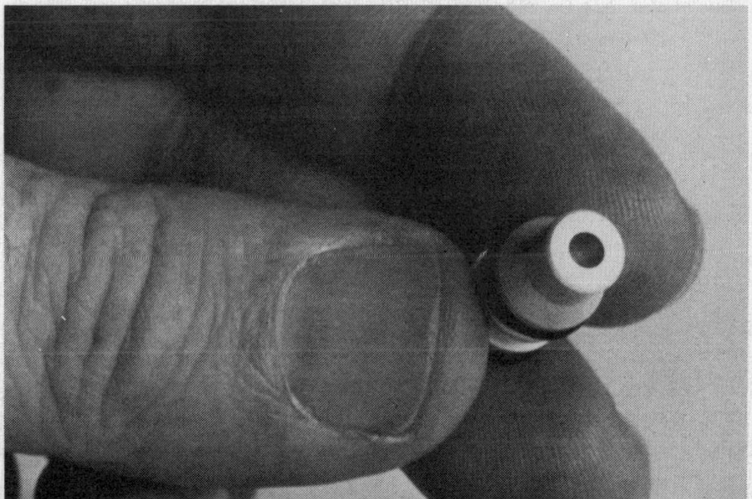

While the "Ringed Rockets" are available separately, the inventor also sells a complete reloading set up. The investment in this sort of item is reasonable, and, perfect for the man who wants to do a fair amount of indoor shooting.

seating dowel, and a small charge cup (also supplied with the kit) is used to dip a 0.6-grain charge of W-W 630 propellant from a reservoir in the tool base. This propellant is then poured into the empty *primer pocket,* where it runs through the flash hole and into the cavity in the base of the bullet—*voilà!* Then, and only then, with the case still impaled upon the bullet-seating dowel, a primer is placed over the pocket and seated home by a short punch and the pivoted handle of the tool. At this stage, the cartridge is completed and ready to shoot.

W-W 630 is rather slow-burning and would not ordinarily be considered suitable for an ultra-light load in 9mm/.38 caliber; however, in this instance, the powder space is actually that tiny cavity in the base of the bullet, so loading density is quite high in spite of miniscule charge weight and volume. In relation to this small volume and high loading density, bullet weight becomes rather large, so the slow-burning 630 *does* perform efficiently. We don't know what velocity this charge produces, simply because as of this writing, we haven't had time to check it through the chronograph. However, accuracy is quite good. The designer, Hoffman, claims that at a range of 25 feet, all shots will "score on a ⅜-inch bullseye." That translates into a group size of just under 1-inch, and that is a good deal better than one can expect *consistently* from wax or plastic bullets of even the best types. Actually , we've been able to obtain quite a few careful groups that measured a good bit less than 1-inch. Further, accuracy remains of about the same order out to 50 feet; that's much better than plastic or wax bullets.

It is stated that these bullets and loading techniques can be used in any caliber which normally shoots bullets of .354- to .360-inch diameter. That is true only if the walls of the case do not taper too sharply to a greater thickness near the base. The .357 Magnum and some .38 Special cases have walls too thick near the head to allow the bullet to be seated deep enough to perform in the manner intended. Nine millimeter and .38 Super cases also share this problem, which can be eliminated by running a neck reamer into the case deep enough to allow the reduced rear portion of the bullet to be seated against the head. Of course, once such cases have been so altered, they'll be too much weakened at the junction of web and walls to be fired with heavy loads, or even full charge loads in some instances. *Such cases should be clearly labeled and reserved for use* only *with Hoffman's bullet and load.*

Like other ultra-light loads, these may be used without any difficulty in revolvers, but require hand chambering and action cycling in autoloaders.

As we pointed out in the beginning of this chapter, ultra-light loads can offer you a wealth of shooting right at home when and where no other facilities are available. Even if you've got a 50-yard range next door, there'll still be times when it would be awfully nice if you could pop off 20 or 30 rounds in the basement. Well, if you get into ultra-light loads, you can do just that.

CHAPTER 17
Blanks: Noise and Smoke Makers

I DON'T SUPPOSE there is hardly a pistolero alive who either hasn't wished for a few blank cartridges or who hasn't been asked by some friend or acquaintance if he can supply some for some athletic or theatrical event. The last request I got was from the director of a marching band through one of my twin, teen-age daughters. The story was that their starting pistol wasn't working right, and they needed something to fire a few signal shots. Spectators and participants accustomed to the feeble "pop" of a .22 starting gun were surprised, to say the least, when my drum majorette daughter hauled out that Colt .38 Special and cut loose with full-case, black powder blanks!

This was another one of those instances where factory-loaded blanks are generally available in the caliber desired, but couldn't be obtained locally, and there wasn't time to special-order them in from a distributor. On top of that, I didn't have a single loaded blank in the shop that would fit a gun suitable for the occasion. Partly because it was the quickest and easiest way, and partly because I pictured in my mind's eye the consternation that would result when about 25 grains of black powder thundered forth in the midst of the band, I quickly made up black powder blanks for the occasion. There were resized and primed .38 Special cases handy on the shelf, so I simply settled them level full of all the FFg they'd hold, then compressed the charge *modestly* with a stiff card wad and crimped it in place in an ordinary seat/crimp die. The result was more than satisfactory.

Factory-loaded blanks are currently available in very few handgun calibers; but, unfortunately, not one gunshop in a hundred will stock them unless there is a fairly high local demand and that situation seldom exists. Some sports-equipment shops stock .22 rimfire blanks for starting guns, but that's about all you can expect to find in the average community. The other calibers available, .32 S&W, .38 S&W, and .38 Special, are usually special-order items, and if you don't make arrangements quite a few weeks in advance, it's unlikely they'll be there when you need them.

If you want to beat this availability problem in center fire blanks, you'll have to handload for the calibers just mentioned, and if you want *any* blanks in other calibers, you'll have no choice except to load them as well.

Above — Commercially available blanks come in (from left to right) .22, .32 S&W and .38 — if they aren't available to you locally you can (in the case of the center fires) roll your own.

Below — Here's the Remington 5-in-1 blank that makes Hollywood roar. On the left is a loaded round; on the right, one that's been fired in a .45 Colt. In most cases the blanks and empty, primed cases have to be special ordered. A handloader will only need the cases to produce suitable blanks.

There are some foreign-made blanks produced for military use. Some of them are quite sophisticated, intended to produce full-auto and semi-auto functioning in appropriate guns when proper adapters are installed. The U.S. military establishment has its standard .45 ACP blank load, but none of these items are generally available to the shooting public.

The military .45 blank load is quite conventional, with special blank powder sealed inside a coned metallic case by a lacquered card wad. The foreign makers show a good deal more imagination, using compound cases and plastic bullets, and in some instances, employing a case and bullet made in a single piece from plastic. The .45 blank cases can be reloaded if you ever run across a supply, but the plastic foreign numbers can't.

There is another blank load produced in this country which is not regularly available on the open market, being made primarily for the entertainment industry as mentioned earlier. This is the Remington "Five-In-One-Blank" long-case type intended for use in .38-40, .44-40, and .45 Colt chambers. It conforms generally to .38-40 external dimensions and has an extended case 1.565-inches long, the forward .410-inch diameter portion serving the guidance purpose of a bullet to insure feeding in repeating rifles. Though not normally available, this empty case can be special-ordered from Remington dealers.

Below — Remington's 5-in-1 blank was designed to be used in a number of firearms. The 5-in-1 on the far left is suitable for use in guns chambered for the .38-40; .44-40 and .45 Colt on the right.

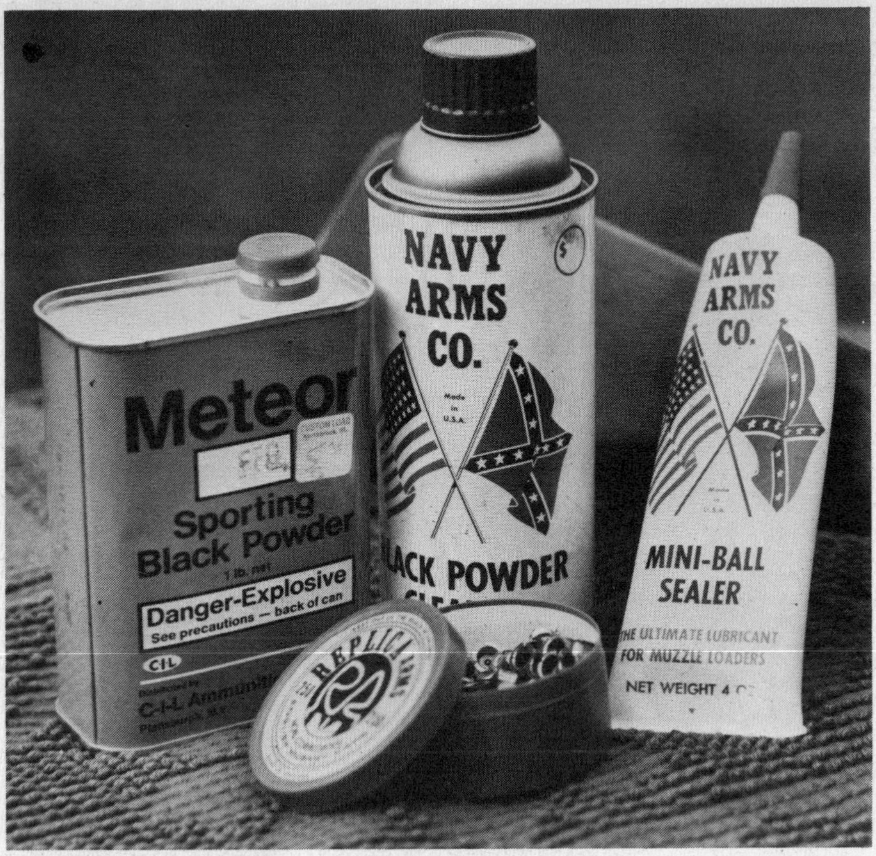

Left — Today, black powder is more available than it was as little as 10 years ago. Growing interest in black powder shooting means that your local gun shop will probably stock a healthy supply — it's just the ticket for blanks.

Above — Once your cases have been primed and loaded with black powder, make up a batch of 1/16-thick cardboard wads. These will be used to seal the mouth of the case.

Loading blanks isn't difficult. You can make them up in any center fire caliber a lot easier than you can load ball ammunition—the only problem is that while they will function perfectly in revolvers, you aren't going to get normal semi-auto functioning in autoloaders unless the gun is extensively altered, often to the point that it will no longer function with standard ammunition. You *can* load blanks that will feed from most autoloader magazines, but only when the slide is cycled manually. There is one minor problem—blanks loaded with powders most likely to do the job safely will foul the gun considerably for the simple reason that it's old-style *black powder*. I call this a minor inconvenience, because it simply means that you'll have to clean the gun *thoroughly* after *each* firing.

"Why no smokeless blanks?" you ask. None of the propellants commercially available are suitable for loading blanks. Bullseye is our fastest-burning pistol propellant, and even it requires substantial confinement by a bullet to burn properly. There is no way you could provide the required confinement in a safe and practical manner. A heavy enough wad will do the job, but if it meets that requirement, it will be as dangerous as a bullet for quite some distance ahead of the muzzle. Most propellant manufacturers produce special blank powders of the smokeless type. They burn so rapidly as to be virtually explosive—as opposed to "flammable solid" as federal government agencies have classified conventional smokeless propellants. The burning rate is far too fast for use with any projectile whatever. For these and other reasons, the manufacturers simply do not sell blank powders in the consumer market. There have been times when I have found this inconvenient, but as a practical matter, especially when I consider the wild projects some handloaders attempt, I'm convinced that they have made the right decision. They sell blank powders only to legitimate loading plants who they *know* will use them only as intended. Looking at this from another point of view, I don't think they would ever sell enough blank powder on the handloading market to pay for distributing it in the first place.

As with any other ammunition, loading blanks begins with selection and preparation of cases. If cases will chamber freely in your gun, resizing isn't required. Cases should be in good condition throughout. They should be primed properly, and belled at the mouth just ever so slightly. Make up some stiff cardboard wads, not more than 1/16-inch thick, which will fit *tightly* in the case mouth. (Chapter 14 covers the making of the wads.) Determine the charge of FFFg black powder—FFg works just as well—that will fill the case to within 1/16-inch of the mouth when well settled. This may produce more noise and smoke than you really want, but the only way you'll find out is to load about three or four rounds and fire them. Okay, with the powder in place; seat a card wad against the powder inside the case and then apply a heavy roll crimp to secure the wad. This can be done easily in a

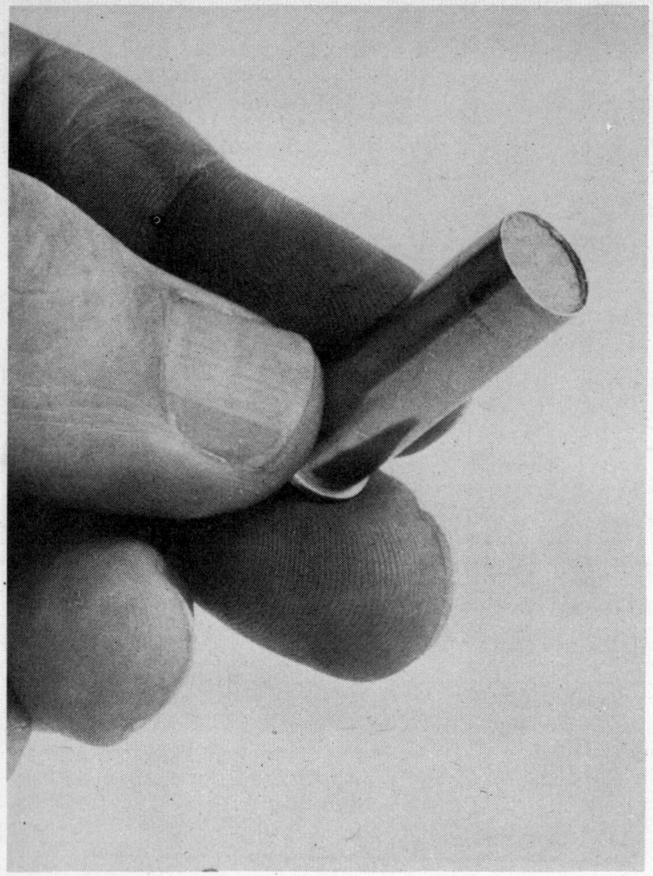

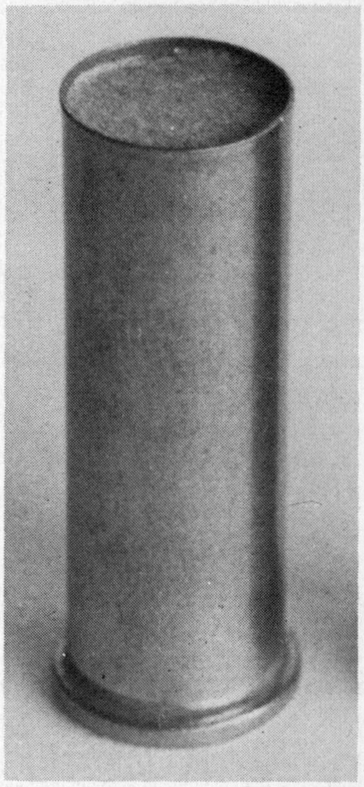

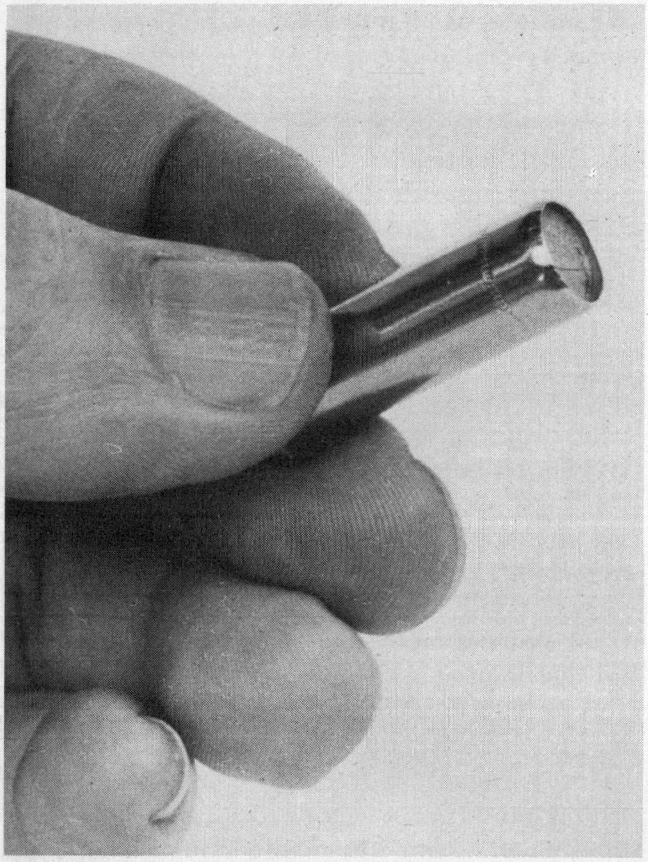

After you've inserted the wad (left, above) make sure the wad itself is just slightly below the mouth of the case (right). Once the wad has been inserted and crimped in place (left, below) you might want to apply a small amount of clear lacquer to seal the load.

roll-crimp die which has been stripped of any internal parts so that a dowel can be inserted from above and held snugly against the wad to keep it in position during crimping. If there is any question about the security of the wad or if you want the load waterproofed, apply clear lacquer fairly heavily to the wad and the edges of the crimp with a small, pointed brush.

Fire two or three of these full-case loads, and if you can live with less report, reduce the charge. When using a reduced charge, use two wads, the first seated fully down on the powder and sealed with lacquer; place the second in the case mouth, apply a *light* crimp, and also seal with lacquer.

Above — Autoloaders, like the MMC-sighted S&W Model 39, function reliably — but not with blanks. The three rounds on the left are standard 9mm H.P. reloads; the ones on the right are blanks. When shooting blanks (in any auto) you'll probably find it easier to single load each round.

Below — When you shoot black powder blanks, be prepared to clean the gun thoroughly. A soap and hot water scrubbing is the place to start — once dry, clean the gun with conventional solvents and oil it with a quality aerosol lubricant.

The purpose of the top wad is simply to close the case mouth so it won't gather up debris which might be fired and become a minor but dangerous projectile. If you have difficulty in keeping the top wad in place, roll a cannelure about 3/32- to 1/8-inch below the case mouth, seat the wad against it, and then apply the crimp. Rolling a cannelure that close to the open end of the case might prove difficult unless the mouth is supported by a piece of dowel or a rejected bullet inserted into the mouth about 1/16-inch.

Once you've developed the black powder load that suits your particular needs, then record the pertinent data, and you'll be able to duplicate it any time you like.

We did mention earlier that you can load blanks which will feed through the action of an autoloading pistol. This requires that you make up cases that include a bullet-like forward extension to guide the load from magazine to chamber. If you'll refer to Chapter 14 on shot loads, we described in detail the procedures for making such an extra-length case for use in the .45 Government Model pistol. Make the case just as described there, then cone the mouth the minimum amount that will allow it to feed with reasonable reliability, then load as just described.

Don't get the idea that because you can make up blanks that will feed in an autoloader that you can simply "plug the barrel," as casually referred to in some literature, and produce autoloading functioning with safety. While this is basically the approach used in *blowback* pistols devoid of any mechanical locking system, more likely than not you'll produce excessive pressures which will damage the gun and, perhaps, your tender body as well. When it comes to locked-breech autoloaders, the problem becomes much more complex, and extensive gun alterations are required, the nature and degree of which depends upon the model and caliber. I learned this the hard way many years ago and wrecked a couple of good autoloaders in the process of discovering that the job is far more complex than it might appear on the surface. It takes a good deal of expertise to rework a locked-breech autoloader so that it will reliably stutter out rapid-fire bursts of the type you see Starsky (of Starsky & Hutch) delivering with his semi-auto on the television.

It is possible for you as a handloader to assemble excellent blanks—just be sure to stick with *black powder*.

CHAPTER 18

Chamber Pressure: How and Why

IT SEEMS THAT almost any conversation between handloaders—neophyte and expert alike—very quickly turns to the subject of chamber pressure. In reading magazine articles, we run into much the same thing, with almost continuous references being made to pressure as if it were something to be avoided. Far too many handloaders appear to stand in awe or fear of chamber pressure, almost as if it were always lurking in some dark corner waiting to spring out and claw the unsuspecting into oblivion.

I've really become quite fed up with chamber pressure being talked (and written) about as if it were something inherently evil and dangerous. Perhaps if we gave it a more truly descriptive name it would seem less fearsome. After all, "chamber pressure" is a rather meaningless term; the chamber certainly doesn't create the pressure directly, and the pressure with which we are dealing is by no means confined to the limits of the chamber. The term "barrel pressure" has been used a few times in the past, but even though it is a bit more descriptive of what takes place, it's not sufficiently so to be worthwhile. Actually, in this scribe's considered opinion, the term "gas pressure" is far more descriptive and more easily understood, especially by the neophyte.

If we use the term gas pressure, we've moved a step in the right direction, but we must still destroy the myth that pressure inside the chamber and barrel of a firearm is inherently evil and dangerous. When some people talk about it, a casual observer might get the impression that they would feel entirely safe only if pressure was eliminated from their guns. In reality, gas pressure is a part and a parcel of any firearms system. In fact, without pressure of quite substantial levels, a firearm simply wouldn't function. Pressure in one form or another is absolutely essential to drive the bullet out of the cartridge case, through the bore, and to accelerate it to sufficient velocity to enable it to do work. To produce this pressure, we ignite and burn a solid fuel inside the cartridge case, and that burning or combustion generates a relatively large quantity of gas at considerable pressure. In other words, we deliberately generate gas at high pressure to enable the firearm to function; then we turn right around and complain about that same pressure.

Let's look at it from another viewpoint. We're all at least passively familiar with the internal-

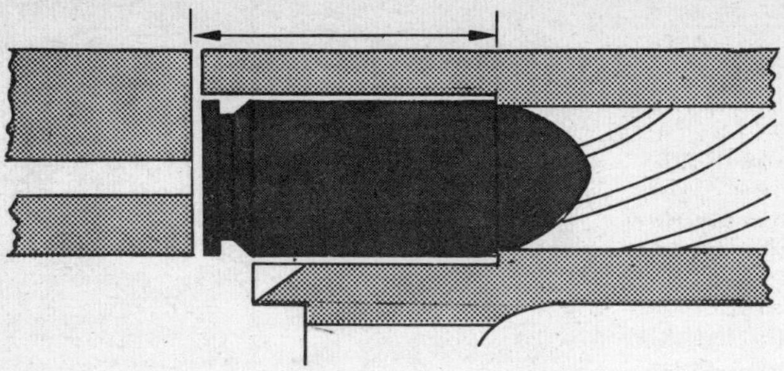

Left—A firearm can be likened to the gasoline engine in that the primer serves as a spark plug to ignite the powder which in turn moves the bullet down the barrel. Actually, a firearm is a "single-cylinder, expendable-piston" heat engine.

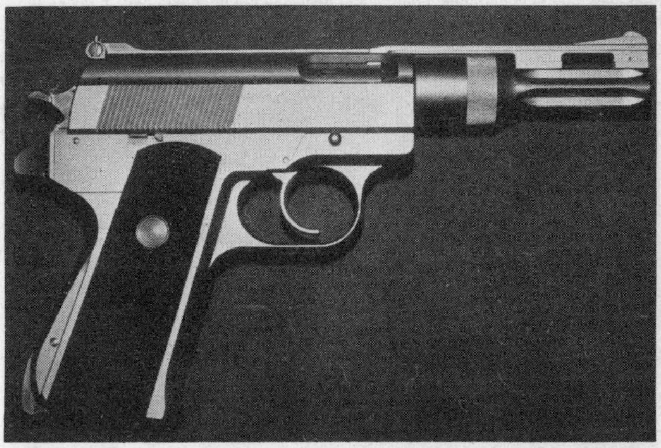

When a new gun and cartridge are being designed concurrently — as in the case of the Wildey seen here (above) — the pressure level necessary to produce the desired cartridge performance is established through testing and/or calculation. The gun is then made to withstand that pressure plus a comfortable safety margin. Below: the Wildey .45 Magnum on the left next to a .45 ACP; the next cartridge is the Wildey 9mm Magnum next to a 9mm Parabellum.

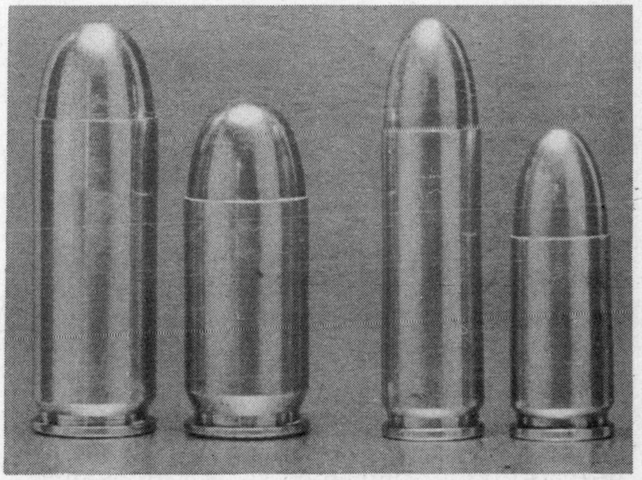

combustion gasoline engine that powers our automobiles, boats, aircraft and the like. Whether the engine is a classic V-16 Cadillac of the early 1930's or a stumbling, one-lung, outboard, it remains a *heat engine*. Using gasoline and oxygen (air) for fuel, it functions through heat and pressure generated by burning the gasoline fuel in an enclosed space. In short, it's a "heat engine," and so is any firearm.

A comparison: In the gasoline engine we have a cylinder closed at one end and containing a piston positioned near that end, and in the firearm we have a barrel closed at one end (by the cartridge case) with a bullet (piston) positioned near the end. In the engine, we have a mixture of gasoline and air enclosed between the end of the cylinder and the piston, and in the firearm we have propellant (which, by virtue of its chemical composition, contains its own oxygen) enclosed between the case and the bullet. In the engine, an electrical spark plug ignites the gasoline/air mixture which begins to burn and generates large volumes of high-temperature gas; in the firearm a primer flame ignites the propellant which burns and generates large volumes of hot gas. In both the engine and the firearm, the gasses formed by burning fuel build up pressure very rapidly, and since both the piston in the engine and the bullet in the firearm are generally free to move, that pressure soon overcomes their inertia and frictional resistance and moves them down the cylinder and barrel respectively. As the gasses continue to grow and expand, the piston and the bullet are accelerated very rapidly.

Up to this point, aside from using different fuels and a different means of ignition, the engine and the firearm function in exactly the same manner. However *at* this point the two slip apart in that the the engine performs its work by converting the linear motion of the piston to rotary motion through a connecting rod and crankshaft, the power eventually being applied to the driving wheels which move the vehicle. The firearm performs its work by that gas pressure forcing the bullet clear of the barrel with sufficient velocity that it travels to a distant point to strike a target and transfer at least a portion of its remaining energy to that target.

In short, then, any firearm is a single-cylinder, expendable-piston, heat engine. The heat and pressure of expanding propellant gasses—and heat and pressure invariably go hand-in-hand—accelerate the bullet to the velocity required to

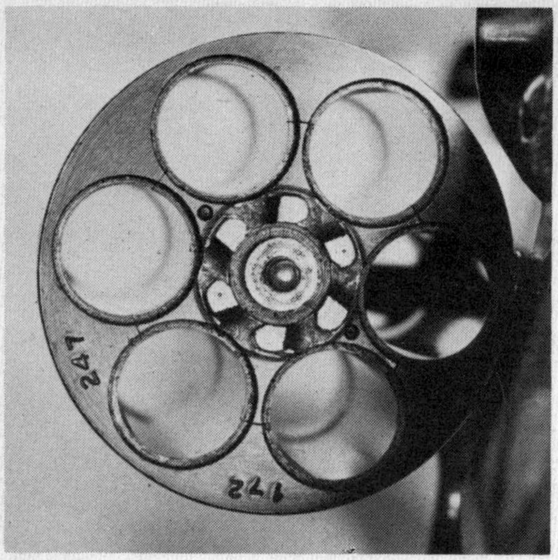

On the left we have a S&W M29 N-frame chambered for the .44 Magnum. To the right, a S&W Highway Patrolman N-frame chambered for the .357 Magnum. Both cylinders share the same diameter; however, note the difference in cylinder wall thickness. While both guns are quite "safe," the thicker cylinder walls on the .357 indicate that it can probably take more pressure and stress.

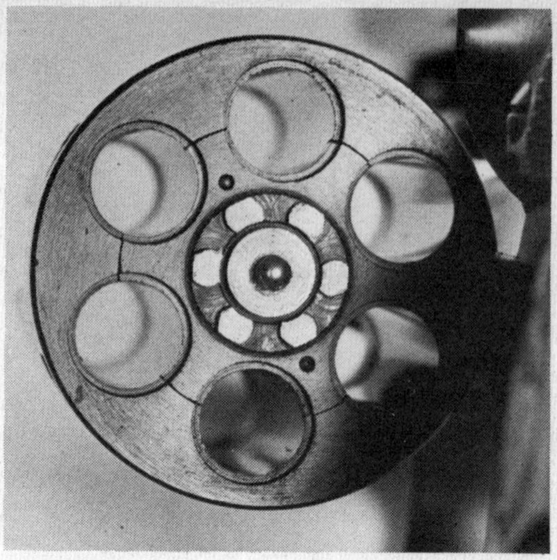

enable it to do its assigned work.

What this all amounts to is that the firearm is a heat engine and that considerable pressure in the chamber and barrel is *absolutely essential* to its function. Now, various firearms are designed to withstand particular levels of pressure. In the case of a gun being designed to handle existing cartridge/load combinations, that pressure level will be established sufficiently above the highest pressure of the ammunition to insure a reasonable safety margin. If a new cartridge and gun are being designed concurrently, as happened with the Auto-Mag and Wildey autoloaders, then the pressure level necessary to produce the desired cartridge performance is established (either empirically or by calculation), and the gun is then designed to withstand that pressure plus a comfortable safety margin. Theoretically, the gun is designed so that when the practical realities of material and manufacturing variations are taken into consideration, along with those of the ammunition, the *weakest* gun that can be expected to come off the production line will withstand the *highest* pressures that might be expected from production ammunition with all its variables.

However, there are other factors that must be considered. For example, there is a wide variation in the diameter of handgun cartridge cases for which a given gun might be chambered. In, for example, the N-frame S&W revolvers, we have a set cylinder diameter; yet that same cylinder might be chambered for cartridges ranging from nominally .375-inch diameter up to .475- to .480-inch diameter. Obviously, the chamber walls in a cylinder chambered for the smaller-diameter case will be much thicker and therefore stronger. A comparable situation is encountered in the Colt GM autoloader where the basic gun is dimensioned around the .45 ACP cartridge. In this instance, chamber walls are nominally .105-inch thick; and when the same gun is chambered for .38 or 9mm caliber, chamber walls are nearly .050-inch thicker and thus stronger. As a consequence of this, we find that maximum allowable pressures of the smaller calibers in this gun are 30-33,000 CUP, while in the big .45 the practical limit is 18-20,000 CUP. So, it can easily be seen that the ability of a particular gun to withstand the pressure in one caliber does not necessarily reflect its ability to do so in another caliber.

Actually, regardless of gun design and materials, the weakest link in the firearm system is the *soft brass* cartridge case. With modern solid-head cases, revolvers of normal headspace present no problems in this respect. The solid brass head of the case enters the rear of the revolver chamber to a sufficient depth that there is no area of the case not supported solidly at the time of discharge. Autoloaders are another matter entirely in that the clearances between the parts necessary for smooth, reliable functioning and the path provided for chambering cartridges from the magazine leave certain areas of the case rather poorly supported. Thus, in most autoloaders, the cartridge case becomes the weakest link in the firearms system. Therefore it and not the gun limits the amounts of pressure that can be withstood. Elsewhere in this volume you'll see references to "feed ramp bulge," a deformation that occurs upon firing when pressures are high enough to distort the brass in the feed-ramp area where the case is not solidly supported.

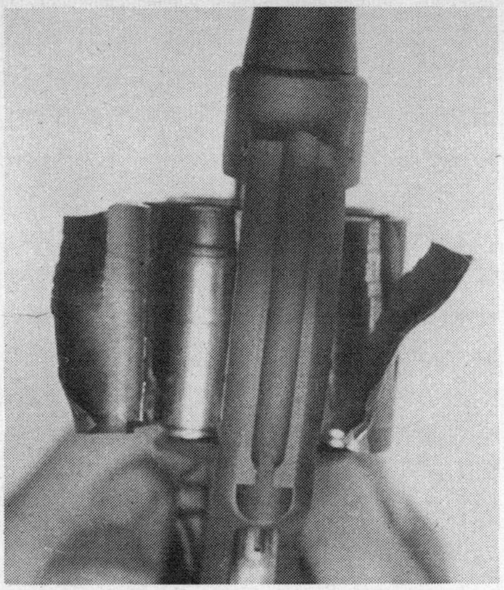

The author's old S&W Model 20 shown, which was originally chambered for .38 Spl., was converted to handle .45 Colt. The first round from a batch of heavy, but not unsafe loads, ruined the gun. When the cylinder was recut to handle the .45 Colt the remaining metal in the bolt-stop areas of the cylinder was unintentionally made paper thin — the result speaks for itself.

Once the brass of the cartridge case begins to deform under gas pressure—and this may range from as little as 15-20,000 CUP up to around 70,000 CUP, depending upon both case and gun design—the deformity is accelerated rapidly and soon ruptures at even the smallest unsupported area. Once it ruptures, the high-pressure gas escapes at a very high velocity (several thousand feet per second) and both melts and erodes brass away, making the opening even larger. This jet of gas can even melt and erode steel in its immediate path and carries with it incandescent particles of metal which can cause painful injuries to the shooter aside from those produced by the velocity and temperature of the escaping gas. Fortunately, conventional handguns are normally held and fired in such a manner that escaping gasses can't come in contact with any major portion of the body; the most one usually suffers is a burned finger or two. However, the damage that gas escaping from a ruptured cartridge case can do to a gun is considerably greater. Almost invariably the magazine catch will be overridden and the magazine blown from the gun; the grips (whether of wood or plastic) will be split or even shattered and can sometimes lacerate the shooting hand; the thin side walls of the slide may be bulged; and the extractor may be bent or broken. Gas will find its way into almost every nook and cranny of the gun, leaving behind propellant fouling and often minute particles or thin layers of brass.

Aside from split or shattered grips, the outward appearance of an autoloader in which a case has failed at high pressure is generally unaltered; but revolvers often present a different picture. Before me as I write is an S&W M20, originally manufactured in .38 Special caliber, but eventually converted to .45 Colt by the installation of a M1917 barrel and rechambering the original cylinder. I detail the conversion so that you'll be fully aware that Smith & Wesson was not in any way responsible for the failure of the gun. In any event, immediately after conversion, this gun was loaded with a relatively heavy (but not normally dangerous) lead-bullet handload. As the first shot was fired, the thin outer cylinder wall of the firing chamber split longitudinally, directly under the top strap, the split apparently originating in the bottom of the bolt cut where the metal was almost paper-thin. As the chamber split, the walls opened up to right and left, and the escaping gas forced the top strap up nearly ½-inch, bending it into a graceful bow and stretching it without breaking it. Naturally, the cartridge case was destroyed in the process. As a result, the ruined cylinder and the apparently undamaged crane are jammed tightly in the frame, and the gun cannot be opened except by hammer blows strong enough to damage it further. Consequently, this gun continues to do service on my desk as a paperweight with five loaded cartridges remaining in the chambers. That's by no means dangerous inasmuch as it is mechanically impossible to rotate any of them in line with the hammer. In this instance the solid web of the failed cartridge case extended approximately ⅛-inch into the chamber, and there was no deformation of the case head. The case was well supported, and it was the converted cylinder wall that failed first, *not* the cartridge case. This is generally true when a revolver fails because of excessive pressure. On the other hand, when an autoloader fails, that failure is *usually* initiated by a rupture of the case, with damage to the gun resulting from the expanding gasses. Of course, in the instance of grossly excessive pressures (such as might be encountered in a double or triple charge of fast-burning propellant) both case and gun might fail simultaneously and also more catastrophically than my converted S&W M20. In the instance cited, the cylinder was apparently defective after conversion, and the pressures of the particular handload involved were sufficient to wreck the gun even though that same load had been used for nearly 20 years in other guns without any problems. In this particular failure, no fragments of cylinder or frame—or case for that matter—were thrown off. But, often in failures of this type, fragments of the chamber walls and top strap are hurled great distances at high velocities. This will not normally harm the shooter since the fragments are thrown upward and outward at roughtly 90 degrees to the barrel, but can certainly be quite dangerous to bystanders. In this failure, the only immediate evidence of what had happened was an unusually loud report and a flash of flame above the cylinder where none should be. This caused me, with ringing ears, to look at the gun whereupon the damage became quite evident. The case was missing from the ruptured chamber, and was found lying on the ground directly ahead of my feet. Looking at the gun made it appear impossible for the case to have escaped in one piece, nevertheless it had, the head intact even though the body had split and unfolded into a single, almost-flat, sheet of brass. Incidentally, the bullet did exit the bore, but it isn't surprising that it missed the target entirely.

Pressure inside the chamber and barrels of a firearm is depicted graphically by what is called the "time-pressure trace." This trace is a curved line shown upon a graph, with the horizontal axis representing time, beginning at the left with ignition of the primer and ending at the right when the bullet exits the muzzle. In essence, the horizontal length of the trace represents the total length of elapsed time from primer ignition until the bullet leaves the barrel—this is generally known as "barrel time," depicted in milliseconds or increments of one thousandth of one second. The vertical axis of the time-pressure trace represents pressure in increments of (usually) one thousand CUP. In any firearm but a revolver, this trace rises very sharply from zero to maximum pressure in a very short period of time, then drops off not quite so rapidly in the beginning, with the rate of decrease slowing as the bullet moves down the bore, and then becomes nearly horizontal by the time the bullet exits the muzzle. This characteristic form of trace is produced by rifles, shotguns, and autoloading and single-shot pistols. Though the trace is generally represented as a smooth flowing curved line when recorded by the most sophisticated instrumentation, it isn't as smooth as it looks. Both the rising and falling portions are sometimes marred by high-frequency fluctuations and even spikes.

By virtue of its unique design, the revolver doesn't allow the time-pressure curve to follow exactly the same pattern. In other firearms the bullet moves from the case directly into the rifling and passes through an uninterrupted barrel. In a revolver, however, the bullet moves from the case through an oversized (usually), smooth-bored throat in the cylinder, then must jump the barrel/cylinder gap and enter the funneled "forcing cone" of the barrel breech, and then finally enter the rifling. As soon as the bullet base exits the cylinder, gas at high pressure and velocity escapes through the gap and continues to escape there until the bullet exits the muzzle. Where in other firearms the resistance the bullet offers to the expanding gasses increases until it is fully into the rifling (and then decreases steadily as it accelerates down the bore), the bullet in a revolver accelerates rapidly in the chamber throat, then is restrained sharply as it enters the forcing cone, thus greatly increasing its resistance to the gasses. It then enters the rifling and begins accelerating again, while gas is being bled off behind it through the barrel/cylinder gap. All this is aggravated by the fact that more often than not the chamber throat is larger in diameter than the groove diameter of the barrel. This causes the bullet to upset to full throat diameter during its initial acceleration, making its external dimensions slightly larger than the internal dimensions of the barrel. When this oversized bullet enters the forcing cone (which is seldom *precisely* aligned with the chamber) the bullet must be swaged back down to barrel groove diameter as it is forced into the rifling. Then, once that is accomplished, the bullet can *begin* accelerating down the bore (beginning about the time the barrel/cylinder gap is uncovered and gas begins

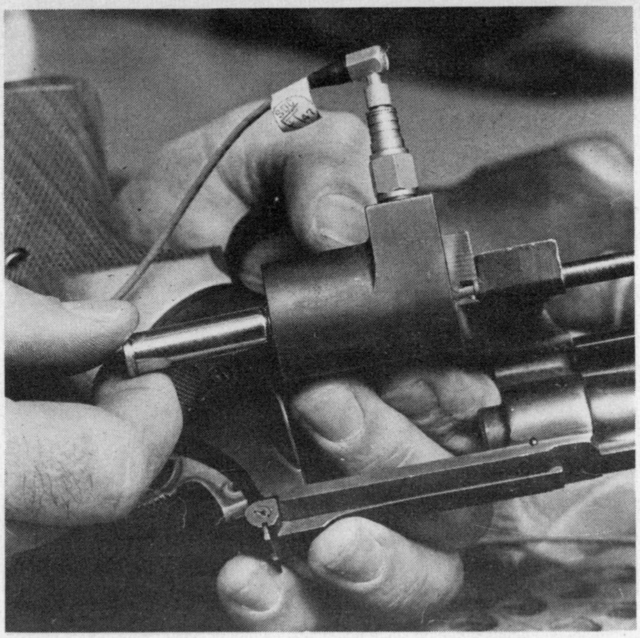

A gun's internal pressure at the time of firing — i.e., from the moment of primer detonation until the projectile exits the barrel or cylinder — can be measured with sensitive, modern equipment. This measurement will result in a graphic "time-pressure trace." In the photo to the left a special transducer-mounted cylinder for a S&W revolver is being loaded for test firing. To the right: the testing process itself. The trace will show up on the oscilloscope to the right of the shooter. (Photos courtesy of Speer Bullets.)

to escape there) and then moving unimpeded until it exits the bore.

From this we can see that there are interruptions to the bullet's movement and interruptions to gas flow as well. Ordinarily, we expect the ascending side of the time-pressure trace to be essentially the same in a revolver as in a autoloader, assuming similar cartridge and load. However, almost immediately after the pressure peak is reached in a revolver, the upset and rapidly accelerating bullet runs into the forcing cone almost as if it hit a brick wall and is decelerated rapidly as it is swaged down to barrel diameter. We have no accurate measurements of the amount of deceleration produced nor of the precise effect this has on gas pressure. Nevertheless, it is obvious that this sudden deceleration and increased resistance to the expanding gasses is going to reduce the rate at which pressure is falling off and, in some instances, actually cause a secondary jump in pressure. At normal revolver cartridge pressures of 10,000-20,000 CUP, the effect is probably not very great; however, in the magnum loadings approaching 40,000 CUP, the effect should be much more pronounced. Few ballisticians have ever explored this peculiar area of ammunition behavior in revolvers, probably because there doesn't seem to be any particular concern with it so long as the basic revolver design functions safely. Nevertheless, handloaders working with the heaviest loads in magnum calibers should be aware of it. In loads nudging the pressure-handling limits of a particular gun, the effect might well be significant. Speer products, the well-known bullet manufacturers, have constructed special pressure-measuring equipment and have conducted some testing in this area. To date, apparently, the testing is not entirely conclusive, but hopefully, in the future the Speer ballisticians will be able to enlighten us on the effect these characteristics of revolver design have upon pressures in general and the time-pressure trace in particular.

Until recent years, published "chamber-pressure data" was stated simply in pounds-per-square-inch (PSI) and accepted as a more-or-less absolute value by laymen even though ballisticians were well aware that the value given did not in fact represent true pressure measured in pounds-per-square-inch. The values were stated in "copper units of pressure" (CUP) which did not and do not correspond precisely with pounds-per-square-inch. Then, a few years back, the firearms industry dispensed with the erroneous PSI label and began listing pressure value in CUP. Virtually all recent loading data lists pressures in CUP, with the exception of those which occasionally are corrected mathematically to PSI. This is all of only academic interest to even the most ardent handloader, since the sophisticated and expensive instruments for measuring pressure are found only in ballistics laboratories.

Laboratory measurements of pressure are taken in special test barrels. In the past, these

Here's a special transducer-mounted test gun which features a universal receiver. Unlike the transducer-mounted revolver shown on the previous page, this gun has no cylinder gap to contend with, hence the time-pressure trace will show a single rise and fall on the oscilliscope. A revolver trace would tend to "spike" unevenly due to the cylinder gap the bullet would have to bridge in flight.

barrels for both autoloader and revolver cartridges have been of one-piece construction, therefore the revolver-caliber pressure barrels did not duplicate the barrel/cylinder gap of functional revolvers. That such a barrel did not produce the same time-pressure trace as a revolver was recognized, but was not considered particularly important. On the other hand, a one-piece barrel did and does closely duplicate the time-pressure trace of an autoloader barrel. There is another difference in autoloader and revolver test barrels: The stated length of an autoloader barrel (5 inches, in the case of the Colt .45 GM) is the true length of the barrel and consequently the pressure-test barrel is made the same length. In a revolver, though, the stated length (6 inches, for example) is the length of only the barrel proper and does not take into consideration the fact that the *effective* barrel length is that of the cylinder and barrel combined. Consequently, a test barrel designed to duplicate the performance of a 6-inch revolver barrel should be as much as 7.640 inches long. Realizing this, SAAMI, (Sporting Arms and Ammunition Manufacturers Institute), long ago established a standard test-barrel length for each revolver caliber in order to eliminate minor variations that might otherwise exist between manufacturers.

Incidentally, the different methods of measuring barrel lengths in revolvers and autoloaders make it impossible to compare cartridge performance between the two based upon published ballastic tables. Revolver lengths quoted in those tables are for the barrel *only* and thus not comparable to the true length of autoloader barrels indicated in the same tables. The tables may well show a revolver load in a stated 4-inch barrel exceeding the performance in a 4-inch autoloader barrel, but they do so only because of the extra 1½-inch or so of *effective* barrel length provided by the cylinder. If both guns possessed the *same effective* barrel length, autoloader performance would probably be superior because of the gas lost at the barrel/cylinder gap of the revolver.

After all that, perhaps we should take a fairly close look at the various factors that affect chamber pressure or gas pressure. As a practical matter, the factors are the same in both revolvers and autoloaders; however, the magnitude of the effect of some factors will differ between the two gun types as there are gun factors present in the revolver which do not exist in the auto.

Pressure-affecting factors should be separated into two categories: Those contained within the gun and those contained within the ammunition.

In both guns, we have basically firing-pin blow, chamber size, and bore size; and in revolvers, specifically, we have a chamber-throat size and forcing cone size and shape. In addition to those factors, anything which tends to impede the bullet's smooth movement out of the case mouth and through the chamber throat—such as a coating of rust or lead—will tend to increase pressures. The effect of the firing-pin blow is slight, but generally an unusually light blow or short protrusion will ignite the primer inadequately, resulting in inadequate or insufficient ignition of the propellant, and this will tend to reduce pressure. By the same token, an excessive firing-pin blow or protrusion may increase pressures. Chamber size, and therefore chamber volume, controls case volume (by limiting case expansion) which in turn affects loading density. As soon as propellant begins to burn (to generate gas and pressure), the case expands to the limit allowed by the chamber walls—the greater the amount of this expansion, the lower the chamber pressures will be. Obviously, if the bore is undersized, the bullet will offer more resistance to expanding gases, and pressures will be raised. On the other hand, an oversized bore will reduce that resistance and thus reduce pressures.

All ammo manufacturers use special pressure test guns to determine the pressure various cartridges and loads generate. On the left, one of these guns has been loaded and the operator is inserting a "copper crusher" pellet. To the right: the fired cartridge case. The hole has been caused by pressure forcing its way upward, through the case, to crush the copper pellet. The amount of deformation to the pellet will determine the amount of registered pressure.

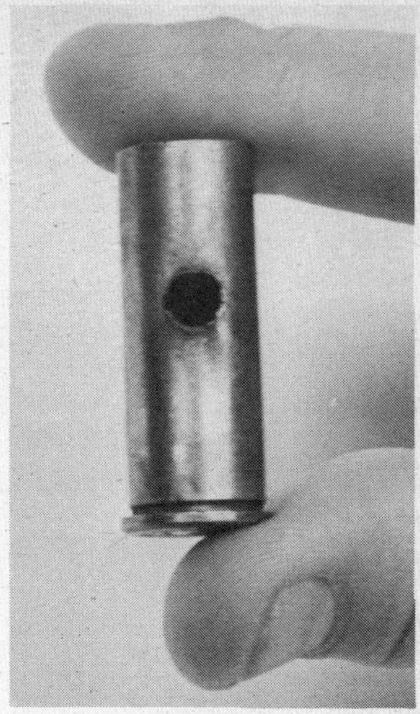

Rust, lead, or other foreign material coating the bore will certainly have the effect of making it undersized and thereby increasing pressure. In revolvers, an oversized chamber throat causes the bullet to be upset to a greater diameter, thereby increasing its oversized relationship to the bore and the forcing cone, creating greater resistance as the bullet passes from chamber to barrel. Naturally, this increases that component of pressure related to that portion of the bullet's travel. Misalignment of the chamber with the barrel will also increase the pressure by increasing the resistance to the bullet's entry into the barrel. The configuration and finish of a forcing cone can also have a substantial effect on the pressure, *any* variation that increases resistance to the bullet's passage increases pressure.

The cartridge contains more pressure factors than the gun. And, while there isn't much the handloader can do about the gun factors, he can vary almost all of the cartridge factors by interchanging components and varying loading techniques.

Primer selection must be made to match the propellant ignition characteristics. If a standard primer is employed with a hard-to-ignite propellant such as H110, then pressure and velocity will tend to be reduced by poor propellant ignition and combustion. The manner in which the primer is seated can also have an effect on pressure in that if a primer is seated with too much force, the compound pellet may be fractured or crumbled and consequently ignite the propellant inefficiently. If the primer is not seated sufficiently deep, the effect may be the same by virtue of a portion of the firing-pin energy being absorbed in moving the primer forward so that the remainder of the energy is not great enough to ignite the primer, and subsequently the propellant, correctly. The cartridge case is also an important factor. A case with thick walls and head will have reduced volume, and this will have the effect of increasing loading density and increasing pressure; the reverse is also true. In heavier loads in particular, bullet pull becomes important to efficient ignition and combustion of propellant; and, if the physical properties of the case mouth are not up to maintaining a tight grip and stiff crimp on the bullet, pressure will be reduced. Variations in case length can affect pressure indirectly: An abnormally short case will not be adequately crimped on the bullet and will also have the effect of reducing ignition efficiency in an autoloader; while a longer case may cause the crimp to be greater than desired and thus increase pressure.

The bullet affects pressure in many ways. If undersized for the bore, pressure will go down; if oversized, pressure will go up. If the bullet is unusually hard, pressure will increase, and if it is unusually soft, pressure will also increase. The former is due to the fact that the very hard bullet offers more resistance to the expanding gases as it is forced into the rifling; the latter is true because a very soft bullet will upset more under

Those tables and manuals often base their performance figures on barrel lengths *alone* not taking into account (in the case of revolvers) the additional 1½-inches of cylinder length. Assuming these two guns were chambered for the same cartridge and had identical barrel lengths, the revolver would decidedly out-perform the auto due to its extra cylinder length.

initial acceleration and thus increase the area of its bearing surface which in turn increases its resistance to entering the rifling. Bullet weight has a substantial effect—the heavier the bullet, the higher the pressure. The length of the bullet shank or actual bearing surface causes pressure to vary directly with the length of that surface. But keep in mind that the effective bearing surface may be substantially greater than it appears on the unfired bullet because of upset during initial acceleration or, in the case of revolvers, passage from chamber throat to barrel. Bullet construction also has its effect: A jacketed bullet causes pressure to be higher than a lead bullet of identical configuration and weight because a jacketed bullet resists entering the rifling more.

Propellant selection is probably the single most important factor affecting pressure. Assuming that a propellant of optimum burning rate is available for a particular case/bullet/gun combination, then a faster-burning propellant will produce higher pressure and a slower-burning variety will produce a lower pressure. And, within reasonable limitations, reducing the amount of propellant will reduce pressure, with the reverse being also true. Actually these factors are covered in much greater detail in Chapter 7 on propellant selection.

The mechanical aspects of loading, resizing, expanding, and crimping probably have a greater affect on pressure than anything else. As already mentioned, if the bullet is not held tightly by the case and if the case is not crimped securely on the bullet (at least in the heavier loads), pressures will be reduced. On the other side of the fence, an excessively tight bullet/case assembly can increase pressure. Bullet-seating depth is also a significant factor. Generally speaking, seating the bullet deeper increases pressure and seating it farther out brings about a corresponding decrease. However, this is by no means a hard and fast rule. At some point with each load in a given gun of a given caliber, the pressure change brought about by varying seating depths may be cancelled out or even reversed by the change in the relationship between the position of the bullet and the origin of the rifling. For instance: Seating the bullet deeper increases the amount of free travel before it engages the rifling, and this "free bore" (often called "leade") *reduces* pressure when all other factors remain equal. Therefore, we have two opposing factors operating at the same time—the deeper-seated bullet tends to increase pressure while the increase in free travel tends to decrease it. Over a decade ago I had pressure tests run on a .357 Magnum load assembled with the 146-grain Speer JHP bullet, Remington cases, 15 grains of 2400 powder and SAC primers. Beginning with an overall cartridge length of 1.66 inches, (all the cylinder allowed) pressure averaged just over 21,000 CUP. As seating depth was increased, pressures rose to 23,400 CUP at an overall cartridge length of 1.58 inches (the normal or recommended length for that bullet). At that point, pressure leveled off sharply and did not increase further even though bullets were seated as deep as compression of the propellant charge allowed. Simply put, with this particular load, increasing seating depth *did not* increase pressure measurably while decreasing seating depth to the maximum amount cylinder length would allow brought about a decrease of approximately 2,400 CUP in chamber pressure. For another bullet and another load, particularly light bullet, high-velocity loads, the situation might be entirely different. It is interesting to note that the velocity changes brought about by

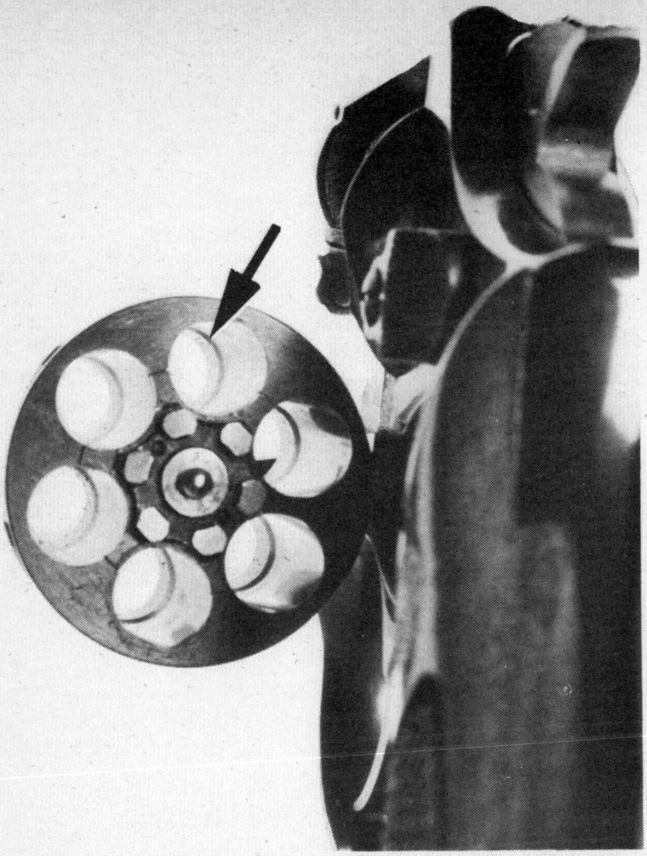

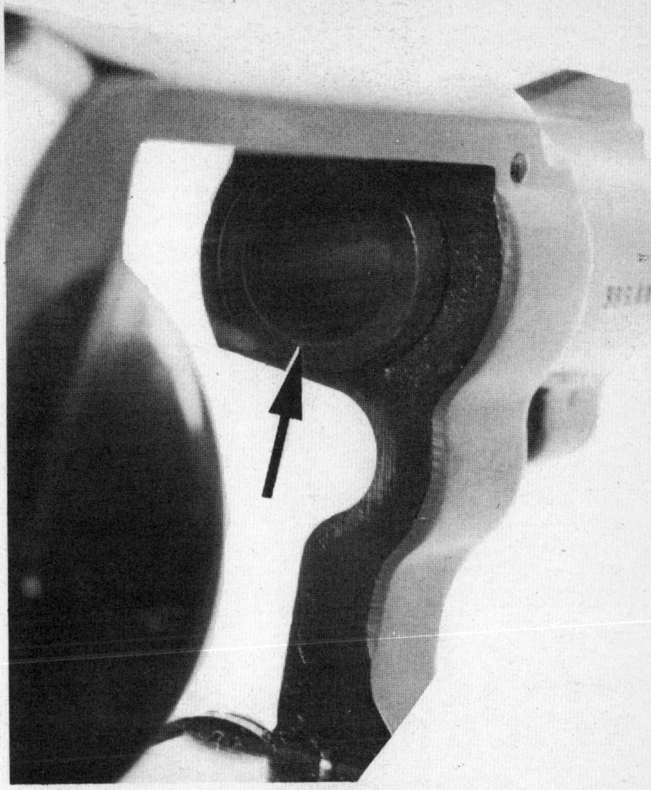

Anything that impedes the bullet's travel such as the chamber throat (left, arrow) or the forcing cone on the barrel (right, arrow) will increase pressure. The pressure that results from the above can be handled safely, assuming a safe load. If, however, you add a little rust, some heavy leading or just plain dirt, pressures can go up — sometimes dangerously.

variations in seating depth *did not* parallel those of pressure. The initial increase in seating depth produced a drop in velocity but further increases in seating depth brought about a sharp rise in velocity to 1465 fps at the normal 1.58-inch overall cartridge length. Remember at this point further seating depth increase produced no measurable pressure increase, yet velocity continued to climb, peaking out at 1505 fps at the greatest seating depth. This effect, too, might very well turn out to be entirely different with another case/propellant/bullet/gun combination. The most important thing to remember about the effect of bullet seating depth is that from the limited amount of testing that has been done in this area, it is impossible to predict whether pressure will go up, will go down, or remain essentially the same; but normally, it *will* change.

There are other minor factors which can produce small effects upon pressure. However, they really aren't worth delving into at this time, even if we had sufficient test data on them, which we don't. Likewise, while sufficient testing has been done to establish the effects and their directions on pressure of the many major variables we've mentioned here, it's impossible to predict the magnitude of the change that will be brought about. We cannot say, for example, that loading a jacketed bullet .003-inch oversize in the .45 ACP cartridge will produce a *specific* increase in pressure—but we can most assuredly predict that it will produce a *measurable* increase. Handguns and their cartridges have working pressure ranges from less than 10,000 CUP up to as much as 40,000 CUP, with some of the unconventional designs (such as the Auto Mag) working at levels considerably higher.

Pressure, per se, is an essential component of the firearms system, and it becomes hazardous only when we employ it to a great magnitude for a particular gun and cartridge. If we keep that in mind and do not attempt to "improve" upon the thousands of loading data combinations published in reputable manuals and references, then, pressure won't get us into trouble. You can't get rid of it, so just learn to understand it.

CHAPTER 19

Wildcats and Case Forming

RIFLEMEN HAVE HAD a field day with so-called "wildcat" cartridges almost from the beginning of the smokeless-propellant era. There have invariably been long-gun enthusiasts who have felt that the factories either couldn't or wouldn't give them the cartridges they wanted, or felt that they were better qualified than the factories to develop cartridges. The result has been that they have necked cartridge cases up, necked them down, shortened them, lengthened them, and otherwise performed virtually every alteration that could be imagined on a cartridge case and still have it suitable for use in a conventional gun. Pistoleros haven't gotten quite that carried away, in fact, handgun wildcats have been relatively scarce.

WILDCATS

As with rifles, wildcatting efforts for handguns have been aimed primarily at producing either greater velocity or greater power—or both. The earliest example I've noticed in shooting literature (and it's likely there may have been others before it) is the .22 Adolph Long Range Pistol of 1913-14, a long, slender, bottle-necked number adapted only to single-shot pistols. Naturally, it didn't go very far, but considering the limitations of the propellants available at the time, and estimating its performance capability from case size, it should have been able to develop performance nearly equal to some post-WWII efforts. In the 1920's, Pop Eimer of Joplin, Missouri, developed an especially potent revolver wildcat which, with today's propellants, very nearly duplicates the .41 Magnum. It was designed for .38-40 and .41 Colt barrels and thus used bullets of nominally .401- to .403-inch diameter in weights up to 200 grains. It employed the .401 Winchester Self-Loading Rifle cartridge case because of its suitability for that diameter of bullet. Eimer shortened this semi-rimmed case to a length of 1.25 inches, this being the maximum length that could easily be accommodated by the Single-Action Army Colt cylinder (the most common gun of the times carrying the proper barrel) with a bullet seated normally. As a practical matter, Eimer produced the first of the magnum revolver cartridges in this, his ".401 Eimer Special," a full decade before the .357 Magnum was introduced, and nearly 40 years before the factory .41 Magnum came on the scene. I recall clearly that when I received my first sample .41

Magnum S&W revolver in 1964 that I was forced to resurrect a batch of .401 Eimer cases and handload them because factory ammunition wasn't yet available. The first several hundred rounds I fired from that gun, out in the Mojave Desert, were loaded in those cases with the mouths expanded to accept the slightly larger .410-inch diameter bullet. Eimer developed other revolver wildcats, including his ".43 Special," based on the .30-06 rifle case (used with half-moon clips in side-swing guns), driving a .44 bullet through a .44 Special or .44-40 barrel. This cartridge's purpose was a bit different, being devised primarily to allow minimum-cost conversion of .45 M1917 revolvers to a more powerful .44 cartridge by simply installing a new barrel. Another was his ".25 Eimer," consisting of a shortened .25 caliber rifle case, loaded with an 86-grain lead bullet at quite high velocity. This was about 1928 and employed the .25-25 Stevens case shortened to about 1.265 inches long.

Very few if any other handgun wildcats achieved any particular notice for the next few decades, though there is reference to a ".22 Lowenstein Luger," which was the 7.65mm Parabellum (.30 Luger) case necked down to .22 caliber.

The post-WWII years saw more activity in this field, and the first wildcat to achieve wide notice was the ".22 K-Chuck,"—a .22 Hornet case shortened slightly and blown out to minimum body taper with a sharp shoulder. It was developed by the late Jim Harvey for use in converted S&W K-22 revolvers. It produced the highest velocity of any revolver load developed until the coming of the factory-loaded .22 Remington Jet in 1961. In fact, I suspect very strongly that the modest success of the .22 K-Chuck was responsible for the Smith & Wesson/Remington joint venture that produced the Jet. The .22 Jet, incidentally, was not a practical success (though the guns sold rather well) because its sharply-tapered, case body caused functional problems in revolvers. On the other hand, because of its near-cylindrical body, the wildcat K-Chuck did not produce this trouble.

The .22 Jet generated its own "improved" wildcat with the case body blown out to near cylindrical form and thus reduced the functional problems of the original cartridge. It did not achieve any particular following for revolver use but was employed in a few custom-built, single-

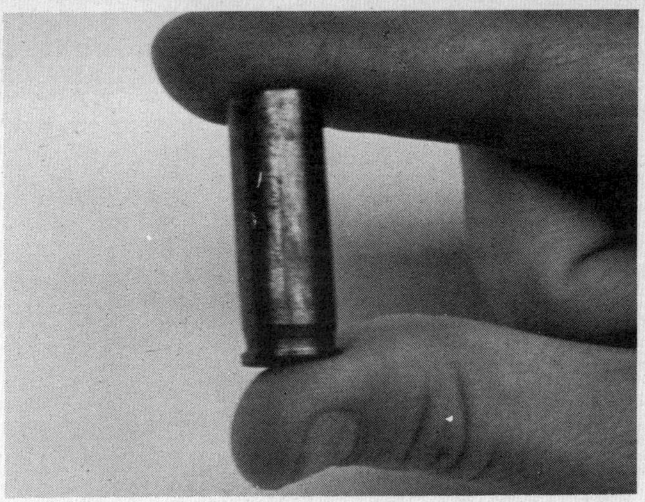

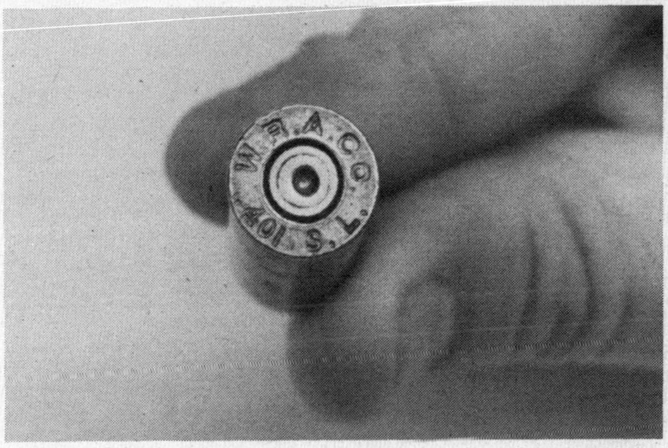

Here (above) we have the .401 Eimer case that's made from .401 Winchester brass (below). This is one of the earlier attempts at wildcatting handgun cartridges which was done in the 1920's by "Pop" Eimer of Joplin, Missouri.

One of the more popular wildcat pistol chamberings was the .22 K-Chuck as developed by the late Jim Harvey. On the left is a regular .22 Hornet, to the right is the K-Hornet. Basically, the .22 K-Chuck is a shortened version of the K-Hornet cartridge.

In 1961 Smith & Wesson introduced their Model 53 revolver chambered for the .22 Remington Jet (left). This was the first commercial offering of a high-performance, small caliber handgun/cartridge combination.

shot rifles and pistols.

Cases sharply *bottle-necked* or *tapered* have always caused functional problems in revolvers which depend upon clearance between the case heads and recoil shield in order that the cylinder be able to revolve freely. (However when a typical, *nearly-cylindrical* case expands on firing in a revolver, it is also thrust hard back against the recoil shield. Then, as chamber pressure drops, the case contracts and reduces sufficiently in diameter that it is free in the chamber, thus its head does not bind on the recoil shield to interfere with cylinder rotation.) When a case is tapered or bottle-necked to any significant degree, it functions as a wedge in the chamber as it expands. It contracts after firing, but since contraction is not all the way back down to original dimensions, it remains wedged rearward in the chamber, with the head bearing against the recoil shield. Because of its expanded, wedge shape, it is prevented from moving forward freely in the chamber, and friction or binding between the case head and recoil shield makes cylinder rotation difficult; in some instances it will prevent rotation entirely.

For this reason the .22 Jet has not been a functional success, even though it has been popular among handgun hunters who can live with single shots and difficult operation.

The .256 Winchester Magnum was a contemporary of the .22 Jet, but because of the problems just outlined, Colt and Ruger—both of whom intended to produce revolvers chambered for it—cancelled and never made production guns. Thus the .256 Magnum survived only in the single-shot Ruger Hawkeye and the much later

Ruger's Hawkeye represents that firm's only attempt in the single shot handgun field. Chambered for the .256 Winchester Magnum, the Hawkeye has a fair amount of collector appeal; and, as a result, shooters have a hard time finding one. To the right, we have the necked-down .256 Magnum compared to the straight-cased .357 Magnum — the cartridge from which the .256 originated.

If you own an Auto-Mag, you're going to have to handload unless you're lucky enough to have a *good* supply of the now scarce commercial brass. While there have been few attempts to wildcat autoloaders, the Auto-Mag stands out as a very powerful exception. The top gun is a first production prototype while the lower gun is the first Auto-Mag ever put together by inventor Harry Sanford. Less than 10,000 were produced.

T/C Contender, and a Marlin lever-action rifle.

The coming of the large-capacity .44 Magnum case brought about numerous attempts to produce high-velocity wildcats by necking the .44 down to various calibers. Though everything from .22 caliber upward was tried on this case, the only one to achieve any measure of success was the .357/.44 introduced (and still offered) by the Bain and Davis Gun Shop. This wildcat avoided the problems mentioned in reference to the .22 Jet by using a near-cylindrical body and a rather sharp-angled shoulder. This cartridge produced very impressive ballistics. With the lighter, .357, jacketed bullets, some loads approach or exceed velocities of the .22 K-Chuck and .22 Jet. Still, it's not been what we might call a howling success.

Autoloader wildcats have been less common. This is understandable inasmuch as neither the cartridge cases or the guns are nearly so adaptable to modification. The best-known autoloader wildcats are several AMP cartridges developed for the Auto-Mag pistols. They were originally designed for factory production, but only one lot of .44 AMP ammunition was ever produced, and it must now be considered a wildcat, along with the smaller-caliber AMP cartridges based upon the same case. (Note: In early 1978, Norma announced it would supply factory-loaded .44 AMP cartridges.) The entire range includes this case necked to .22, 6mm, .25, .357, .41, the basic .44, and a .45 variation. With less than 10,000 of the Auto-Mag pistols produced (and apparently no more to be expected), there is little likelihood that these calibers will ever step out of the wildcat class. They all produce, by a most substantial margin, velocity and energy far exceeding all other autoloader cartridges of comparable bullet diameters.

One other autoloader wildcat obtained a modest following in the early 1960s, and retains some popularity yet today. This is the .38/.45, which is simply the .45 ACP case necked down sharply to accept a wide variety of .38/.357 bullets. The original purpose of this wildcat was not to produce a very high level of ballistic performance, but to enable the then cheap and plentiful .45 GM pistol to be adapted easily to a light, .38/.357 caliber target load. This was the time when a great deal of effort was being devoted to development of the basic GM design to handle the standard .38 Special, mid-range, wadcutter, target

One autoloader wildcat that enjoyed modest success — a conversion is still available from Whitney Sales — is the .38/45 which is simply a .45 case (left) necked down to accommodate a variety of .357 caliber bullets. One of the benefits of this particular wildcat is the fact that the shooter of a G.M. .45 need only purchase a .38/45 barrel, .38 Super bushing and .38 Super recoil spring to make the conversion.

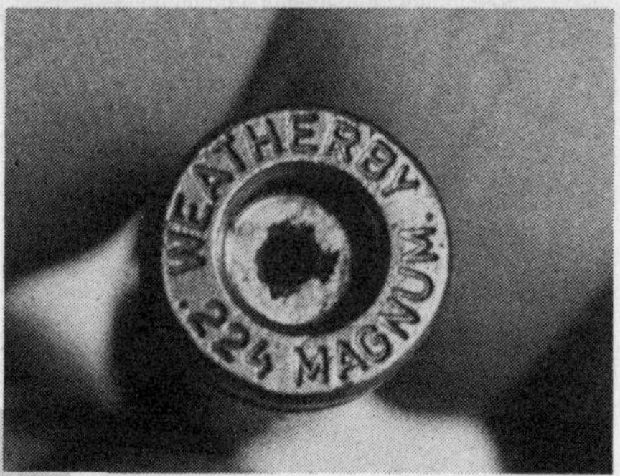

A recent attempt at autoloader wildcatting is the .40 G&A *(Guns & Ammo)*. The gun used was a Browning HP (far left) converted by Bar-Sto and Dr. Ralph C. Glaze to handle a .40 caliber slug. Center: on the left is a .40 G&A compared to a 9mm Luger cartridge (the round the Browning HP is usually chambered for). Above: the basis for .40 G&A round is the .224 Weatherby Magnum case which features a belted rim.

load. Bo Clerke felt the need for replacing the expensive and complex conversions to the rimmed case by using the original rimless case and merely necking it down to handle the desired diameter of bullet. The cartridge was technically successful in this respect, but simply was not popular for that purpose. Instead, handloaders attempted to "magnumize" it to exceed the performance of the .38 Super cartridge. Unfortunately, these efforts were not particularly successful. The design and dimensions of the GM barrel are such that while the .38 Super can be safely loaded to pressures exceeding 30,000 CUP, the much larger-diameter .45 case cannot because of reduced barrel wall thickness. Consequently, the .38/.45 cannot exceed .38 Super performance by any worthwhile margin within safe pressure limits. In spite of this, the .38/.45 still enjoys modest sales success, and this is probably partly due to the fact that its use requires no permanent or irrevocable modifications to the gun. A conversion kit in this caliber is available at a relatively low cost from Whitney Sales. It consists of a rechambered .38 Super barrel and matching barrel bushing, and a recoil spring. The standard .45 GM (whether of military or commercial origin) is converted to .38/.45 by simply replacing the comparable, original parts with those in the kit. And, it is just as easily and quickly converted back to .45. No other wildcat conversion—either revolver or autoloader—offers that advantage.

Other autoloader wildcats have appeared very briefly. One of the more recent ones is the .40 G&A, developed by the staff of *Guns & Ammo Magazine* and first written up by Dr. Ralph C. Glaze. The cartridge (which featured a .400 jacketed bullet) was made from shortened and reamed .224 Weatherby brass and adapted for use in a Bar-Sto modified Browning HP pistol—the goal being to gain greater power than the 9mm Parabellum, for which the gun is normally chambered, while retaining the other advantages of compactness and large magazine capacity. This scribe devised a similar cartridge with different goals—a rimless .41 (.410-inch bullet) made by shortening the .41 Magnum case to the maximum length that could be accommodated in the Colt GM action when loaded with a 130- to 140-grain, truncated-cone bullet. Of course, the case was converted to rimless form, and it was (is) hoped that in this cartridge we could combine the best features of the .45 and .38 ACP cartridges—the velocity range of the latter and the big-bullet smash of the former. Thus far only a couple guns have been made up in two slightly differing variations of this cartridge, and we're still working with them now and then after several years. Autoloader cartridges in the .40-.41 caliber range are by no means new; there have been factory attempts to develop such cartridges going back more than 60 years, but none have ever achieved production status.

Doubtless over the years, there have been other handgun wildcats. After all, anyone who can afford the price of a chambering reamer can "design" a cartridge and modify a single gun to handle it. Once you have a chamber, dies can be jury-rigged to make cases that can be effectively fire-formed. Once you've done this, the loading-tool manufacturers will be happy to make up a set of dies from the fired cases, that's all it takes. For a bit over a hundred bucks (a lot more if you want) you can have your very own special wildcat cartridge.

The handgun field is no different than that of rifles in that countless of thousands of revolvers and autos exist chambered for cartridges which are no longer available. Some, but by no means all of them, are military arms long since discarded for one reason or another, such as the Austrian M12 Steyr and the several variations of the Japanese Nambu pistols. Even though cartridges haven't been made for them in decades, many of these guns still exist in large quantities and in good mechanical condition. Then, there are calibers which are simply too difficult or too costly to obtain, or, perhaps, which are obtainable only in Berdan-primed cases with their attendant reloading problems. Examples of these calibers might be the .455 Webley which is indifferently available and usually Berdan primed, or the Soviet 9mm Makarov which simply isn't available outside the Soviet Block, even though an occasional example of the gun filters through the Iron Curtain.

CASE FORMING

"Case forming" can come to the rescue in many instances where you either can't obtain what you need, can't afford it, or aren't satisfied with what is available. With very few exceptions, virtually any handgun-caliber case can be made from some other case. The job may be as simple as resizing the .44-40 to .38-40, trimming the .44 Special to make .44 Russian; or, a complex process involving trimming, resizing, reaming and even turning of the head. In essence, case forming consists simply of changing the shape and/or dimensions of a case of one caliber to function correctly and *safely* in the chamber of another caliber.

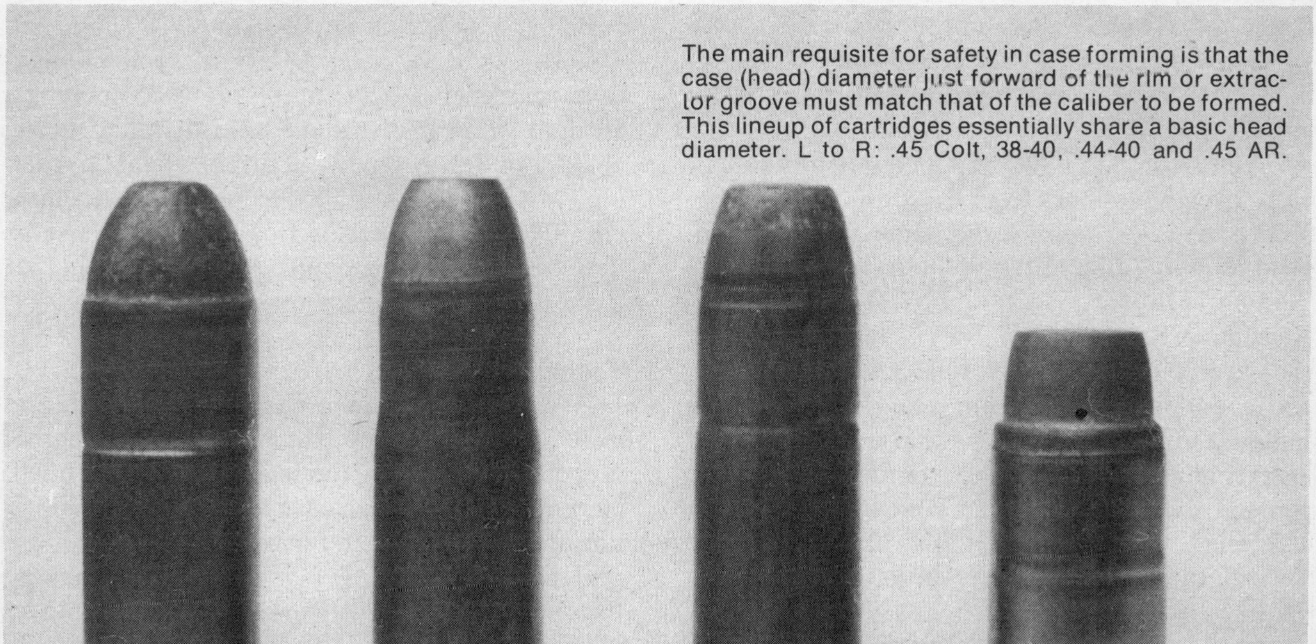

The main requisite for safety in case forming is that the case (head) diameter just forward of the rim or extractor groove must match that of the caliber to be formed. This lineup of cartridges essentially share a basic head diameter. L to R: .45 Colt, 38-40, .44-40 and .45 AR.

Here's another "family" of rounds that share basic head diameter. L to R: 9mm Mauser, 7.63 Mauser (same as 7.62 Soviet pistol). 9mm Bergmann-Bayard and 9mm Steyr.

The main requisite for safety in case forming is that the case diameter—head diameter—just forward of the rim or extractor groove must match that of the caliber to be formed. Actually, that doesn't present as much of a problem as you might think—there are really only a few basic head diameters upon which virtually all calibers are based. The .38 class embraces the .38 Short and .38 Long Colt, .38 Special, and .357 Magnum; the .45 size revolver includes the .38-40, .44-40, .45 S&W, .455 Webley, .450 Revolver, .45 Auto-Rim, .45 Colt, and a couple of uncommon foreign numbers from the 19th century; the 9mm size is *essentially* the same for the 9mm Parabellum, 7.65mm Parabellum, 9mm Mauser, 7.63mm Mauser, 7.62mm Soviet pistol, 9mm Bergmann-Bayard, 9mm Steyr, and a few more that aren't quite so common. The .38 ACP is close enough to the 9mm size—though semi-rimmed instead of rimless—that a fair degree of interchangeability is possible, and it matches the 9mm Browning Long head almost exactly. The .38 Smith & Wesson, though a rimmed revolver cartridge, virtually matches the 9mm Parabellum in head diameter. There are deviations, of course, especially among foreign calibers. The 7.65 French Long and both 7mm and 8mm Japanese Nambu cartridges have had dimensions unique unto themselves. We've included a dimensional chart which, if it doesn't list every center fire handgun cartridge ever made commercially, it at least lists all of those you are likely to encounter.

We'll run through each of the basic case forming operations in some detail.

Resizing

This doesn't differ in the least from ordinary resizing, except that it will change the shape and dimensions of the case. As a case forming operation, resizing *generally* applies only to necking a straight or nearly-straight case down to use a smaller bullet. A typical example would be resizing .44-40 cases in a .38-40 die. Be particularly careful to avoid over-lubrication of the case so that there won't be any oil dents; aside from that it's a perfectly ordinary resizing operation.

Trimming

Again, trimming as a case forming operation differs from ordinary trimming only in degree. Normally only a few thousandths of an inch will be removed from each case to clean a batch up to uniform length, but in forming, ¼-inch or more may have to come off. For example, shortening the .44 Magnum to .44 Russian requires removal of ⅓-inch of brass. The typical hand-operated case trimmer turns that into a very onerous chore. Trimming only a half-dozen cases can blister your hand unless you're already well endowed with calluses. A common, hardware store tubing cutter can come to the rescue here,

When forming some varieties of brass, you'll have to do some healthy trimming, as in this example — shortening .44 Magnum brass (right) by ⅓-inch to make .44 Russian brass (left). A tubing cutter can speed up this sort of operation.

being used in its intended fashion to lop off the excess to within about 1/32-inch of the desired length. Then, it isn't too much of a job to clean off the rest with a conventional trimmer. If a large quantity of cases is involved, then a power-driven trimmer such as those offered by Forester and Lyman for use in a drill press, will allow the largest amount to be cut off to final length very quickly. As many as several hundred cases per hour can be brought to finished length when using this technique. Though, ordinary, fired cases don't really require deburring the inside of the mouth, cases that have been trimmed do need this treatment; just don't overdo it.

When both trimming and resizing are involved in a forming job, trimming should be done last. If it is done first, there's always the possibility that the mouth may be pulled slightly out of square by the friction of resizing.

Neck Expansion

This is seldom required, mainly because it isn't often that a case will be reformed to utilize a larger bullet than for which it was intended. However, it is a practical solution to some problems. If you happen to have available a supply of 7.63mm Mauser cases and need 9mm Mauser cases, expanding the mouth or neck of the former will do the job. This amount of expansion cannot be done in a single operation without distortion, but a satisfactory job can be accomplished in two or three small increments, using expander rods or balls of successively larger sizes. The major problem is that the case may be foreshortened by friction, and also that the mouth will be pulled down more on one side than the other and thus be out of square. Since there is no excess length to allow trimming to square up the new mouth, this can be a bit of a problem. Because of this, it's more practical to avoid large expansion of bottle-neck cases where possible.

Head Turning

There are two reasons for turning the head of a case; the first is to reduce rim thickness and/or diameter, and the second is to convert a rimmed or semi-rimmed case to rimless form. The most common examples are thinning the rim of the .45 Auto-Rim or .45 Colt to produce .455 or .450 cases. The excess thickness must always be removed from the forward face of the rim, simply because if it's taken off the rear, the primer pocket will then be too shallow; and, while the pocket can be deepened, the bottom will then become so thin that it blows out after a very few firings. This operation is by far best accomplished on a bench lathe, though the economical miniature lathes widely advertised today are quite adequate for the job. However, if only a few cases need be processed, very careful use of a safe-edge file will get the job done. It just requires perseverance and patience. If you've a real sure

If you need brass for 9mm Mauser (left), you can neck up 7.63 Mauser to do the job; however, you can't count on doing it in a single operation without some distortion. Neck up in, say, 2 or 3 increments to avoid distortion.

hand with a file, the case can be pressed over a mandrel chucked in a drill press, and the rim filed while it's spinning. This also serves very nicely to reduce rim diameter. Conversion of a rimmed case to rimless form simply can't be done accurately or satisfactorily without a lathe. Further, it requires a tool bit carefully ground to produce the correct extractor-groove profile. Again, this can be done on a miniature lathe, but it requires at least modest skill in operating a lathe. If you lack these skills, then it's far better to farm the job out. No matter who does the job, particular care must be taken that the extractor groove is not cut either deeper or longer than necessary to insure proper functioning in the gun. Simply duplicating the extractor groove of the caliber you're forming can be a mistake if the wall and web thickness of the case being used are not as great as the case being formed. Under those circumstances, cutting the groove full size and full depth may weaken the head to the extent that it will fail. If there's any doubt whatsoever, it's best to sacrifice an *original* case of the caliber you're forming by sectioning it, then sectioning the case you'll be using and comparing the amount of metal in the head.

Fire Forming

Generally speaking, this will be required only when making up wildcat cases such as the .22 K-Chuck and "improved" .22 Remington Jet. In the latter instance, nothing more is required than firing either a factory load or handload in the improved chamber. Gas pressure inside the case will expand its walls into firm contact with the chamber, the case thus being reshaped. The K-Chuck is a different matter in that the case must first be trimmed to the correct length, then handloaded, and *then* fired in the new chamber. You might even encounter a wildcat caliber which would require resizing as well as trimming before it can be loaded for fire-forming. While it isn't common in handgun calibers, some lots of brass may produce a few split cases. The amount of expansion at the largest diameter may be more than the brass can stand without exceeding its normal elasticity. A split case cannot be salvaged, but if the case survives the initial fire-forming without splitting, it is unlikely to do so thereafter.

Annealing

Seldom required in handgun calibers, annealing makes use of heat and very rapid cooling to "soften" the brass of the case and make it more ductile so that it is less likely to split or crack from the stresses placed upon it by fire-forming and/or repeated loading and firing. The problem in annealing handgun cases is their very short length; it's difficult to apply sufficient heat to the mouth without the head also becoming hot enough to be softened. And if the head becomes too soft, it will deform under the stresses of firing and perhaps rupture to damage the gun. Annealing is best done by immersing the front half of the case in molten lead—the same lead from which you cast bullets. The typical handgun case is too short to be held in the fingers while doing this, so I have made up a simple set of aluminum tongs. They hold the case head securely, and so long as they are closely fitted to it, they function as an effective heat sink and thus prevent overheating of the case head. The case is prepared for annealing by first decapping, then dipping it in light oil or pow-

One of the more common examples of head turning is taking a .45 Colt (left) or .45 AR (center) and turning the head down to make a .455 case (right).

dered graphite, making certain that the interior is coated as well as the outside. It is then grasped with the tongs, and the forward half is immersed in the lead vertically for approximately 5 seconds; it is then quickly withdrawn, given a quick flip of the wrist to throw off any molten lead still clinging inside, and then dropped instantly into a container of cold water. This treatment will not make the mouth dead soft, but will soften it enough to give the desired effect. As a matter of fact, we don't want the mouth dead-soft.

Neck Reaming

Neck reaming is another operation which is seldom required in handgun calibers, and then only when a case is being necked down a substantial amount. The only occasion I have had for reaming the necks of handgun cases is when forming the AMP series from .308 Winchester brass, and when forming 7.63mm Mauser cases from .222 or .223 Remington rifle *cases*. In both instances, the neck of the case being formed falls well back on the body of the case being used, at a point where the walls have become much thicker than they are at the original mouth. Attempting to use such formed cases without reaming results in the neck diameter being far too great over the seated bullet to enter the chamber properly. While accuracy-minded rifle shooters would consider the method inaccurate and imprecise, I have found the neck-reaming attachments for the Forester case trimmer to be quite adequate for all handgun uses except the AMP calibers. For them nothing less than the special neck-reaming die set made by RCBS should be considered. It is also possible to form short 9mm Parabellum and even .380 ACP cases from .222 or .223 Remington brass. If this job is undertaken, reaming will also be required, and for the same reason, even though the cases are straight. The walls are so thick under these conditions that the operation is more counterboring than reaming, but with care it can still be done on the Forester trimmer/reamer setup with a standard .35 caliber reamer.

Swaging

This is a rather advanced operation, and one requiring special dies. Unless you simply want to, there is no need to do the job yourself, because there are a few shops around the country who will do the job for you if you just need a supply of cases to shoot in a particular gun. It involves forcing the case into a hardened and polished ring die to reduce its diameter full-length except for the rim. Two examples of this that come to mind are forming the old, black powder .44 S&W American or .44 Colt cases which are smaller in diameter than today's .44 Special and .44 Magnum. These modern cases measure nominally .455-inch just ahead of the rim, and also have a rim thicker than that of those old .44s which have a head diameter of about .440-inch. After the rim thickness is reduced and a V-shaped clearance groove is cut on a lathe or with a file, the cases are forced completely into a die which will reduce them to the correct diameter. Since much more force is required here than when resizing, and also because the case must be forced entirely into the die, we can't pull it back out by the rim. Instead, we'll have to extract it from the die by means of a punch inserted from the open end. Since we can't use a shell holder, then, a simple unthreaded die can be used in a heavy-duty bench vise.

There is another form of swaging, (used to reduce rim thickness in altering .45 Auto-Rim cases to .455 Webley caliber), that I believe I

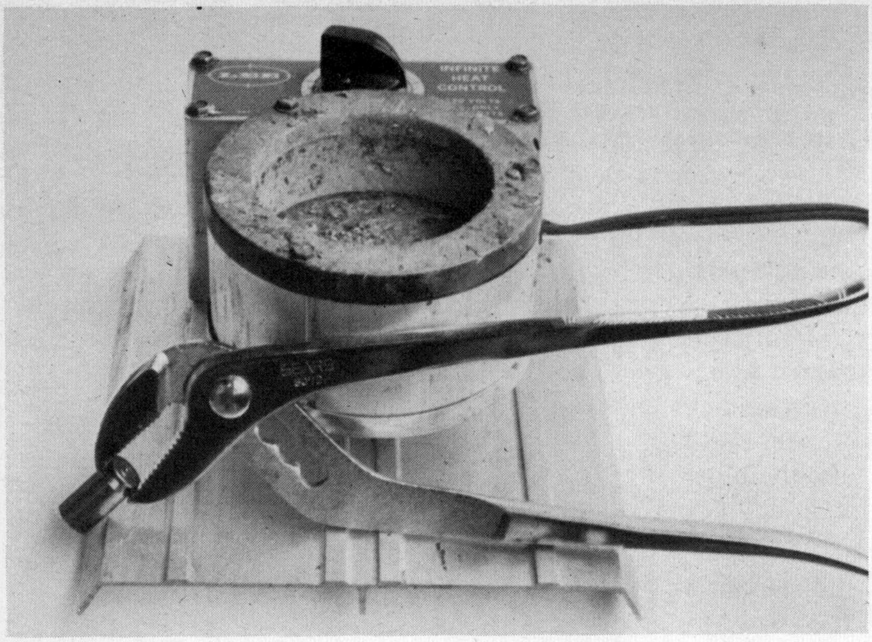

Your bullet casting pot and a pair of pliers are all you need to anneal cases. Turn the pot on, dip the case into light oil or powdered graphite and dip the case *mouth* into the molten lead for 5 seconds, remove, shake off any lead, and drop the case into a can of cold water.

originated some years back. This consists of making up a punch and die, dropping the case into the die, then inserting the punch and driving it down solidly against the stop with a hammer. This forces the brass of the thick A-R rim to flow outward as it is reduced in thickness. Once the die and punch set is made up, this is by far the quickest and most convenient method of making .455 or .450 cases. The swaged rim may occasionally be a bit oversize in diameter, but no more than can be accommodated readily in Colt and S&W D-A revolvers, and also not so much that oversized portions can't be removed easily with a file for use in other guns.

Swage-Modified Berdan Primer Pockets

Modifying Berdan primer pockets to accept domestic Boxer primers was not covered in our section on priming simply because it is a more complex swaging operation which properly belongs here. A hardened, steel punch is made up, and then a rod or anvil is also made up and affixed to a very solid metal base. The rod must fit closely inside the case, making firm contact with the head around its perimeter, leaving room in the concave end for the bottom of the existing primer pocket to be driven forward. With the case slipped over the anvil, the pocket-forming punch is placed in the existing primer pocket, carefully aligned, and then struck smartly with a hammer until the Berdan anvil is flattened, and the pocket bottom is lowered sufficiently to accept a Boxer primer. Most Berdan handgun cases, upon which you might want to perform this operation, use the standard small-pistol size primer, nominally .177-inch in diameter—and that is .002-inch larger than our small pistol Boxer size. The shape of that portion of the punch which contacts the case head—and it must be made accurately—will close the mouth of the primer pocket sufficiently to make the Boxer primer a tight fit.

A few years back there was quite a large quantity of Berdan-primed .45 ACP ammunition sold cheaply in this country. If you've shot this ammunition and were foresighted enough to save the cases, they can be easily converted to use a readily-available domestic primer. This time, though, we're going to use standard, 209-size, battery-cup shotshell primers, not the usual pistol type. The job is best done on a lathe, with the case first resized so that it can be held concentrically in a collet; but it can be done on a drill press if you rig some sort of simple fixture for holding the case head-up on the press table. Leave the fired primer in place and secure the case in the collet; chuck a very short 15/64-inch drill (preferably ground for cutting brass) in the tailstock, and drill through the case head. Use the shortest possible drill to eliminate walking. Back the drill out and cut a 45-degree chamfer a bit less than 1/16-inch wide at the mouth of the hole with a scraper or hand-held countersink. After this has been done to all of the cases, replace the drill bit in the tailstock with a $^{5}/_{16}$-inch end mill. Again place the case in the collet, and advance the end mill to cut a counterbore .025-inch deep.

Once all this has been done to the case, repriming is simply a matter of seating a battery-cup, shotshell primer in the normal fashion; making certain that the entire battery cup is flush with or slightly below the surface of the case head. For subsequent reloading, decap by hand with a ⅛-inch punch and light mallet. My own reloading for these shotshell primed .45 ACP cases has *never exceeded* recommended service-type loads found in current manuals. Stick with these—or target loads—and you should experience no difficulty.

So much for the various case forming operations. The accompanying charts list the dimensions for the many handgun calibers and also contain information as to the basic case and forming operations which must be accomplished to make what you need from what you've got. With these at hand and a little attention to detail, there aren't many calibers for which you can't make conventional cases and load ammunition if you're willing to work at it.

MAKING PIN FIRE CASES

As we mentioned earlier while discussing the historical development of handgun ammunition, the pin fire cartridge was popular—more so abroad than here—during the transition period from loose powder and ball to the rim fire and center fire cartridge. The pin fire type was widely known by the period of the American Civil War, and while no such guns appear to have ever been manufactured here, they were imported from that time on for several decades. Cheapness of both the guns and the ammunition were their principal virutes after the early 1870s. Manufacture of pin fire revolvers continued for a number of years in Europe, but pin fire ammunition continued to be made up until WWII, and possibly for a short time afterward. The fact remains that there are enough pin fire revolvers still around that a good many people would like to shoot them. Without ammunition, that can't be done.

It isn't really all that difficult to *make* pin fire cases that can be reloaded a number of times. All we need do is come up with a thin-walled copper or brass body of the proper diameter for the

Pin fire cases are hard to come by, however, by following the author's directions you can make your own. This pin fire cartridge is for a .38 caliber gun and was made in the U.S. well prior to WWII. The external stem, when hit by the hammer, strikes an internally-situated priming pellet which in turn detonates the black powder charge.

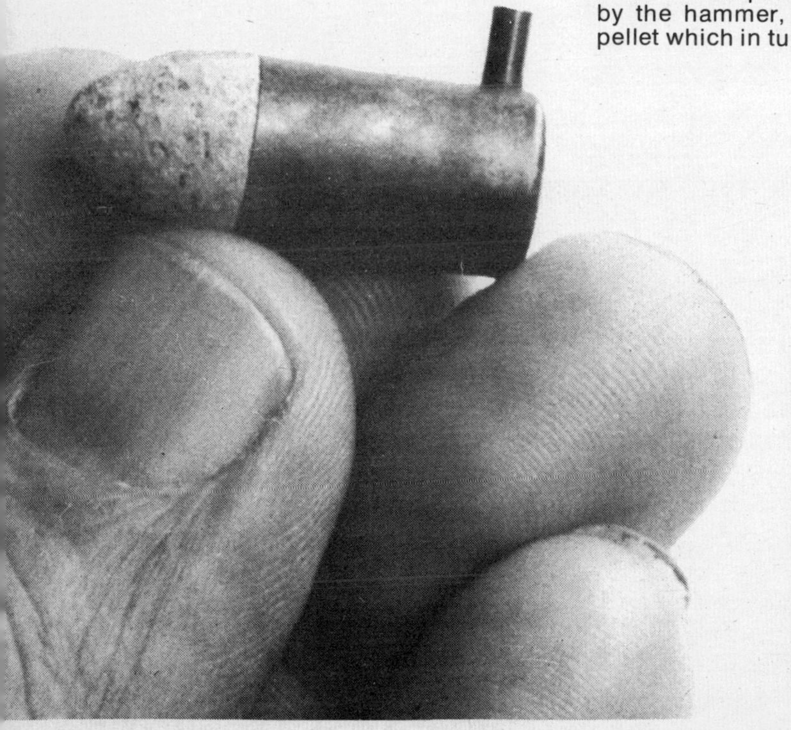

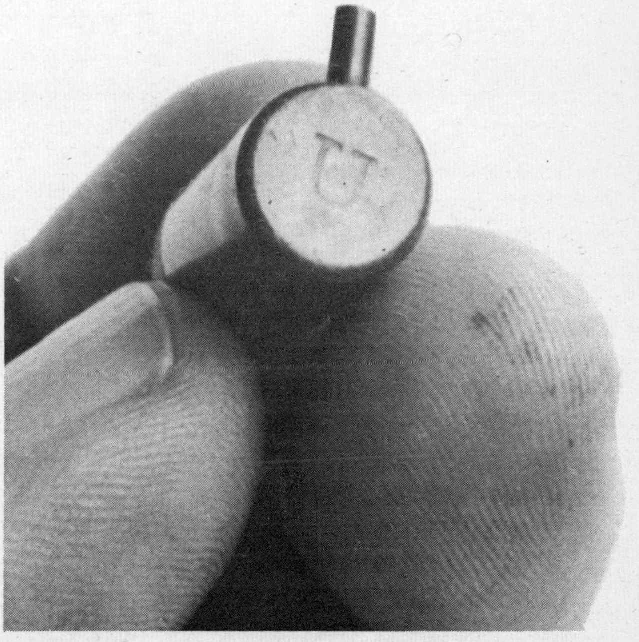

HANDGUN CARTRIDGES
Current and Obsolete Centerfire — Black Powder and Smokeless
Dimensional Data

Cartridge	Case type	Bullet dia.	Neck dia.	Shoulder dia.	Base dia.	Rim dia.	Case length	Ctge. length	Twist	Primer
2.7mm Kolibri	D	.107	.139	—	.140	.140	0.37	0.43		B
3mm Kolibri	D	.120	.150	—	.150	.150	0.32	0.43		B
4.25mm Liliput	D	.167	.198	—	.198	.198	0.41	0.56		B
5mm Clement	C	.202	.223	.277	.281	.281	0.71	1.01		B
5mm Bergmann	D	.203	.230	—	.273	.274	0.59	0.96		B
5.5mm Velo Dog	B	.225	.248	—	.253	.308	1.12	1.35	8.2	S-B
22 Remington Jet	A	.223	.247	.350	.376	.440	1.28	1.58	10	S
221 Fire Ball	C	.224	.251	.355	.375	.375	1.40	1.82	14	S
25 ACP	D	.251	.276	—	.277	.298	0.62	0.91	16	S
256 Winchester Mag.	A	.257	.277	.378	.378	.440	1.30	1.53	14	S
6.5mm Bergmann	C	.264	.289	.325	.367	.370	0.87	1.23		B
7mm Nambu	C	.280	.296	.337	.351	.359	0.78	1.06	12.5	B
7.62mm Nagant (Russian)	B	.295	.286	—	.335	.388	1.53	1.53	9.5	B
7.62mm Tokarev	C	.307	.330	.370	.380	.390	0.97	1.35	10	B
7.63mm Mauser	C	.308	.332	.370	.381	.390	0.99	1.36	7.9	S-B
30 Borchardt	C	.307	.331	.370	.385	.390	0.99	1.34		S-B
30 Carbine	D	.308	.336	—	.356	.360	1.290	—	18¾	S
7.63mm (7.65) Mannlicher	D	.308	.331	—	.332	.334	0.84	1.12	10	B
7.65mm (30) Luger	C	.308	.322	.374	.388	.391	0.75	1.15	9.8	S-B
7.65mm MAS (French)	D	.309	.336	—	.337	.337	0.78	1.19		B
7.65mm Roth-Sauer	D	.301	.332	—	.335	.335	0.51	0.84	14.2	B
32 ACP	H	.309	.336	—	.336	.354	0.68	1.03	16	S
32 S&W	B	.312	.334	—	.335	.375	0.61	0.92	16-18	S
32 S&W Long	B	.312	.335	—	.335	.375	0.93	1.27	16-18	S
320 Revolver	B	.317	.320	—	.322	.350	0.62	0.90	22	B
32 Colt	B	.313	.318	—	.318	.374	0.92	1.26	16	S
32-20	A	.312	.326	.338	.353	.405	1.32	1.59	16	S
35 S&W Auto	D	.309	.345	—	.346	.348	0.67	0.97	12	S
7.5mm Swiss Army	B	.317	.335	—	.345	.407	0.89	1.29		B
7.5mm Nagant (Swedish)	B	.325	.328	—	.350	.406	0.89	1.35	18	B
8 mm Nambu	G	.320	.338	.388	.408	.413	0.86	1.25	11	B
8mm Lebel Revolver	B	.323	.350	—	.384	.400	1.07	1.44	9.5	B
8mm Roth-Steyr	D	.329	.353	—	.355	.356	0.74	1.14	10	B
9mm (9x18) Ultra	I	.355	.375	—	.382	.374	.705	1.00		S
9mm Glisenti	D	.355	.380	—	.392	.393	0.75	1.15	10	B
9mm Luger	D	.355	.380	—	.392	.393	0.76	1.16	9.8	S-B
9mm Wildey Mag.	D	.356	.380	—	.393	.393	—	—		S
9mm Bayard	D	.355	.375	—	.390	.392	0.91	1.32		B
9mm Steyr	D	.355	.380	—	.380	.381	0.90	1.30		B
9mm Browning Long	D	.355	.376	—	.384	.404	0.80	1.10	12-16	B
9mm Makarov	D	.363	.384	—	.389	.396	0.71	0.97		B
357 S&W Magnum	B	.357	.379	—	.379	.440	1.29	1.51	18.8	S
380 Revolver	B	.357	.377	—	.380	.426	0.70	1.10	15.1	S-B
38 Long Colt	B	.357	.377	—	.378	.433	1.03	1.32	16	S
38 Special	B	.357	.379	—	.379	.440	1.16	1.55	16-18	S
38 S&W	B	.359	.386	—	.386	.433	0.78	1.20	16-18	S
38 Colt ACP & Super Auto	H	.358	.382	—	.383	.405	0.90	1.28	16	S
380 ACP (9mm Browning Short)	D	.356	.373	—	.373	.374	0.68	0.98	12-16	S
38-40	A	.401	.416	.438	.465	.520	1.30	1.59	16	L
41 Long Colt	B	.386	.404	—	.405	.430	1.13	1.39	16	S
41 S&W Magnum	B	.410	.432	—	.433	.488	1.28	1.58	18	L
10.4mm Italian	B	.422	.444	—	.451	.505	0.89	1.25	10	B
44 AMP	D	.429	.458	—	.470	.473	1.298	—	20	L
44 S&W American	B	.434	.438	—	.440	.506	0.91	1.44	20	L
44 S&W Russian	B	.429	.457	—	.457	.515	0.97	1.43	20	L
11mm German Service	B	.426	.449	—	.453	.509	0.96	1.21	23	B
11mm French Ordnance	B	.425	.449	—	.460	.491	0.71	1.18	16.2	B
44 S&W Special	B	.429	.457	—	.457	.514	1.16	1.62	20	L
44 S&W Magnum	B	.429	.457	—	.457	.514	1.29	1.61	20	L
44 Colt	B	.443	.450	—	.456	.483	1.10	1.50	16	L
44 Webley	B	.436	.470	—	.472	.503	0.69	1.10	20	L-B
44 Bull Dog	B	.440	.470	—	.473	.503	0.57	0.95	21	S-B
44-40	A	.427	.443	—	.471	.525	1.31	1.55	16-20	L
11.75mm Montenegrin	B	.445	.472	—	.490	.555	1.40	1.73		B
45 Colt ACP	D	.452	.476	—	.476	.476	0.898	1.17	16	L
45 Auto-Rim	B	.452	.472	—	.476	.516	0.898	1.28	15-16	L
45 Colt	B	.454	.476	—	.480	.512	1.29	1.60	16	L
45 S&W Schofield	B	.454	.477	—	.476	.522	1.10	1.43	24	L

Cartridge	Case type	Bullet dia.	Neck dia.	Shoulder dia.	Base dia.	Rim dia.	Case length	Ctge. length	Twist	Primer
45 Webley	B	.452	.471	—	.471	.504	0.82	1.15		L-B
45 Wildey Mag.	D	.451	.476	—	.476	.476	—	—		L
450 Revolver	B	.455	.475	—	.477	.510	0.69	1.10	16	L-B
455 Enfield (455 Colt)	B	.455	.473	—	.478	.530	0.87	1.35		L-B
455 Webley Revolver Mk-II	B	.454	.476	—	.480	.535	0.77	1.23	16-20	L-B
455 Webley Auto	H	.455	.473	—	.474	.500	0.93	1.23	10	B
476 Enfield	B	.472	.474	—	.478	.530	0.87	1.33		B
50 Remington Army	A	.508	.532	.563	.565	.665	0.57	1.24		L

Notes on handgun primers: Magnum pistol cartridges are usually loaded with special Magnum primers and the .22 Remington Jet and .256 Winchester are sometimes loaded with small rifle primers. During WW 1, Frankford Arsenal made .45 ACP cases with special #70 primer of .204″ diameter instead of the standard .210″.

Case types:

A/Rim, bottleneck
B/Rim, straight
C/Rimless, bottleneck
D/Rimless, straight
G/Semi-rimmed, bottleneck
H/Semi-rimmed, straight
I/Rebated rim, Straight

Primer: S—Small (.175″). L—Large (.210″). B—Berdan type

Unless otherwise noted, all dimensions are in inches.

CASE FORMING SUGGESTIONS

Case	Forming
.22 Rem. Jet	Resize .357 Magnum or .256 Win.; plain brass cases form better than nickeled cases.
.221 Fireball	Resize and trim .222, .223, .222 Mag. Rem.
.256 Win. Mag	Resize .357 Magnum (see .22 Jet)
7mm Nambu	Resize and trim .32 S&W Long; turn head to rimless; ream neck
7.62mm Tokarev	Use 7.63mm Mauser
7.63mm Mauser	Resize and trim .222 Rem.; ream neck
7.65mm Parabellum	Resize and trim .38 ACP
7.65mm French	Resize and trim .32 S&W Long; turn head to rimless
7.5mm Swiss Rev	Resize and trim .32-20
8mm Lebel Rev	Resize and trim .38 Spcl; turn down rim
9mm Glisenti	Use 9mm Parabellum
9mm Parabellum	Resize and trim .38 ACP
9mm Bayard & Steyr	Use .38 ACP; reduce rim diameter if necessary
9mm Browning Long	Resize and trim .38 ACP
9mm Makarov	Resize, expand and trim .38 ACP
.38 Long & Short Colt	Resize and trim .38 Spcl.
.38 ACP	Resize and trim .357 Mag.; turn head to semi-rim shape
.380 ACP	Resize and trim .38 Spcl.; turn head to rimless
9mm Ultra	Trim 9mm P.; turn head to rebated rim
.38-40	Resize .44-40 or .45 Colt
.44 AMP	Resize and trim .308 Win.; ream neck
11mm German Rev	Resize and trim .44 Spcl.
.45 Schofield	Trim .45 Colt
.450 Revolver	Trim .455 Webley
.455 Webley/Enfield	Trim .45 Colt or .45 A-R; reduce rim thickness
.45 Wildey Mag	Resize and trim .308 Win.; ream neck

NOTE: Above list is by no means complete; check dimensions data table for numerous other possibilities.

chamber, make it gas-tight at the rear, and provide it with an internal primer and an ignition pin. While the smaller calibers present technical problems in doing this, those above .38 can be handled without too much difficulty.

If thin-walled .010- to .015-inch wall thickness (brass or copper) tubing cannot be found of the proper diameter for the case body, you can probably make do with a section of a conventional cartridge. The rear of the body is then closed with a plug, and ideally this can be made from a disc of sheet copper about .03-inch thick formed into a cup with a simple punch and ring die. Lacking that, a copper or brass plug can be turned on a miniature lathe to the proper shape. Either type of plug is then coated with low-temperature solder paste, pressed into the end of the body tube and heated just to the point where the solder flows freely. A 1/16- to 3/32-inch hole is then drilled through the body and the forward-extending part of the plug, at right angles to the case, allowing the drill to go through far enough to just barely mark the inner surface of the plug directly opposite the hole. The dimple formed there by the drill point serves only to aid in properly locating the primer for loading; it *must not be deep enough* to weaken the case. A firing pin or ignition pin is then made from copper, brass, or aluminum rod to fit snugly in the hole just drilled. The case is now complete.

Original pin fire cartridges were primed with the common percussion cap used on muzzle-loading guns. You may still do that, or you may substitute a conventional small-pistol size primer. In either instance, a dab of quick-setting adhesive is placed on the closed end of the primer or cap which is then carefully positioned inside the case, mouth-up, over the dimple opposite the ignition-pin hole. Once the adhesive has set, insert the ignition pin in its hole, (it should be a snug fit), and position it so that its inner end is seated solidly against the priming compound pellet in the percussion cap, or the center of the exposed anvil in the primer. It should be tight enough in its hole that it will remain in this position during normal handling, but to make sure, apply a drop of Lok-Tite or cyanoacrylate adhesive where the pin passes through the case wall. This will not only secure the pin against handling displacement, but will waterproof the joint.

You now have a primed, pin fire case. It is a *weak* case, as were the originals, and it must be kept in mind that the pin fire revolvers were also of relatively weak construction and materials, and designed for *black powder only*. With the primed case at hand, charge it with FFg powder—*black powder only*—the amount being regulated to permit *slight* compression of the charge when the bullet is seated. Tap the case *gently* (preferably not over the primer) after the powder is added so that it will settle in around the primer and ignition pin and help retain them in their proper postion. I prefer to load such pin fire cases with round balls, mainly because most of them would require a special-diameter bullet for which a production mold is not available. Round-ball molds are available in such a wide variety of sizes that no difficulty should be encountered in obtaining one to suit almost any pin fire revolver. Since balls contain no provision for lubricant, I use a disc of bullet lubricant between the ball and powder as described in Chapter 16. The ball is then seated half or very slightly more than half its diameter into the case, and if it isn't tight enough to remain there during normal handling, two or three light stab crimps or a drop of cyanoacrylate adhesive will secure it solidly in place.

It is also sometimes possible to make pin fire cases from ordinary center fire cases if the outside diameter is correct. This involves simply facing off the head to eliminate the primer pocket, then plugging the flash hole and installing the primer and ignition pin as already described. If there were sufficient demand, I'm sure that quite satisfactory pin fire cases could be injection-molded from polyethelene, as is done in making cases from some transition-period, outside-primed, breech-loading rifles of the Civil War era.

The fact remains that if you do possess a big-bore pin fire revolver, you *can* make cases and load ammunition for it, and it need no longer be relegated to a hidden corner of the closet.

CHAPTER 20

Hand-Tool and No-Tool Loading

PERHAPS WE HAVE GIVEN you the impression that one must spend a good deal of money for the tools, equipment and accessories necessary to get set up for loading even a single caliber. Well, you *can* spend a lot of money if you are simply inclined to do so, or in the event your requirements are such that you'll need the more costly and sophisticated equipment. On the other hand, it is entirely possible to load superbly accurate ammunition that is functionally perfect (at least for some uses) with tools costing only a handful of dollars. It is also possible to assemble some particular types of loads without the use of any store-bought tools whatever.

NO-TOOLS LOADING

Let's take the latter first. Let's say you have a bunch of fired cases in a straight-case revolver caliber such as .38 Special or .45 Colt and, further, that you're willing to settle for a relatively light lead-bullet load. Decapping the cases requires only a stick of wood (to be used as a hammer) and some form of punch that will enter the flash hole. A 1/16-inch pin punch will do the job, if you happen to have one, but alternatively just file down the end of a spike/nail or other piece of steel to fit. The reduced portion should enter the flash hole freely and needs to be about 5/16-inch long. Drill a hole in a piece of wood that the fired primer will pass through (¼-inch hole is big enough), set the case over the hole, start the punch in the flash hole through the case mouth, and tap it smartly to eject the old primer. You can clean the primer pocket with a scraper or brush if you like, but otherwise the job is finished.

Seating a new primer may seem like a rather complex job if you've never done it except with a priming tool. However, it's really quite simple. Lay out a smooth, flat piece of metal that is fairly thick (a piece of thick plate glass will also do nicely) and set a primer on it, anvil-up. Carefully place the primer pocket over the primer (you'll be able to feel the primer enter as far as the pocket radius), making certain the case stands vertically. Slip a piece of dowel inside the case to bear against the head, and tap *gently* on the dowel to drive the case down over the primer. When the case head makes firm and complete contact with the plate, the primer is seated flush with the head. I realize that this seating method is somewhat in contradiction to what we recommend elsewhere in this book in that the primer does not make con-

"No-tools loading" could also be called "limited tools loading." Decapping requires only a stick (or mallet) and a shaped piece of metal (a nail ground to proper dimensions will do). Here we've used a mallet and a Lee decapping rod and universal shell base. The Lee equipment costs only a shade over $2, and the mallet was on hand.

tact with the bottom of its pocket. Nevertheless, for light loads, this is an entirely satisfactory method of priming even though it may not be ideal. At first you might think driving the case down over the primer by means of hammer blows creates a danger of primer detonation—*not* true. If only *light taps* are applied, the primer cup will not be in the least deformed, and since there is no support for the anvil, the compound pellet cannot be crushed. (As long as it isn't crushed violently, the primer will not fire.) I suppose it *is* possible to set off a primer in this fashion if the case were dealt a really fierce blow, but I don't think, we're dumb enough to do that. After all, as you'll learn from experience, several very lights taps on the dowel will seat the primer fully.

We now have a primed case, though admittedly it has not been resized. Sizing isn't really necessary so long as the case will chamber freely in the gun for which it is intended, and we must presume that you checked this out in the very beginning by trying all the cases and setting aside any which did not chamber easily.

Adding propellant is the next step, and there is no reason why we can't use a simple charge cup for the job. A .22 Short empty *fired* case will normally hold 2.0 grains of Bullseye or 1.9 grains of Unique. An empty, fired .22 LR case will hold 2.9 grains and 2.8 grains respectively of the same propellants. Thus, all we need to do is epoxy or solder a paper-clip handle on a fired case and use it to dip propellant charges. Start with a coffee cup or similar container about ¾-full of propellant, and submerge the charge cup in it vertically so that propellant flows in to fill it. Don't scoop the cup through the propellant more or less horizontally or it will pack in unevenly, producing wider variations in the charge. Lift the dipper straight up and with a knife blade or similar instrument scrape the excess off flush with the mouth. Pour the propellant powder carefully into the case to avoid spillage, and another operation is complete.

At this point, you have a choice to make in order to avoid the costs of bullet-casting equipment—either settle for round lead balls which may be purchased cheaply, or buy cast bullets from one of the numerous custom-casters which advertise in the various shooting periodicals. If you require only a light, short-range load, the round ball is certainly the cheapest and most convenient; however, if you want something with a bit more range and power, you'll have to buy conventional bullets. Actually, this really isn't bad at all, for as this is written, bullets can be purchased for as little as $15-18 per 500. At that price, you can buy a lot of bullets for what a mold, melting pot and ladle will cost. On the other hand, if you plan to do a lot of shooting with these handloads, say more than a 1,000-1,500 rounds over the next few months, you'll save money in the end by breaking down and buying a low-cost Lee bullet mold, along with a melting pot and dipper.

Having decided upon bullets, it's now necessary to seat them and secure them in the primed and charged cases. The round ball is easy: Place a disc of bullet lubricant in the case mouth, then tap the ball in half its depth. Preparing the lubricants is easy, just flatten a lump of it between your hands, then press it over the case mouth to cut out the disc which will remain in place. If the

Left — Seating requires a mallet, a length of dowel and a flat surface. Right — Lacking an expensive powder measure, you can make your own using a fired .22 Short case and a length of paper clip. The fired .22 Short case will hold 2.0 grains of Bullseye or 1.9 grains of Unique.

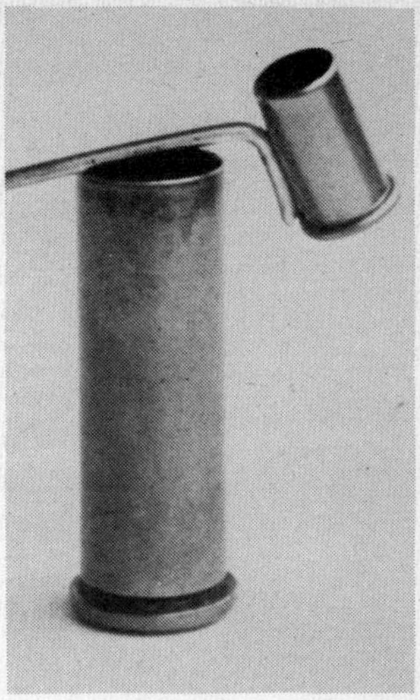

ball fits tightly in the case, no further operation is necessary. However, if it isn't all that tight, press the blade of a small screwdriver into the case mouth, indenting it slightly to hold the ball more securely. If you're using store-bought lead bullets which have been sized and lubricated, they may well be a loose fit in the unsized case mouth. If so, simply press the bullet in with your fingers to the proper depth, then indent the mouth as just described to hold it there. If perchance the bullet is snug, tap it in to the proper depth where friction and the tackiness of the lubricant will hold it. If you have popped for a mold and have cast your own bullets, plan on using them *unsized,* just as they come from the mold. Set them on their bases in a pie tin or similar container and melt bullet lubricant over *low* heat (be sure it's *low*) to a depth that covers the lubricating grooves. Allow the lubricant to harden, then using a snug-fitting piece of tubing or an expanded fired case with its head cut off, press down over each bullet to cut it free from the cake of lube. This will leave lubricant in the grooves while scraping the bearing surface clean. If you don't want to make all this mess, simply rub lubricant from the stick into the grooves of each bullet. Now, start the bullet in the case, and if it fits too tightly to be forced in by hand, tap it in to the proper depth with a stick or light mallet. These as-cast (unsized) bullets will normally be large enough in diameter to fit snugly if not tightly in the case and thus no crimping of any sort is required.

With light loads, cases may be reloaded a number of times without any resizing. The shallow stab crimps described may be reduced very slightly by repeated firing, but generally remain. Upon subsequent reloadings they may form shallow grooves in the bullet's bearing surface, but this does not appear to affect accuracy to any measurable extent; and, in fact, they will often grip subsequent bullets tightly enough that no further crimping is necessary. It if does become necessary to crimp in subsequent loadings, always deepen the existing stab crimps (with the same instrument) rather than impressing more upon the case.

Even with the lightest loads (that is, anything heavier than ultra-light loads), cases will eventually expand after several firings so that they no longer chamber or extract freely. When that happens, they must be resized if they are to continue in service. You can locate a friend or neighbor with a resizing setup and have him do the job; or you can simply switch to another batch of cases and set the expanded ones aside; or, as a last resort, you can purchase a simple, hand-type resizing die. This is nothing more or less than a length of round steel stock containing a smoothly-polished resizing cavity. As originally sold, dies of this type were intended to be used by using a hammer and punch to drive the case in, then the hammer and a "knockout rod" to remove it. Old Lyman/Ideal handbooks illustrate this operation, but these days, many households contain a bench vise in the basement or garage. It's a lot easier to simply start the lubricated case in the hand die, then place the assembly between the vise jaws and crank the handle until the case is forced home; then, if the vise has enough jaw spread, a wood spacer containing a hole to admit the case will allow the vise to be used with the knock-out rod to force out the case. Incidentally,

Left — Depending on your pocketbook, you can choose between cast slugs, round balls or swaged commercial slugs for your shooting. Actually you might want to consider all three for separate uses.

Right — Crimping a bullet into place can be easily done with a screwdriver and a light tap with a hammer. Note the "stab" crimp; just below the mouth of this round ball, .38 Special load.

one summer many years ago in west Texas (El Paso Del Norte), I shot all season with less than a hundred .45 Colt cases which I resized in this fashion so they could be loaded up in preparation for the coming weekend of shooting. Hand-type, full-length resizing dies aren't as common as they once were, but thousands of used ones are floating around, as well as those of current manufacture in a relatively few calibers. With a little diligent effort you'll be able to find one.

HAND-TOOL LOADING

Progressing upward from "no-tools" loading, the next step is the hand-tool. Hand-tools have been made in an almost unbelievable array of designs and types and are the oldest types of handloading equipment. They appeared on the scene immediately following the reloadable center fire metallic cartridge and have been with us ever since. Back in the last quarter of the 19th century virtually every major (and plenty of minor) firearms manufacturers cataloged loading tools right along with their guns and supplied them routinely upon request. They appeared in a wide variety of designs, but most consisted essentially of some sort of frame housing a sliding ram connected to a small lever. With the proper punch or die in place, operation of the lever decapped, recapped, sized case necks, and seated and crimped bullets. To the best of my knowledge none of these lever-actuated tools had the capability of resizing cases full-length, but there was at least one screw-type design which could do so. It employed a ram rotating in the frame by means of a coarse screw thread, and when the threaded portion was turned by a means of a folding handle, it could exert enough force to resize the case. Separate bullet molds were often supplied with the tools, but some designs actually incorporated the mold in the tool proper. The most notable of these was the original Ideal Tong Tool which remains in production today, albeit in considerably modified form. A continuous production run of nearly 100 years for the same basic design isn't doing too badly.

Ideal and Lyman Tong Tools

The tong tool is typical of all of the older designs and a description of the current Lyman #310 form will show how simple it is to reload cartridges with a complete tool costing only $33. The tool proper consists of a pair of handles pivoted together at their forward end and resembling nothing so much as an old-style nutcracker gone pregnant. A boss on the bottom handle is threaded to accept interchangeable dies and accessories, while a spring-loaded hook is let into the upper handle to engage case rims. To begin, a "decapping chamber" containing a decapping stem and pin is screwed into the lower handle from the underside; it is adjusted so that the decapping pin extends about ⅛-inch above the upper surface of the handle. The handles are opened wide and a fired case dropped down over the pin; closing the handles forces the case down over the decapping pin and the fired primer is pressed out. The handles are then opened and the hook on the upper handle pulls the case clear. In some calibers, the decapping chamber may also serve to resize the mouth (only) of the case. In others a separate "muzzle resizer" must be screwed into the handle and employed in the same manner to reduce the mouth of the case so it will hold a fresh bullet securely. The case is reprimed in a unique attachment which also threads into the lower handle. Half of its lower portion is cut away to permit the case to enter it sideways, just as a case is placed in a shell holder. A primer-seating punch slides in the upper portion of the attachment, to be forced downward to seat a primer by the upper handle. With this "priming chamber" properly adjusted, a case is placed in the lower part, making certain the rim seats fully in its groove, a primer is dropped into the channel ahead of the primer-seating punch, and the tool handles are then closed to force the primer into the pocket. Expansion of the case mouth—and flaring as well—is accomplished by a simple threaded tube screwed into the lower handle and carrying inside its lower end a short conventional expander plug. Again, the case is dropped into the hole in the upper surface of the lower handle, after which the handles are closed to force the case down over the expander plug.

Older tong tools will frequently be encountered in which the new primer is seated by a short stud in the underside of the upper handle; a separate hole in the lower handle receives the case, a primer is then placed over the pocket, and the handles closed to seat it. For rimmed cases there is simply a hole in the handle, but a U-shaped steel clip is added to hold rimless cases.

With the tong tool, no provision is made for measuring the proper charge of propellant. Any form of charge cup (holding the proper amount of propellant) may be used, even those already described. However, a far more useful and versatile means is the graduated set of molded plastic charge cups made by Lee. More about them later.

A bullet-seating chamber is supplied with the tong tool and is screwed into the lower handle in the usual fashion. It is adjusted to seat the bullet to the proper depth when the handles are fully closed. To use it you first dump the propellant charge in the case, then start the bullet in the case mouth. The tool is then, inverted (to guard against the possibility of the bullet falling out and dumping the propellant) and the case is placed in the chamber bullet-up. Closing the handles completes the job, seating the bullet to the proper depth. Chambers are available in some calibers

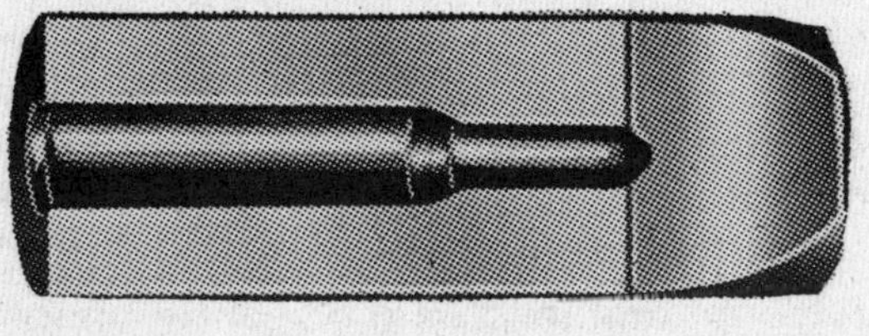

Above — Hand-type resizing dies are relatively inexpensive and can do the job quite nicely. This Lyman die is for a rifle cartridge and has been sectioned to show its inner surfaces.

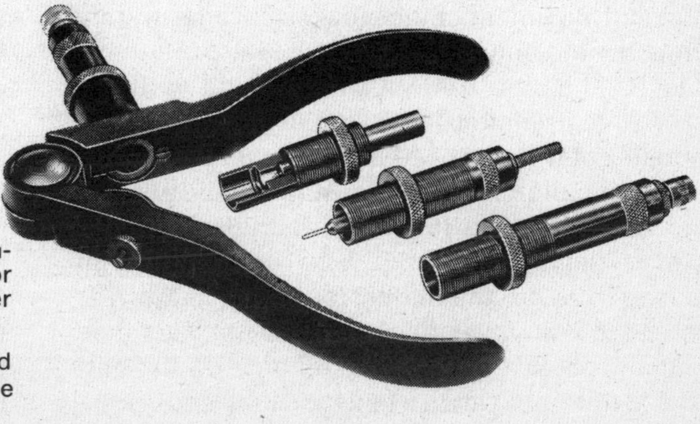

Right — One of the oldest designs in the hand-tool field is the Lyman 310 tong-type. It costs under $35 complete with a set of dies.

to crimp the case mouth as well.

A "bullet-sizing" chamber is also available as an option for the tong tool. It consists of nothing more than a threaded tubular section which fits into the lower handle with a passage at its bottom which is the diameter to which the bullet is to be sized. A separate push rod is supplied. In use, the chamber is threaded into the handles, a bullet dropped into the chamber, the push rod placed on top of it, and the handles closed. The rod pushes the bullet through the sizing portion of the chamber, ejects it, and it is ready to load. Opening the handles causes the rod to be lifted by means of a flange on its upper end upon which the extractor hook acts. I've always found it faster and more convenient, though, to simply roll the tool inverted as the handles are opened, allowing the rod to fall out.

Incidentally, operation of the tong tool in the bullet-sizing mode can be speeded up a good bit by clamping the lower handle in a vise and placing a container underneath to catch the sized bullets. Bullets can then be sized almost as fast as with a bench tool, operating the handle with one hand and feeding bullets in with the other. Most of the other loading operations can also be speeded up substantially if the lower handle of the tong tool is held in a vise or in some other manner clamped solidly to the bench top. Priming is especially speeded up if a small piece of steel angle is screwed to the right side of the lower handle (for a right hander) so the tool can be clamped solidly in a vise lying on its side.

Incidentally, should you encounter one of the oldest Lyman or Ideal tong tools which incorporate the bullet mold, you may try using the mold if you like, but don't expect too much from it in the way of bullet quality. In those tools, the mold blocks are integral with the handles (each handle/block being machined from a single casting) and alignment of the blocks depends entirely upon the tool hinge pin. The blocks are relatively small and have a habit of warping a bit with any extensive use; after they warp, it's impossible to obtain good bullets. Also, the handles quickly become quite hot during casting, and most uncomfortable, even if you're wearing the heavy gloves that you should. You might also run across one of the older tools not made for threaded individual dies or chambers. This type simply contains a large integral boss on the bottom handle into which a bullet-seating and crimping cavity has been machined. These tools were supplied with a separate decapping stem and pin which was dropped into this cavity, followed by the case for decapping; then the case was reprimed via the hole and punch in the handles, after which the case was charged with powder and the bullet seated and crimped without any sizing of the case.

Lee Loader

There have been numerous more recent handtool designs. The best known today is the "Lee Loader." In fact, this tool has become tremendously popular in the last two decades and has certainly been more widely sold than the Lyman Tong Tool. It consists of a heavy metal base into which various individual dies and accessories fit loosely, with the power for sizing, priming, bullet-seating, etc., being applied by punches and a mallet or hammer. At first, some people might think this method of operation harmful or even dangerous for priming, it really isn't. As mentioned before, seating new primers by lightly tapping the case down over them is a practice almost as old as handloading and perfectly safe unless one strikes the punch so hard as to deform the case head and thus crush the primer. The beauty of the basic Lee design is that the individual dies and parts of the set require no adjustment; and, further, that they are of very simple design for screw-machine manufacture and can thus be produced at very low cost. If in considering its price the Lee Loader can be said to have any real disadvantages, they are in its very slow operation and the noise of the hammer blows attendant to its use. Whereas most other conventional handloading tools operate in almost complete silence, the hammering necessary to use the Lee tool can hardly be muffled. The use of a rawhide or rubber mallet reduces the noise somewhat, but not entirely. Though any form of measure or scale can be used, Lee supplies a simple molded-plastic charge cup with which to regulate the amount of propellant used. One is best advised to obtain a complete set of the graduated-capacity Lee charge cups and then he'll be set for almost any propellant/size/caliber combination. There is also the advantage that Lee supplies a table listing the capacity (dipped properly and struck level) of each cup with virtually every propellant that is readily available. This offers a considerable advantage over a homemade cup which you'll

The Lee loader is, perhaps, the best known hand-type loading tool. It consists of: (1) priming chamber; (2) lock nut and bullet seater; (3) body and stop collar; (4) flaring tool; (5) decapping chamber; (6) decapper; (7) priming rod; and (8) powder measure.

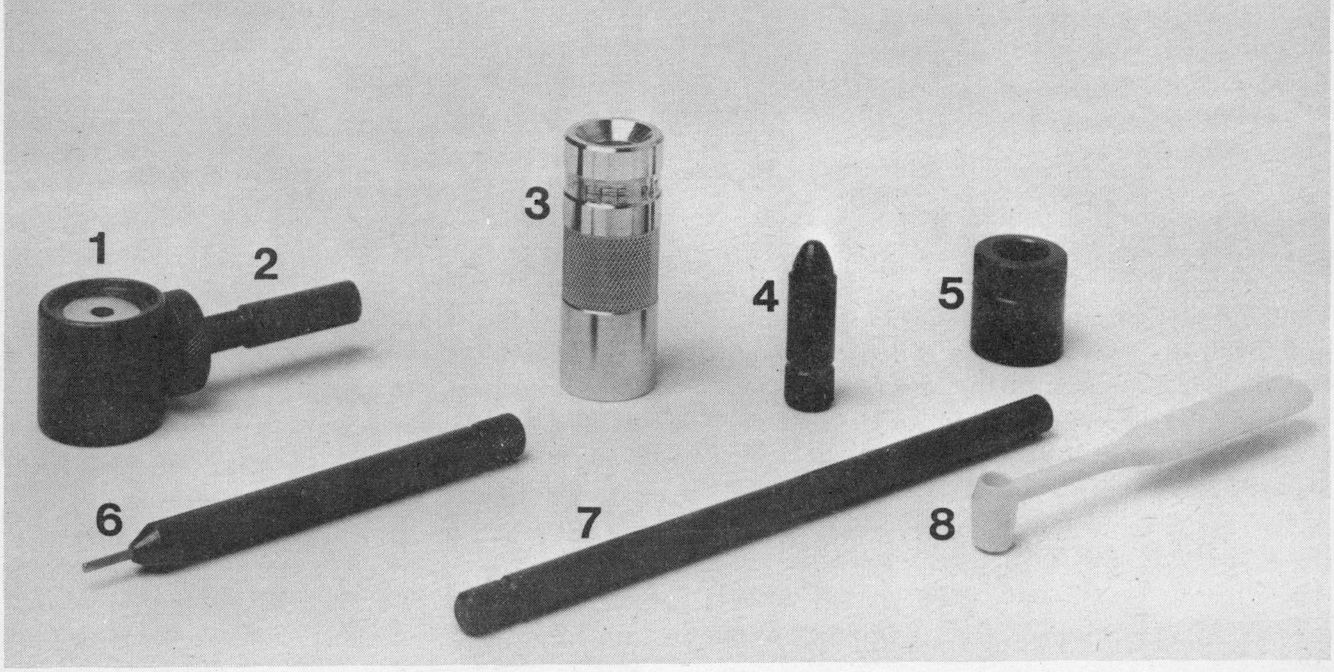

have to check against an accurate scale before knowing exactly what charge of Bullseye, Unique, 2400, etc., it will throw.

Lee also offers a broad line of incredibly simple and innovative handloading tools and accessories. For example, the deburring tool is nothing more than a simple hollow steel cone slotted across its peak, then squeezed into a slightly oval cross-section so that the sides of the slot are displaced to form cutting edges, both inside and outside. All of these accessories are intended for individual hand use and include primer pocket reamers, pocket cleaners, case trimmer, an especially handy priming tool, and numerous other items. Hardly any of these little tools follow traditional handloading design practices. Aside from simple and easy operation, their principal features are compactness, light weight, portability, and low cost.

Hopeful, would-be handloaders with little money to spend usually went the tong-tool route before the 1960's; but, since then, the Lee Loader seems to have become the favorite.

While both the Lyman Tong Tool and the Lee Loader are technically quite portable, they aren't really designed or packaged for *maximum* portablility. Of course, when the tong tool began its long history, it was considered portable enough to carry along in saddlebag, bedroll, or whatever so that ammunition could be loaded in camp. In fact, inclusion of the bullet mold in the original design made it virtually a one-piece tool (with the exception of the decapping stem) and thus reduced the probability of loss in the field. Technically the Lee outfit is quite portable, but its numerous parts and pieces (to which a hammer must be added) and the need for a solid working surface makes it somewhat impractical for use in the field.

Pak-Tool

On the other hand, we have the "Pak-Tool," especially designed and manufactured by Bill English for a high degree of portability and low probability of parts loss. The complete Pak-Tool comes packaged in a tough, tubular, plastic container, carrying tight-fitting end caps that, if not completely waterproof, certainly provide a high degree of protection from damage and from the elements. This carrying case is 2³/₁₆ inches in diameter and 10⅛ inches long and, will fit almost anywhere in one's duffle. The entire outfit

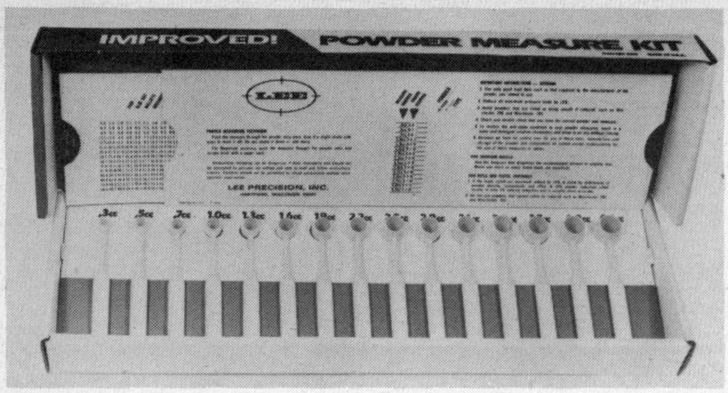

Lee also offers a complete powder-measure kit which consists of 15 different charge cups — the price is more than reasonable.

weighs a mere 1¾ pounds.

The Pak-Tool will perform all loading operations except full-length resizing. It consists of a tubular body, cut away for a part of its diameter and for nearly ⅔ of its length, with a removable shell holder threaded into the lower end. A short ram slides in the upper end, to which is pinned a second-class lever handle whose short end is secured to the body by two steel links. Raising and lowering the handle moves the ram vertically inside the body (Bill English calls it a carrier rod). The lower end of the ram carries female threads to accept a decapping rod, expander plug, primer-seating punch, or bullet-seating die. These are interchanged as necessary and are positioned longitudinally in the tool by different length carrier rods or carrier-rod extensions, depending upon the caliber. With the shell holder and any one of the dies in place and the handle closed, the complete tool measures only about 1¼ inches across and a hair over 9 inches long—extremely compact. In fact, there is ample room inside the plastic case to store extra dies and shell holders for a couple additional calibers.

Operation of the Pak-Tool is quite simple, both ends of the shell holder are slotted to accept the case rim. The longest carrier rod is pinned into the handle, and the decapping stem screwed into the rod. The handle is opened fully, retracting the decapping pin, and the case is slipped into the upper end of the shell holder inside the open portion of the body; the handle is closed, and the decapping pin pushes out the fired primer in the usual fashion. The decapping rod is replaced by the primer-seating punch, the handle is partially closed, and a fresh primer is placed in the hole in the shell holder; the case is then placed in the outer end of the shell holder, and the tool handle closed to seat the primer. Jam nuts are provided on the carrier rod to limit primer punch travel and regulate seating depth, but I find it better to "feel" the primer into position with the nuts backed off. The neck-sizing die is then screwed into the carrier rod, the handle opened all the way, and case placed in the inner end of the shell holder. Closing the handle forces the die down over the case and resizes its neck; jam nuts are provided to limit die travel as desired. The expander plug is installed next, the handle opened, the case placed in the inner end of the shell holder, and the handle closed to force the plug into the case mouth. Plug travel is adjusted by jam nuts on the carrier rod. Bullet-seating is carried out the same way, by first screwing the seating die into the carrier rod, then placing the charged case in the shell holder and closing the handle. A separate seating stem is threaded inside the seating die and is adjustable for seating depth. It is a minor inconvenience that the die must be removed from the carrier rod to permit this adjustment to be made; the slotted end of the seating screw is enclosed by the carrier rod when the die is in position. However, once the adjustment is made, it is locked with the jam nut, and no further problems will be encountered. Again, though, die travel should be adjusted by the jam nuts on the carrier rod.

Because of its configuration, the Pak-Tool can be held and operated in almost any position except for seating primers; for that operation it needs to be held with the shell holder up, and in this position it can be operated with one hand inasmuch as only a very small amount of handle travel is necessary. So long as bullets are started firmly in the case, that operation may be performed with the tool in any position; but, if there's any chance of the bullet falling out, the tool should be held shell-holder down so that the cartridge remains upright and the bullet remains in place until it is seated.

Of the three hand-type tools just described, I find the Lyman Tong Tool the most convenient for use at home. It can certainly be used on the range or in the field, and with careful manipulation, it will turn out first-class ammunition. On the other hand, though, I lean toward the Pak-Tool for use in the field, probably more because of its compactness and carrying case than any mechanical superiority it might have over the other two. I suppose it should be pointed out at this time that while many people buy hand-type loading tools with the idea of using them on the range or on hunting trips, they very seldom actually do so. Loading afield may seem attractive while one is reading about it, and it was certainly important to market hunters and buffalo hunters in the 19th century. However, today's cartridges are relatively light and compact, and it is usually far simpler to carry a sufficient supply of ammo along rather than a tool with primers, propellant and bullets. If one were to be away from civilization and sources of supply for a long time it would probably be simpler to pack in components for 1,000 rounds or so than it would be to carry that much loaded ammunition. However, except for my number two son who wants to hide away in the northern Canadian wilderness, damn few of us are ever likely to find ourselves in that position.

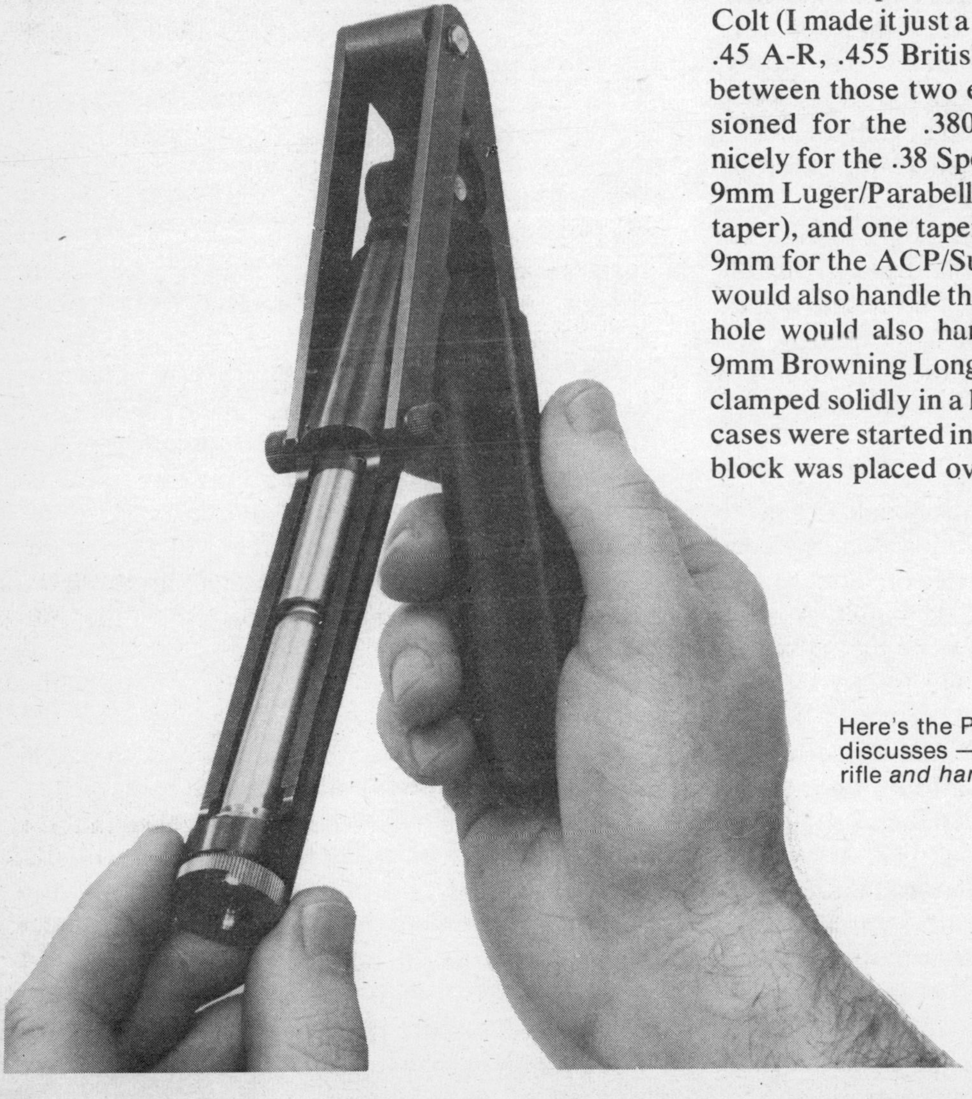

Here's the Pak-Tool the author discusses — it's available for both rifle *and handgun* calibers.

IMPROVISED TOOLS

Resizing Die

We shouldn't leave the subject of this chapter without some discussion of improvised tools. Take full-length resizing, for example, an operation which we normally associate with a loading press and bench (or at least a heavy-duty vise) and a store-bought die. Well, a resizing die is nothing but a smoothly-polished hole in a piece of metal, and if the need is strong enough, we can usually produce that. Just to prove it could be done, I once drilled a row of different-sized holes in a bar of steel, then reamed and polished each of them to the proper diameter for resizing a different size cartridge case. The smallest was for the .32 S&W Long and could also be used to resize the .32 S&W and .32 ACP, not to mention the .32 Colt New Police (which is nothing more than the .32 S&W Long). The largest hole was for the .45 ACP, and it performed equally as well on the .45 Colt (I made it just a tad oversize for this reason), .45 A-R, .455 British and the .450 Revolver. In between those two extremes were holes dimensioned for the .380 ACP (which also served nicely for the .38 Special and .357 Magnum), the 9mm Luger/Parabellum (this one required a bit of taper), and one tapered not quite so much as the 9mm for the ACP/Super. Either of the latter two would also handle the .38 S&W, and the .38 ACP hole would also handle the Bayard, Steyr and 9mm Browning Long cases. For use, this bar was clamped solidly in a heavy bench vise, lubricated cases were started in the proper hole, a hardwood block was placed over the case head and struck with a hammer to drive the case home. Rimless cases could be driven in without any problem until their heads were flush with the bar, but those with rims required a bit of care, else the rims might be distorted against the bar. Cases were removed from their respective resizing holes with appropriate-diameter lengths of steel rod used as knockout punches. This could be simplified a great deal with the two truly rimless/straight cases in the lot, the .380 and .45. These two numbers could be started *head-first* in their respective holes, then driven completely through via the knockout rod to fall free, completely resized.

All the foregoing paragraph tells us is that in an emergency, or in the absence of any supply of tools, any reasonably round, highly-polished hole of the right diameter in a piece of metal may be used for resizing cases full-length. As a matter of fact, for moderate use, the hole need not be in steel. When I obtained my first .401 Eimer Special revolver conversion, no dies whatever were available. New cases were loaded without tools, but after firing, the mouths were too large to grip fresh bullets. I drilled a ⅜-inch hole in a piece of brass bar stock, then polished it out carefully until it would squeeze those .401 case mouths down just enough to produce a tight bullet/case assembly. I only had to use that brass sizing die for a couple hundred rounds, and it still will do the job. It was made up over 25 years ago, and unlike many other gadgets of those days, is still with me.

Bullet Seating Die

A bullet-seating die is also just a hole in a piece of metal, large enough to barely accept the case mouth, and fitted with a plunger of some sort to exert pressure on the nose of the bullet. A piece of ⅜-inch inside-diameter pipe or tubing with a piece of dowel pinned or glued inside will seat .38/.357 bullets smoothly and correctly. Back in the old days, when tools were scarce and hard to come by, straight-line seating tools were sometimes made out of the chamber end of a shot-out rifle barrel. I own one of these (picked up somewhere in a trade) made from the rear 4 inches of a Krag rifle barrel and fitted with a threaded cap over the breech end and a plunger sliding in the bore from which the rifling has been reamed. The same could be done with an autoloader barrel, the charged case being placed in the chamber, the bullet dropped down the bore, and a plunger and mallet used to seat the proper depth. A revolver cylinder of the proper caliber could be used the same way. With the ratchet and/or extractor removed, it could be set down over a charged case, a bullet dropped into the chamber throat, and a short length of dowel used to force the bullet into the case.

Some reloading tools can be easily improvised. A length of pipe of appropriate diameter (properly belled) can be used for crimping.

Expander Die

Expanding case mouths to the proper diameter is easily done without store-bought tools. File a U-shaped notch in a piece of flat steel so that the case head just forward of the rim or the extractor groove fits snugly in it then screw this piece of steel to the bench top with a spacer slightly thicker than the rim beneath it so that cases may be slipped easily into place. Then polish a piece of steel rod down to the proper diameter, giving it a slight taper at the lower end as well. Slip a case under the holder, tap the rod into the mouth, and pull it out. If removing the rod is difficult, place a block of wood about an inch thicker than the case is long nearby, and use pliers clamped on the rod as a lever, with the wood block as a fulcrum. Opposing notches filed on the rod to accept the snout of the pliers will simplify the matter.

Priming Seaters

We've already described decapping and the necessary filed-down nail or punch, and recapping is nearly as simple. However, for more rapid and convenient seating of primers than allowed with the punch and plate method, make a tool out of a pair of cheap (or old) pliers. Cut one jaw off and then silver-solder or braze in place a metal block containing a hole just large enough to admit the case body. This alone will serve to support rimmed cases, and for rimless cases, take a piece of 1/32-inch thick steel and file a notch into which the extractor groove of the case will fit snugly. Slip the case in the notch and then into the hole, and the notched plate will support it solidly. Making certain it will be centered over the primer pocket of the case in the lower jaw, drill a small hole in the upper jaw and press or solder into it a short length of round rod of approximately primer diameter. File this "seating punch" as short as possible but long enough to force a primer fully home when the handles of the pliers are closed completely. An even simpler—and very crude in appearance but effective—priming tool can be made by drilling a hole in one leaf of a strap hinge and locating the seating punch centered over it in the other limb. In doing this, it's better to use a rather small hinge and attach wood handles to facilitate operation; a large hinge can be operated without adding handles, but it is a clumsy, heavy rig at best.

Crimping Die

When a proper crimping or crimp/seat die is not available, things become a bit more difficult, In the unlikely event that a lathe of some sort is available, a satisfactory die for producing a roll crimp can be made from almost any metal by the use of boring bar. On the other hand, a taper crimp die isn't all that hard to make. Or, you might be able in some instances to turn out a satisfactory crimp by running the case and seated bullet only a very short distance into the mouth of a resizing die of the same caliber. The mouth of the die is radiused and very slightly tapered to facilitate starting of the expanded, oversized, fired case. Once the bullet is seated in a resized case, this oversized portion of the sizing die will turn the case mouth in snugly on the bullet if the two are brought together with just the right amount of pressure. It takes only a little practice to develop the feel that will tell you that the case mouth has been turned into the bullet without excessive deformation. But, back to the taper-crimp die. A hole in the metal block of slightly less than case diameter can be reamed (if a ordinary taper reamer is available) or polished to sufficient taper that forcing the case into it a short distance will squeeze the mouth inward just as in a commercial taper-crimp die.

If you must have a fairly heavy crimp and no die is available, the use of a small screwdriver bit or blunt punch as described in Chapter 16 on ultra-light loads will work reasonably well if applied vigorously. Unfortunately, driving a punch or screwdriver into the case deeply enough to secure the bullet against heavy recoil is likely to cause the case to bulge and distort around the stab crimp produced. This can make chambering difficult, or in some instances can even prevent it. To avoid this, lay the case with

Need a cannelure? The cannelure on the case of this .38 Special was applied with the *dull* knife shown — just find a flat surface and roll the edge of blade on the body of the case.

seated bullet down on a smooth flat surface and then place the *dulled* edge of a long knife blade at right angles to the case over one of the lubricating grooves in the bullet; press down moderately and *roll* the case under the knife blade. With a little practice, you'll be able to apply just enough pressure to form a cannelure in the case which enters the lubricating groove and thus secures the bullet against fore-and-aft movement. The deeper the cannelure, the more secure the bullet, but if it's made too deep, the case will be seriously weakened for any further use. In addition, if the blade is too sharp and the cannelure made too deep, the case mouth forward of it may tear off and go down the barrel with the bullet.

Doubtless other improvisations will occur if you delve more deeply into this area of handloading. To a large extent, they depend upon your mechanical ability and the tools and the materials that happen to be available.

DIE INTERCHANGEABILITY

One aspect of improvisation opens all sorts of other possibilities. It doesn't involve doing without dies in general, but rather with using a die or dies of one caliber to load cases of another—and in doing so in such a manner that the ammunition produced will perform both correctly and safely in properly-chambered guns. We call this simply "die interchangeability" and it is a subject little noted except among the most advanced handloaders. Interchangeability is not something to which we resort only when we don't have the proper dies—actually, it is a characteristic of loading dies that we can and should exploit regularly, even to the extent of making our initial purchases with interchangeability in mind. It is rather well known that dies in .38 Special caliber may be used to load .357 Magnum and that the same relationship exists between .44 Special and .44 Magnum; beyond that, though, most hull-stuffers are unaware of the tremendous amount of interchangeability that can be accomplished among many calibers.

Resizing dies are the most critical to interchangeability and one should begin by purchasing good quality dies. The better dies have closer dimensional control, and their material, finish and heat treatment are better suited to the extra loads some interchangeability problems will place upon them. A relatively soft and rough .38 ACP die may not work well with the slightly larger base of the .38 S&W case, while a first-class die will handle the job quite well. In the reverse, a cheaply-made .38 S&W die may be too much oversize to reduce the .38 ACP adequately, even though it works well enough in its intended caliber.

Resizing dies are generally made of a good grade of tool steel with the resizing cavity drilled, reamed, and polished to shape and finish, then heat-treated or hardened, and again polished to the final finish. At least one manufacturer produces the cavities differently, using instead a closely-controlled boring process rather than reaming. Most important of all are the shape and dimensions of the resizing cavity which must reduce fired cases sufficiently to insure that they will enter freely in any industry-standard chamber, and yet at the same time not reduce the case *too* much so that it is excessively worked to shorten its reloading life. Almost equally important to the die's function is the quality of the finish inside the cavity, which must be smooth enough to avoid marring the case surface, and yet not subject it to too many expensive smoothing operations which would raise the cost. Final polishing of most resizing dies is done by hand, and all should know by now that skilled individual handwork is the highest of all labor costs.

With all this in mind, and with today's dies running as high as $45 for a four-die pistol-caliber set, it behooves one to select them carefully—not only for a reasonable balance of quality and price, but for a design and type that best suits one's particular needs, and for maximum utility.

Restricting our discussion for the time to resizing dies, let's take those factors in order. Quality is something that the majority of handloaders simply can't evaluate except from the appearance and function of the die. If it shows good workmanship throughout, both inside and outside, then it's probably a good die. If it resizes a clean and properly lubricated (meaning sparingly) fired case with only modest effort on the tool handle and produces a smoothly-burnished, unmarred finish on that case, then the surface finish and hardness inside the cavity are at least adequate. If the resized case will pass muster in a case gauge (either the Wilson or micrometer type), chamber freely in guns of the proper caliber, and not show excessive expansion after further firing, then the dimensions of the sizing cavity are likely correct if not necessarily ideal.

Below—The tungsten carbide sizing die is made to last a lifetime — the author once sized over a *million* cases using one! Right! You can see the T/C button insert (arrow). Admittedly, T/C dies cost more, but they last forever.

Above — If you need to reload .357 ammo and don't have a set of .357 dies, you can do the job on dies made for .38 Special. The same relationship exists between .44 Special and .44 Magnum.

It can be unnecessarily costly to buy dies of a type and design not needed for either the ammunition you wish to assemble or the use to which the ammunition will be put. For example, if you'll be loading a particular pistol caliber only occasionally and in relatively small quantities, it seems rather foolish to spend an extra $15-25 just for the more efficient, longer-lasting, and technically superior carbide-insert resizing die. Likewise, the extra $8 or so for a four-die pistol set with its separate crimping die is hardly worthwhile unless your requirements are for the very highest accuracy with lead bullets. Much the same can be said about some rifle calibers in that, unless you actually require maximum accuracy, there is no need to purchase the more costly and complex bullet-seating dies containing the spring-loaded alignment sleeve. On the other hand, if you do legitimately require maximum accuracy, it would be counter-productive and false economy to buy the simplest basic die set just to save a few dollars. You won't be satisfied and will eventually have to buy the better dies anyway; so, you'll be better off and less out of pocket in the end to buy what you *really* need to start.

Achieving maximum utility from your dies means employing varying degrees of interchangeability between calibers. With handgun cartridges in particular, much more interchangeability exists than the average handloader realizes. Surely most pistoleros recognize the fact that .38 Special dies can be simply readjusted to handle the .357 Magnum perfectly; and the same situation exists in regard to the .44 Special and .44 Magnum. However, the reverse is not true, and in order to take advantage of this interchangeability among cases of the same diameter but different lengths, one must always purchase his die set for the *shortest* of those cartridges. Then, it will be possible to adjust dies higher in the press and accommodate the cases of greater length. It *is* possible to resize and expand all lengths of a given basic case in the same die; length only becomes a problem in crimping the case mouth upon the bullet. Ideally, then, to be able to load the maximum number of calibers in the .38/.357 range, one could purchase a seat/crimp die in .38 Short Colt which would have the additional capability of being adjusted for the .38 Long Colt, .38 Special and .357 Magnum. It would also serve to seat bullets in .38 ACP, 9mm Bergmann-Bayard, 9mm Steyr, and perhaps les-

Recently, Redding introduced a "Titanium Carbide" die that's much smoother than the tungsten variety and requires less force to size and extract cases.

ser known cartridges in the 9mm/.38 class.

However, interchangeability goes farther than that. Cases which are not entirely identical in dimensions can often be resized in the same die. For example, the 9mm Parabellum, .38 ACP, and .38 S&W are all slightly tapered, ranging in length from .725-inch to .900-inch. Their normal bullet diameters cover a spread of only about .005-inch, and their base diameters don't vary much more. Usually, a resizing die in .38 ACP or .38 S&W caliber will handle the other two cases adequately. A resizing die in any of these three calibers will usually also partially resize .38 Special and .357 Magnum to the point that they will rechamber freely and hold a bullet tightly. If the .38/.357 cases are overly expanded near the case mouth, this may not work out, but generally it will. In this same area, while there are minor dimensional differences in the case of the .45 ACP, .45 Auto-Rim, .45 Colt and .455 Webley, a resizing die in any one of those calibers will usually function satisfactorily for the others.

The foregoing are just examples of *practical* die interchangeability that exists. There are many, many other combinations, and we have prepared the accompanying table which in and of itself is by no means complete, simply because if we listed all of the possible combinations, it would occupy far too much space. However, in viewing this table, the additional interchangeabilty possiblities will be readily apparent. If you want to investigate the matter more fully, and this would be particularly applicable if you shoot a variety of obsolete handguns, then your best bet is to go to a source of cartridge dimensions. The best source I have is *Pistol & Revolver Cartridges* by White & Munhall, followed by *Cartridges of the World* by Frank Barnes. If you suddenly feel the urge to load some additional caliber, check out its case dimensions and you may very well find that dies you already own will do the job satisfactorily. We have included a case capacity table and case dimensional drawings of the more well-known cartridges in the appendices that follow.

One additional note on resizing dies: Most ardent handloaders are familiar with the so-called "carbide die," which is in reality an oversize steel die body containing a tungsten carbide ring cemented or soldered in its mouth—the ring is dimensioned to reduce the case body to the proper diameter. Quite a few manufacturers supply this type of die with the carbide ring shaped to resize *straight* cases, and a few supply it with a longer insert somewhat tapered in bore to handle cases such as the 9mm Parabellum and .38 Super with their tapered bodies. Naturally, tungsten-carbide being a very costly material and being very difficult to shape, the price is considerably above that of ordinary all-steel dies. The advantages of the carbide die are well known—resistance to wear and abrasion. Other advantages are its burnishing effect upon the case, its resistance to scoring and virtual refusal to accumulate grit, and its ability to accept cases which are not thoroughly clean and would otherwise scratch or score a conventional resizing die. I once used a single die of this type for over *one million cases*.

Now, Redding-Hunter has introduced a new variation of the carbide die, substituting "titanium carbide" for the tungsten carbide almost universally used until now. Construction of this new die is the same, with a titanium carbide ring installed in the mouth of an oversize body. If you ask, "Why *another* carbide die?" The answer is to be found in the crystalline structure of titanium carbide. Under high magnification, the grain structure of tungsten carbide shows up as more-or-less sharp-edged, angular grains resembling chunks of coal such as we used to see in the basement or in the backyard bin. Thus, no matter how smoothly tungsten carbide is polished, its surface still features a multitude of tiny sharp edges.

Since tungsten carbide's principal use is in cutting tools, this is quite proper. However, this makes it less than ideal for the working surface of a resizing die, even though it possesses many advantages over the conventional all-steel die. On the other hand, titanium carbide is composed of irregular rounded grains, similar in shape to water-worn pebbles one might find in the bed of a fast-running creek. Thus, when the working surface of a titanium carbide sizing die is smoothly polished, it does not possess the millions of tiny teeth just described. Even though some sharp edges result, the surface is measurably (admittedly with expensive instrumentation) smoother and less abrasive than that of tungsten carbide. The result is simply a working surface with all the advantages of tungsten carbide, but even smoother and slippier, and thus requires less force (translated into muscle power) to resize and extract a case.

If you already own a good tungsten carbide die, I doubt it would be worthwhile to replace it with titanium carbide; on the other hand, if you are planning to buy a die of this type, the increased cost of titanium carbide should be well worth while. At the moment, a Redding-Hunter three-die pistol set containing the titanium-carbide resizing die sells for a bit under $45; that's $6-8 more than a corresponding set utilizing tungsten carbide.

Die Interchangeability Table

Die Caliber	Other Calibers
.32 S&W[1]	.32 ACP, .32 S&W Long, .32 Colt New Police, 7.65mm French Long
.32 ACP[1]	.32 S&W[a], .32 S&W Long, .32 Colt New Police, 7.65mm French Long
7.65mm French Long	.32 ACP[a], .32 S&W Long, .32 Colt New Police
.380 ACP	.38 Short Colt, .38 Long Colt, .38 Special, .357 Magnum, .38 AMU
.38 Short Colt	.380 ACP[a], .38 Long Colt, .38 Special, .357 Magnum, .38 AMU
.38 ACP[2]	.38 Super Auto, .38 S&W[a,b], .38/200, 9mm Parabellum[a,b], 9mm Bergmann-Bayard (9mm/38), 9mm Steyr, 9mm Mauser, 9mm Ultra[a,b], 9mm Browning Long.
.38 S&W[2]	.38-200, .38 ACP[c], .38 Colt New Police, .38 Revolver, 9mm Japanese Revolver Type 26, 9mm Parabellum, 9mm Makarov[b], and as for .38 ACP
.38 Special	.38 Long Colt[a], .38 Short Colt[a], .38 AMU, .380 ACP[a], .357 Magnum
9mm Browning Long[4,2]	.38 ACP, 9mm Parabellum, 9mm Bergmann-Bayard, 9mm Steyr, 9mm Mauser, .38 S&W[a,b], .38/200[a], 9mm Ultra[a]
9mm Parabellum[4,2]	.38 S&W[a], .38/200[a], .38 ACP[c], 9mm Ultra[a], and as for .38 ACP[b]
.44 Russian[3]	.44 Special, .44 Magnum
.44 Special[3]	.44 Russian[a], .44 Magnum
.44 Magnum[3]	.44 Russian[a], .44 Special[a]
.45 ACP	.45 Auto-Rim, .45 Colt[b], .455 Colt, .45 S&W, .450 Revolver, .455 Webley/Enfields, .476 Enfield, .455 Webley Auto.
.45 Colt	.45 ACP[a,b], .45 Auto-Rim[a], .455 Colt[a], .45 S&W[a], .450 Revolver[a], .476 Enfield, .455 Webley/Enfield[a]

Quite a few possible combinations are not listed for lack of space. The above will prompt others, as will a study of case dimensions in our appendix.

Notes:

[1] May also be used to neck-size .32-20 and .30 Carbine.
[2] May be used to neck-size and partially body size .38 Special, .357 Magnum, etc.
[3] May be sometimes used to neck-size .44-40 enough to hold bullet.
[4] May be used to seat bullets in .38/.357.
[a] Case too short to be crimped in seat die.
[b] Case *may* be too large for die at head.
[c] Die may not reduce head quite enough.

APPENDICES

Handloader's Glossary

Accuracy—The measure of a gun/ammunition combinations's capability to place its bullets close together on target.

Action—The basic mechanism of any firearm which produces all the mechanical movements necessary for functioning.

Anneal—Softening the mouth area of a cartridge case by application of heat followed quickly by quenching in cool water.

Anvil—In a primer, the conical surface against which the firing pin crushes and ignites the priming compound; may be located in the primer proper (Boxer) or the case head (Berdan) or in a separate cup as in shotshell primers.

Back Thrust—See *Bolt (Back) Thrust*.

Ball—Earliest term applied to a bullet, meaning then and now a lead sphere in that context; also means a full metal jacketed bullet or a cartridge so loaded.

Ball Powder—A small arms propellant developed by Western. It rose to prominence during WWII because of the rapidity with which it could be produced. Kernels are either spherical or spheres flattened to some degree by passing between rollers.

Ballistics—The study of projectiles in motion; *internal* ballistics deals with the bullet while in the barrel; *external* ballistics deals with the bullet's flight through the air. Terminal ballistics applies when the bullet enters the target.

Ballistic Coefficient—Ratio of the bullet's sectional density to its coefficient of form. Indicates bullet's ability to retain its velocity during flight.

Balloon Head—The earliest form of "solid head" case construction wherein the web is too thin to accommodate the primer pocket, so a forward bulge is formed to do so. See also *Solid Head*.

Barrel-Cylinder Gap—In a revolver the clearance between barrel and cylinder; necessary to permit free rotation of cylinder and typically about .006-inch in a new gun.

Battery Cup—The metal cup containing the cap and anvil of a shotshell primer in the proper relationship; the name given to typical American shotshell primers.

Bearing Surface—That cylindrical portion of a bullet behind the curved nose which contacts the bore surface. In soft bullets, this surface may lengthen during acceleration as the nose sets back.

Bell—Flaring or funneling of a case mouth to permit easy starting of a new bullet.

Berdan—Named after U.S. Army Col. Berdan; a type of primer depending upon an anvil formed integrally with the case. The primer proper is simply an open cup filled with priming compound which seats over the anvil.

Bench Rest—A shooting bench or table, very solid, to support a gun to allow achievement of maximum accuracy.

Bevel Base—A bullet shape incorporating an angled or beveled outer edge on the base for easy seating.

Bleed—In swaging bullets, a secondary extrusion through a "bleed" hole in the die of excess lead or core material.

Body—That part of a cylindrical cartridge case that lies between the head and mouth; in a bottle-neck case, the area between head and shoulder.

Bolt (Back) Thrust—The actual pressure applied to the slide of an autoloader or recoil shield of a revolver through the case head by firing.

Bore—The interior of a barrel which guides the bullet and contains the propellant gases so they can act against the bullet base. Bore diameter is the measure of the hole before the rifling grooves are cut or formed.

Boxer—Named after British Col. Boxer, the standard U.S. small arms ammunition primer type which contains its own conical anvil pressed into the mouth of the cup. Now slowly displacing the Berdan primer in the rest of the world because it simplifies case manufacture.

Brass—A term often applied to cartridge cases; a particular copper/zinc alloy known in the trade as "cartridge brass" from which cases are made.

Brisance—The shattering power of an explosion; usually applied only to primers as an indication of their quickness or violence of function.

Bullet—The projectile fired by a firearm; the projectile only, not the complete cartridge.

Bullet Pull—The force necessary to extract a bullet from the cartridge case; a measure of crimp efficiency and of resistance offered to propellant ignition and initial combustion.

Bullet Puller—Any device used to extract a bullet from a loaded cartridge.

Cake Cutter—A tubular tool pressed over a bullet to strip off excess lubricant and to free it from a solidified cake of lubricant.

Caliber—Principally the bore (or land-to-land) diameter of a barrel; *not* actual bullet diameter in modern usage. Also used to refer to the actual designation of a cartridge such as, "caliber .308 Winchester." Also used as a unit measure to describe bullet nose shape, i.e. "4-caliber ogive" or seating depth, "1-caliber seating depth," and in big-gun terms, barrel length in units of bore diameter. A 50-caliber 6-inch gun has a barrel 50 X 6″ or 300 inches (25 feet) long.

Cannelure—A groove around a bullet or case; may be merely for identification, to support the bullet, to receive a crimp, or to contain lubricant.

Cannister Powder/Propellant—Powder blended to standard performance and sold to handloaders. Standard characteristics are necessary since handloaders depend upon published data and cannot test individual lots and batches like ammunition manufacturers do.

Cap—A percussion cap for use with cap-and-ball guns; also the cup and priming-compound pellet of a conventional shotshell primer. Also sometimes used in place of the word "primer."

Cartridge—The complete assembly for firing one shot; the case with primer, propellant and bullet assembled to it. Technically correct for shotshells but usually applied only to rifle and pistol ammunition in this country.

Case—For any cartridge, the metal, paper, or plastic (or combination thereof) container to which all the other components are assembled. Sometimes called "hull," "shell" or "brass."

Case Forming—The reshaping of one cartridge case to fit a different chamber.

Case Trimming—Shortening a case to standard length, or to a new length as part of case forming.

Cast Bullet—Normally a one-piece bullet formed by filling a cavity of the desired shape and size with molten lead alloy which is then removed after it solidifies or "freezes."

Center Fire—Central location of the primer in the head of a cartridge case and essential for practical reloading.

Chamber—That part of the rear of the barrel (or cylinder in a revolver) enlarged to accept the cartridge.

Chamber Cast—The form of a chamber made by filling the chamber with low-temperature alloy or other hardening liquid and then removing it for examination and/or measurement.

Chamber Throat—In a revolver the bullet-diameter extension of the chamber in the cylinder; in a single-shot or auto, that enlarged part of the bore ahead of the chamber to accept the bullet and align it with the rifling.

Chamfer—A small narrow bevel at the case mouth or primer pocket to permit easy entry of the new component.

Charge—The amount, by weight, of propellant (and in a shotshell also the shot) placed in the case; also the act of placing propellant in the case.

Charge Cup—A propellant measuring device used to charge empty, primed cases; usually found in the form of a scoop or cup, designed for a specific amount of a specific propellant.

Chronograph—An electronic instrument for measuring bullet or shot velocity. The bullet breaks a start-and-stop circuit by passing through screens or over light sensors so that the time it takes the bullet to pass is measured and this value is converted to velocity.

Closed Base—A form of bullet jacket where the nose is interrupted and the core is inserted through the open base.

Components—Those items assembled to form a loaded cartridge, case, primer, propellant, bullet; and in a shotshell, wad and shot.

Compressed Charge—A propellant charge bulking greater than the space available for it in the case and therefore compressed by the seating of the bullet.

Copper Crusher—The method of chamber or pressure measurement most commonly used when a soft copper cylinder is deformed by gas pressure acting against it; also the copper cylinder used.

Cordite—An obsolete form of double-base, smokeless, nitrocellulose propellant, made and used in long extruded strings or "cords" and widely used in the British Empire up into the 1950's.

Corrosion—The eating away of metal by rust which may be promoted in a barrel by salts in primer or propellant residue.

Corrosive Primer—A primer containing chlorates which deposit a salt-bearing hygroscopic residue in the bore; the salts attract moisture which, in turn, promotes corrosion if not removed by thorough cleaning.

Crimp—Turning or compression of the case against the bullet or primer to produce a secure assembly.

Crimped Primer—In military ammunition, a primer normally secured in the case by a ring or stab crimp. Crimp must be removed to allow repriming.

Deburr—To scrape or cut away burrs and sharp edges on the case mouth.

Decap—To punch the fired primer out of its pocket in the case head by means of a pin passing through the flash hole. May also be done by other methods in the instance of Berdan primers.

Deterrent—A chemical agent applied in, or as a coating to, propellant granules to slow or control burning rate.

Double-Base Powder/Propellant—A propellant with nitroglycerine added to its basic nitrocellulose.

Drop Tube—The tube (usually clear plastic) on a powder measure which guides discharged powder from the metering chamber into the cartridge case.

Duplex Load—A confusing term meaning either the use of

more than one bullet in a cartridge or the loading of more than one propellant in the case to achieve maximum velocities. The latter is considered quite hazardous and *should not* be attempted.

Energy—The amount of work, measured in foot-pounds, done by the bullet. Can be calculated by squaring the velocity in fps, then multiplying that value by bullet weight in grains and, finally dividing by 450,240. Also, see *Muzzle Energy*.

Engraving—The impression of rifling lands into the bullet, forming grooves in its bearing surface.

Erosion—The wearing and melting away of the bore surface by friction and high-temperature propellant gases; normally occurs mainly just ahead of the chamber.

Expander—A rod or ball pressed into and out of the case neck or mouth to produce the desired inside diameter irrespective of wall thickness or sizing die diameter.

Expansion Ratio—The ratio of case volume to base volume.

Extruded Primer—A primer whose cup has flowed back into the firing pin hole during firing; sometimes called *cratering*.

Feed Ramp—In an autoloader, the surfaces behind and below the chamber to guide cartridges into the chamber.

Feed Ramp Bulge—Bulging of an unsupported segment of the rear of the case where a portion of the chamber wall is cut away to form the feed ramp.

Fire Form—To fire a cartridge in a chamber whose forward portion is larger than the case; gas pressure expands the case to fill the chamber, permanently altering its shape.

Flash Hole—In Boxer cases, a single central hole from primer into the interior to allow the primer flame to reach and ignite the propellant. More than one off-center holes are common to Berdan cases because of the central integral anvil.

Folded Head—A long-obsolete form of case construction where the rim and primer pocket are formed by folding the thin walls back upon themselves; a weak type not suitable for smokeless propellants or high pressure.

Foot-Pound—A measure of kinetic energy used to indicate the "power" of a bullet in flight; the theoretical effort required to raise 1-pound 1-foot vertically.

Forcing Cone—In a revolver the slightly funneled rear of the barrel which permits the bullet to enter more freely when chamber alignment is not precise.

Freebore—The smooth (unrifled) portion of the barrel ahead of the chamber through which the bullet must pass before engraving takes place.

Full Metal Jacket (FMJ)—See *Metal Case*.

Galling—Rubbing off of brass or nickel particles from a case into the surface of a resizing die; caused by excessive friction, usually due to lack of lubricant or an unsuitable lubricant.

Grain, Avoirdupois—The unit of measure used in this country for ammunition components; 7000 grains in 1-pound, 437.5 grains in 1-ounce.

Grease Groove—A shallow recess around a bullet, deeper than the rifling, to contain lubricant.

Groove—See *Cannelure*.

Group—The pattern formed on target by a series of consecutive shots; group diameter is a practical standard measure of accuracy, though "mean radius" is technically more meaningful.

Half Jacket—A short jacket extending only over the bullet bearing surface, and then not quite to the front of that surface; also a bullet with such a jacket.

Handloading—Technically the assembly of other components to a new case, but normally used synonymously with "reloading," (see same).

Hangfire—The circumstance where there is a perceptible delay between the firing pin strike and the report of firing.

Headspace—The distance between the chamber surface which arrests forward movement and the surface which supports the case head at the instant of firing. In revolvers this is the distance between the rear of the cylinder (where the case rim rests) and the recoil shield; in rimless-caliber autos, the distance from the step or shoulder at the front of the chamber to the slide breech face when all recoiling parts are fully in battery.

Heel—Edge of the bullet base.

Heel-Type Bullet—A bullet whose major diameter is the same as the case and with a reduced-diameter section at its base to fit inside the case mouth.

Hollow-Point—A lead bullet containing a hole in its nose to promote expansion. "Jacketed Hollow Point" (JHP) when of jacketed construction.

Ignition—Beginning the burning of the propellant by passage of the primer flame into it via the flash hole or—in the case of rimfires—the rim.

IMR—"Improved Military Rifle" powder series made by Du Pont, some of which are of use in loading certain handgun cartridges.

Inside Lubricated—An obsolete center fire cartridge/bullet design; the bullet is of the same diameter as the case and lubricant is applied over the exposed part of the bullet; bullet is normally of heel and/or hollow base type.

IPSC—International Practical Shooting Confederation.

Jacket—An envelope of hard metal covering a bullet's lead core; normally of copper or copper alloy, though steel and other metals are often used in other countries.

Jump—The upward lift of a handgun muzzle brought about by recoil and the manner in which the gun is held for firing; depending on numerous factors, this can *partially* occur *before* the bullet leaves the barrel.

Keyhole—When a bullet is unstable in flight and travels broadside rather than point-on; named from the appearance of the elongated bullet hole in the target.

Knurling—A pattern of cross-hatched lines or grooves rolled into the bearing surface of a bullet to hold lubricant; of only recent application to handgun bullets for this purpose, but long used in other areas.

Lands—The spiral ribs left standing in the bore when the rifling grooves are cut; the lands bite into the bullet and give it the spin necessary for stability.

Leade (Leed)—See *Freebore*

Leading—A deposit of lead on the bore surface, rubbed off the bullet during its passage.

Line (Angle) of Departure—The angle from the horizontal or the direction in which a bullet departs from the barrel; normally a greater vertical angle than that formed by the bore center line at the instant of firing, due to jump.

Line of Sight—The straight line formed by the front and rear sights to the point of aim.

Loading Block—Any block—normally of wood or man-made materials—containing a series of vertical holes in which cases are placed to facilitiate storage, handling or charging of cases. May also be used to temporarily hold fully reloaded cartridges.

Loading Density—As defined in NRA publications, the ratio of the weight of the propellant charge to the weight of water, the *usable* case volume will hold. Also sometimes stated as the ratio of the propellant charge volume to the usable case volume.

Lubricator-Sizer—A tool for bringing bullets to roundness and desired diameter while simultaneously forcing lubricant (under pressure) into the grease grooves.

Magnum—A revolver name copyrighted by Smith & Wesson in the early 1930's, and also a class of extra-powerful revolver cartridges begun by the .357 Magnum. Generically applied to such cartridges and all guns chambered for them. Also applied to certain rifle and shotgun cartridges and loadings.

Mean Radius—A measure of accuracy; the average radial distance of all shots in a group from the center (center of impact) of that group.

Meplat—The small flat found on the point or nose of nearly all bullets.

Mercuric Primer—A long obsolete type of primer containing fulminate of mercury; discontinued because the mercury released by firing attacked and weakened the case, reducing or destroying its value for reloading. Also, this type of priming promotes bore corrosion.

Metal Case (MC)—A military type bullet with a jacket continuing unbroken over the bearing surface and point.

Metal Fouling—Bits of jacket metal rubbed off the bullet which adhere to the bore surface.

Metal Point—A bullet in which the jacket covers only the nose, leaving the lead bearing surface exposed. Not common today.

Mid-Range Trajectory (MRT)—The maximum height the bullet rises above the line of sight during its flight; maximum height does not actually occur at the mid-point, but about 55 percent of the range from the muzzle.

Minute of Angle (MOA)—An angular unit of measure, 1/60 of one degree actually 1.05 inches per 100 yards but generally approximated at 1-inch per 100 yards. Used as a measurement of accuracy, i.e., a 2-inch group at 200 yards is indicated as "2 MOA accuracy."

Misfire—Failure of the cartridge to fire when the primer is struck; usually due to a primer failure or contaminated propellant.

MOA—See *Minute of Angle*.

Mold Blocks—Two pieces of metal fitting closely together, each containing half the cavity into which molten lead is poured to harden and form a bullet.

Mouth Sealer—An asphalt-base compound applied in liquid form inside the case mouth. It becomes semi-hard and seals the case/bullet joint; also contributes to bullet pull.

MRT—See *Mid-Range Trajectory*.

Multi-Ball—A cartridge loaded with more than one full-caliber projectile.

Mushroom—A term applied to classic bullet expansion in which the nose distorts outward and rearward to resemble a mushroom.

Muzzle—The forward end of the barrel from which the bullet exits.

Muzzle Blast—The exit of hot, expanding propellant gases from the muzzle behind the bullet, producing flash, noise and a temporary pressure differential sometimes painful and damaging to the ears.

Muzzle Energy—The kinetic energy possessed by the bullet as it exits the muzzle; see also *Energy*.

Muzzle Pressure—The pressure of the propellant gases inside the bore the instant the bullet leaves the barrel.

Muzzle Velocity—The velocity of the bullet as it exits the muzzle.

Neck—That forward part of the case which grips the bullet.

Neck Down—To reduce neck diameter.

Neck Expansion—See *Expander*.

Neck Size—To reduce the diameter of the case neck by forcing it into a die.

Neck Ream—To reduce the thickness of case neck walls by cutting away brass inside the neck.

Neck Turn—As above but by removing brass from *outside* the neck.

Neck Up—To increase neck diameter.

Non-Corrosive—A primer not containing any ingredient whose combustion residue promotes corrosion.

Non-Mercuric—A primer not containing any mercuric compound.

NRA—National Rifle Association.

NSL—National Shooters League.

NSSF—National Shooting Sports Foundation.

Oil Dent—A dent in the cartridge case caused by excess lubricant trapped between case and die walls.

Open Base—A form of bullet construction where the rear (the base) of the jacket is not closed.

Outside Lubricated—A bullet or cartridge with lubricant applied to the exposed bullet surface; now obsolete except for .22 rimfire cartridges.

Pattern—The distribution of shot pellets from a shotshell or shot load upon a test target or surface; also applied to projectile distribution of a multi-ball load.

Pierced Primer—A primer whose cup is pierced by a defective firing pin or by blowout of the firing pin indentation; sometimes confused with "extruded primer."

Point of Aim—The point upon which a gun's sights are aligned; normally the point the bullet will strike.

Powder (Propellant)—Smokeless; the fuel burned to cause a cartridge to function; normally a mixture of mainly nitrocellulose (single base) and sometimes nitroglycerine (double base) and other materials needed to facilitate manufacture and to give it the desired performance characteristics. Produced in a wide variety of burning rates for particular applications, each designated by a number or name. Most are extruded in solid or tubular form and cut into short lengths, the form, size and area of the individual granules largely controlling burning rate; the exception to extruded powder is Olin "Ball Powder," (see same). Some European forms are rolled into thin sheets, rather than extruded, and then cut into squarish or diamond-shape flakes. Smokeless powders arrived in the 1870's, and quickly displaced black powder which is a mechanical mixture of charcoal, sulphur and saltpeter. Manufacture of all types is dangerous and very taxing; powder-plant explosions are by no means uncommon.

Powder Bridging—The clumping together of propellant granules in a drop tube, most common with large granules; can cause a missing charge, double charge, or any variation in between.

Powder Measure—A device using a rotating or sliding member containing a cavity to measure propellant by volume and place it in the case; may use a fixed, adjustable, or replaceable cavity. Also, see *Charge Cup*.

Powder Trickler—A small device using a rotating tube to discharge propellant by very small quantities or individual granules into a scale.

Pressure—The force generated by burning propellant to drive the bullet out of the barrel; expressed in CUP (copper units of pressure) which is analagous (but not exactly comparable) to pounds per square inch; in European practice, expressed in terms of "atmospheres" or in metric units of measure.

Pressure Gun—A special action and barrel used in laboratory to measure gas pressure. The most common allows the gas to act upon and deform a copper cylinder. That deformation is translated into CUP (Copper Units of Pressure) by means of a "Tarage Table." Other methods are used, including strain gauges and piezo-electric crystals.

Pressure/Velocity Ratio—The ratio of bullet velocity to gas pressure; meaningful only for a given load and gun.

Primer—The removable ignition device in the head of a case. The primer contains a sensitive explosive which generates flame to ignite the propellant charge. A small metal cup containing a pellet is crushed by the firing pin and ignited by pressure and friction. See also *Boxer Berdan* and *Battery Cup*.

Primer Flipper (Tray)—A two-piece tray that facilitates handling of primers.

Primer Indentation (Dent)—Actually a "firing pin indentation" formed when the firing pin strikes the primer to ignite it.

Primer Leak—Escape of gas between the primer cup and case head; caused by excessive pressure or primer or case defect.

Primer Pocket—The cylindrical hole in the head of a case into which the primer is seated tightly.

Primer Pocket Reamer—A cutting tool for removing military crimp from the mouth of a primer pocket; may also be used to form a larger radius on the mouth of the pocket in some uncrimped cases.

Primer Pocket Swage—A forming punch forced into a primer pocket to remove crimp or to resize or reshape the pocket.

Primer Tool—A special tool for seating primers more precisely as a separate operation.

Progressive Burning—Applied to propellants which burn relatively slowly due to granule shape and area; plus, deterrent ingredients and coatings.

Projectile—A bullet once it has been placed in motion by the force of propellant gases.

Proof (Prove)—Testing a gun by firing in it a cartridge ("proof cartridge") developing 20 percent (or more) greater than normal pressure.

Propellant—See *Powder*.

Protruding Primer—The condition when a primer is not fully seated and extends beyond the case head; may jam revolvers and cause premature firing in autoloaders. If a properly seated primer protrudes after firing, it *may* indicate excess headspace. (See *Headspace*.)

Ram—The large cylindrical rod or plunger of some loading presses which forces the case in and out of dies.

Rebated Head—A rimless case whose extraction flange is substantially smaller in diameter than the case head.

Recoil—The rearward movement of a firearm generated as an opposite reaction to the forward acceleration of the bullet and propellant.

Recoil Buffer—In an autoloader, an additional spring which is compressed during the last part of rearward slide travel to slow it more gradually and thus reduce felt recoil.

Reprime—After a fired primer is removed (decapped) it is "reprimed" with a fresh primer.

Resizing—See *Sizing*.

Rifling—The spiral grooves and ribs formed in a bore to spin the bullet about its longitudinal axis. This stabilizes the bullet for predictable flight beyond the barrel.

Rim—A flange at the rear of a case providing a surface for extraction and for positioning the cartridge in the chamber.

Rimfire—A type of cartridge case in which the priming compound is placed in a hollow rim and ignited as that rim is being crushed by the firing pin; not reloadable.

Rimless—A cartridge case whose extraction flange is of no greater diameter than the head and is formed by a V-shape groove cut circumferentially in the case.

Round—A complete loaded cartridge or, in muzzle-loading or separate-loading arms, all the ammunition components necessary to load and fire a single shot.

Round Nose—The most common bullet shape where the cylindrical bearing surface has a semi-spherical front part.

Rupture—A break in the case wall or head; quite dangerous if at the rear where gas can escape.

SAAMI—Sporting Arms & Ammunition Manufacturers Institute; an industry organization which establishes certain standards for ammunition and components, mainly to insure safety and interchangeability.

Seating Depth—The distance a bullet enters the case mouth when loaded.

Sectional Density—The weight of a bullet in *pounds*, divided by the square of its diameter in inches; a decimal value indicating the bullet's ability to retain velocity and to penetrate targets.

Semi-Rim—As in rimless, except that the flange protrudes *slightly* (perhaps .015-inch) beyond the case head.

Shank—The cylindrical rear portion of a bullet; also see *Bearing Surface*.

Shell—Originally an explosive artillery projectile; by common current usage a loaded shotgun cartridge or a cartridge case.

Shell Holder—A device containing a recess to fit the case/ head closely so it may be forced into and out of dies; attached to the "Ram," (see same) by a T-slot and tenon.

Shock—The physical and nervous effect of bullet striking a live target.

Shot—The firing of a cartridge; the report of a cartridge being fired; the lead-alloy, spherical pellets loaded in shotshells as missiles.

Shoulder—In a cartridge case, the conical transition area from the body to a smaller neck.

Single Base Powder—See *Powder*.

Sizing (Resizing)—Reduction of an expanded, fired case down to its original size so it will enter standard chambers freely and will hold a new bullet tightly; done by forcing the case into a hardened, polished die cavity of the proper shape and dimension(s).

Slugging—The passing of an oversize lead slug through a bore to determine proper bore diameter. Also; see *Upset*.

Soft Point (SP)—A bullet with a closed-base jacket, exposing a part of the lead core at the nose.

Solid Head—Modern case construction where the web is thick enough to accommodate the primer pocket entirely.

Swage—Technically, the bringing of an undersize bullet or other object upwards in size and shape by applying pressure in a die which expands it to the limits of the die; in common usage; also means the reduction of an oversize object by forcing it through a die.

Throat—See *Freebore* and *Forcing Cone*.

Trajectory—The path of a bullet in flight, generally an approximation of a parabolic curve.

Twist—The spiral path of the rifling; the rate of twist expressed in the distance in inches required for one complete turn, as 1:10, meaning 1 turn in 10 inches of barrel.

Upset (Slugging)—The foreshortening and blunting of the front of a bullet, sometimes accompanied by radial expansion, as the bullet accelerates.

Velocity—The speed at which a bullet travels, expressed in FPS (feet per second) here in the U.S.; or, in meters per second under the metric system abroad.

Vernier Caliper—A slide-type measuring instrument much used by handloaders.

Wad—In handgun loads, a disc of paper, fiber, plastic, etc., used to close the case mouth or separate shot and propellant. Sometimes made of or containing lubricant.

Wadcutter—A cylindrical bullet shape most used for light target loads in revolvers; cuts a clean, sharp-edged hole in paper targets and is sometimes used defensively for that reason.

Web—In a cartridge case that thick portion of the head containing the primer pocket.

Wildcat—A reshaped cartridge case intended to produce performance not available from factory-loaded cartridges.

Charges Per Pound of Propellant

Charge wt., gr.	No. Charges* per lb.	Charge wt., gr.	No. Charges* per lb.
1.0	7000	11.0	636
1.5	4666	11.5	608
2.0	3500	12.0	583
2.5	2800	13.0	538
2.7	2592	14.0	500
3.0	2333	15.0	466
3.5	2000	16.0	437
4.0	1750	17.0	411
4.5	1555	18.0	388
5.0	1400	20.0	350
5.5	1272	21.0	333
6.0	1166	22.0	318
6.5	1076	23.0	304
7.0	1000	24.0	291
7.5	933	25.0	280
8.0	875	26.0	269
8.5	823	27.0	259
9.0	777	28.0	250
9.5	736	29.0	241
10.0	700	30.0	233
10.5	666		

**NOTE: Number of charges are rounded off, dropping all fractions, and allows for no spillage; allow about 5% for spillage and other losses.*

Number of Bullets From 1 Lb. Melt

Bullet wt., gr.	Bullets Per Pound of Melt*	Bullet wt., gr.	Bullets per Pound of Melt*
50	140	170	41
55	127	180	38
60	116	190	36
65	107	200	35
70	100	210	33
75	93	220	31
80	87	230	30
85	82	240	29
90	77	250	28
95	73	260	26
100	70	270	25
105	66	280	24
110	63	290	23
115	60	300	23
120	58	310	22
130	53	320	21
140	50	330	21
150	46	340	20
160	43	350	20

**NOTE: Figures given do not allow for any wastage; allow up to 10% when calculating requirements. In number of bullets, all fractions have been dropped.*

Common Firearms Abbreviations

AE: Automatic Ejectors
AP: Armor Piercing in reference to bullets.
API: Armor Piercing Incendiary in reference to bullets.
AR: Automatic Rifle
AS: Automatic Safety

BB: Bevel Base in reference to bullets.
BC: Ballistic Coefficient in reference to bullets.
BC: Battery Cup primer.
BN: Bottle Neck in reference to a cartridge case.
BP: Black Powder
BT: Boat Tail in reference to bullets; Beaver Tail in reference to stocks.

CF: Centerfire
CI: Center of Impact
CT: Copper Tube in a bullet designation.

DA: Double action in a handgun.
DB: Double Barrel in reference to a gun; Double Base in reference to powder.
Den: Headstamp for Denver Ordnance Plant.
DWM: Deutsche Waffenundmunitions Fabrik, used as a headstamp and as a prefix to numerical cartridge designations.

EC: Headstamp for Evansville-Chrysler Ammunition plant.
Ex: Excellent in NRA condition standards.
ECS: Headstamp for Evansville/Chrysler/Sunbeam Ammunition plant.

F: Fair in NRA condition standards.
F: Fire, in reference to a gun safety.
FA: Headstamp for Frankford Arsenal.
FB: Flat Base in reference to bullets.
FC: Full Choke in reference to shotguns; Full Charge in reference to ammunition.
FC or **FCC:** Federal Cartridge Corporation headstamp.
FP: Foot-Pounds in reference to energy.
FPS: Feet per Second in reference to velocity.
FP: Full Patch (FMJ) in reference to bullets.
FMJ: Full Metal Jacket in reference to bullets.
FW: Featherweight.

G: Good in NRA condition standards.
GC: Gas Check in reference to bullets.
GR: G. Roth, used as a headstamp and as a prefix to numerical cartridge designations.

H: Hodgdon (B.E.) as a prefix in a powder designation.
HB: Hollow Base in reference to bullets; Heavy Barrel in reference to barrels.
HE: Hand Ejector (ejection) in reference to a revolver.
H&H: Holland & Holland, old-line British gun-making firm sometimes called "Holland's."
HJ: Half Jacket in reference to bullets.
HP: Hollow Point in reference to bullets; High Power in reference to cartridge or gun designations.
Hy: Hornady Manufacturing Company.

IL: Inside Lubricated in reference to bullets.
Imp: Improved in a cartridge designation.
IMR: Improved Military Powder as a prefix in a powder designation.

JHP: Jacketed Hollow Point in reference to Super Vel bullets.
JSP: Jacketed Soft Point in reference to Super Vel bullets.

L: Lead in reference to bullets.
LC: Headstamp for Lake City Arsenal.
LC: Long Colt, in reference to handgun cartridges.
LGS: Lyman Gun Sight Co.
LP: Large Pistol primer
LR: Large Rifle primer
LW: Light Weight

M: Model, as M-16 rifle.
Mag: Magazine
MC: Metal Case in reference to bullets; Monte-Carlo style comb in reference to stocks.
ME: Muzzle Energy of a bullet in foot pounds.
MH: Martini-Henry rifle or Merwin Hulbert revolver.

MK: Mark, sometimes used instead of M or Model, as in Revolver, .455, MKVI.

MOA: Minute of Angle

MP: Metal Point or Metal Piercing in reference to bullets.

MR: Mean Radius, average shot distance from center of impact.

MRT: Mid-Range Trajectory height

MV: Muzzle Velocity of a bullet in feet per second.

N: Norma Projectilfabrik as a prefix in a powder designation.

NC: Non-Corrosive in reference to primers.

NE: Nitro Express in a cartridge designation.

NM: National Match, in reference to guns, equipment, or ammunition manufactured or selected specifically for the U.S. National Rifle & Pistol Matches.

NMNC: Non-Mercuric, Non-Corrosive in reference to primers.

OL: Outside lubricated in reference to bullets.

OP: Over Powder wad; Open Point in reference to bullets.

OU: Over-under, barrels placed one above the other. Sometimes indicates "Superposed."

OS: Over Shot wad

PB: Plain Base in reference to bullets.

PC or **PCC:** Peters Cartridge Co., in reference to a headstamp or ammunition label.

PG: Pistol Grip

PP: Paper Patch in reference to bullets.

PP: Power-Point (Winchester) or protected point, in reference to bullets; Police Positive in reference to Colt Revolvers; Pistole Polizei (Police Pistol) in reference to Walther pistols.

PPC: Police Pistol Course, a course of fire used in training police in handgun marksmanship.

PSP: Pointed Soft Point in reference to bullets.

QD: Quick Detachable

R: Sturm, Ruger, Inc.

RB: Rolling Block (Remington) rifle or pistol

RE: Recoil Energy

Rem-UMC: Remington-Union Metallic Cartridge Co.

RF: Rapid Fire in reference to target shooting.

RF: Rim Fire in reference to ammunition.

R-P: Remington-Peters

RPM: Rounds per Minute in reference to cyclic rate of fire of an automatic weapon.

RV: Recoil Velocity

S: Safe, in reference to a gun safety.

SA: Savage Arms

SA: Single Action in a handgun; also Springfield Armory.

SAA: Single Action Army, a Colt revolver.

SA: Sierra Products

SB: Single Base Powder

SD: Sectional Density in reference to bullets.

SF: Slow fire

SL: Headstamp for St. Louis Ordnance Plant.

SMG: Submachine Gun

SO: Secant Ogive, a term used by Hornady to describe a particular form of bullet ogive.

SP: Small Pistol primer

Sp: Speer Products

SP: Spire Point or Soft Point in reference to bullets.

Spcl: Special, as in .44 Special, in a cartridge designation

Spitz: Spitzer (pointed) bullet

SR: Semi-Rimmed in reference to a cartridge case.

SS: Single Shot

ST: Silvertip (Winchester-Western) in reference to bullets.

S&W: Smith & Wesson

SWC: Semi Wadcutter in reference to bullets.

S&W/F: Smith & Wesson Fiocci

SV: Super Vel Cartridge Corp.

SX: Super X (Winchester-Western) in cartridge designations; also designates a form of varmint bullet by Hornady.

TW: Headstamp for Twin Cities Arsenal.

TF: Timed Fire

Tr: Tracer in reference to bullets.

UMC: Union Metallic Cartridge Co.

VG: Very Good in NRA condition standards.

VR: Ventilated Rib

WC: Wad Cutter in reference to bullets.

WCC: Headstamp for Western Cartridge Co.

WRA: Winchester Repeating Arms Co.

W-W: Winchester-Western

Some Notes on Components Storage

IF WE HAVE GIVEN the impression that one need not be concerned with the various handloading components except when actually putting them to use, believe me, it was not so intended. Naturally, anyone getting into handloading will very shortly develop a substantial inventory of cases, propellants, primers, bullets, lubricant, and other miscellaneous items. This may result from simply buying in quantity to obtain the best price, or from left-overs as you try different components to obtain the best results. Regardless of how an inventory is generated, none of these items can simply be tossed casually aside into a convenient corner or into a cardboard box and then be expected to somehow pop up again pristinely new and undeteriorated months from now when you suddenly discover a use for them.

CARTRIDGE CASES

You'll most likely acquire a surplus of cases very quickly. Cases are the basic ingredient of handloading, and often, even before one loads his first cartridge, he starts gathering up brass wherever he can find it, in all calibers and conditions. After all, each fired, reloadable case represents roughly half of the cost of a factory-loaded cartridge; and, if you can get it cheap or for free, it seems worthwhile to acquire it and squirrel it away against some future requirement, even if it isn't a caliber you expect to load in the near future. Cases fired with black powder and set away without thoroughly removing the gummy fouling from the interior will most assuredly develop corrosion and eventually be ruined except, perhaps, in a very dry desert atmosphere. Such cases should be thoroughly washed in a strong detergent solution immediately after firing, then dried completely, after which they can be stored indefinitely without deterioration, providing they aren't exposed to excessive moisture. However, those cases really *should* be decapped before washing so that fouling will be removed from the primer pocket and flash hole as well, insuring against corrosion and moisture damage there. Once thoroughly cleaned and dried, cases that have been fired with black powder may be stored like any others.

Cases fired with smokeless powder loads (and that's all most of us will be concerned with) usually aren't given any attention at all before being stashed away in paper bags, tow sacks, cardboard boxes, etc. Given that treatment (or lack thereof) and parked on the floor, in a corner of the garage or basement, they may not fare too well, especially those not plated and thus exposing naked brass to the elements. The concrete floor is by nature damp, and plain brass, in combination with the dirt and other debris it may have picked up on the ground, will slowly but surely accumulate dark stains and greenish corrosion usually called "verdigris." Given enough time this greenish corrosion builds up to substantial thickness and deeply etches the brass beneath. It can become quite difficult to remove; and, even after it is scrubbed away, the case surface is roughened and sometimes even weakened. To a lesser degree, the same thing can happen to plated cases. The initial firing and cycling through the gun often scrapes through the plating in spots, exposing the brass to the ravages of corrosion. Resizing tends to break down the plating even more, so cases used several times (especially if your resizing die isn't quite perfect) may corrode easily.

The problem of corrosion during storage is very easily whipped by washing or cleaning cases to remove fouling and foreign material before storage, and then by placing them in the driest locations possible and in a container which does not attract moisture in and of itself (such as a cardboard box or cloth bag might). Actually, clean cardboard boxes won't cause any trouble in this respect in any but the most humid areas. If garage or basement is the only location available to you, simply stow the boxes well up off the floor—and by that I don't mean just on a couple of bricks or boards on the floor. The higher they are placed, the drier they'll be and the better they'll survive long-term storage.

When I acquire new cases and wish to set them aside for any great length of time, I leave them in the original containers and park them above waist level on shelves in a room equipped with a dehumidifier. Fired cases (whether generated myself or acquired elsewhere) are stored in the same fashion, in clean cardboard boxes, but are first thoroughly cleaned by tumbling. I just looked at some cases on those shelves, and they have deteriorated not in the least, in spite of having been sitting there over 10 years. When storing especially valuable cases, such as some new ones I have in the big British Express calibers, I seal them up in metal boxes with some form of desiccant to extract moisture from the air. Metal military ammunition boxes work beautifully for this purpose and can still often be obtained from various military surplus outlets. Unless damaged, their lids contain a water-proof gasket, and the cases are perfectly protected for an almost indefinite time. One fellow I know stores his surplus cases (after cleaning) in small cardboard boxes placed atop the heating ducts running from his furnace throughout the basement. During most of the damp times of the year the heat is on, and those cases never deteriorate.

PRIMERS AND PROPELLANTS

Both propellants and primers present much the same safety hazards and are also subject to deterioration from the same influences. Consequently, we'll treat them together. Prolonged exposure to excessive heat and dryness will cause moisture to evaporate from both priming compound and propellant; the ultimate result is that individual propellant granules shrink and warp, often lighten slightly in color, and may split and crack. The effect is similar on the priming compound pellet, but not so visible. Once this condition has been reached, stability has been reduced and performance has been adversely affected. If propellant granules are split or cracked, the burning rate may be increased to a danger point. For this reason, primers and propellants should be stored at a temperature of approximately 70 degrees Fahrenheit, but certainly should never be allowed to reach a temperature exceeding 100 degrees for any extended period of time.

Storage in extreme cold has no deleterious effect upon propellants or primers, as witnessed by the fact that vast quantities of these commodities are often stored by the military establishment (as is ammunition) in unheated igloos in some pretty frigid parts of the country. Propellants and primers *fired* while at very low temperatures will produce substandard performance, but will perform normally when brought up to more-or-less standard temperature. So, if you're forced to choose between a hot place or a cold place for propellant and primer storage, by all means go the frigid route.

Excess moisture can cause or contribute to propellant and primer deterioration if the exposure is prolonged. Storing them well above the floor will avoid any problems in this area.

The foregoing may lead one to believe that temporary wetting will render primers or propellant unsuitable for use. If you're unfortunate enough, as I once was, to wake up some morning with water in the basement, lapping over a shelf on which propellants and primers were stored, don't despair. If the propellant is sealed in metal cans, the contents probably won't get wet (check to be sure) anyway; but if it is stored in fiber or cardboard containers, moisture might very well seep through to wet the propellant. Primers are normally shipped in cardboard packages and do tend to get wet under such conditions. In my particular experience, the propellant was spread out thinly on butcher paper and allowed to air-dry. Primers were given the same treatment, but I used an electric fan to speed things up. The primers took longer to dry, mainly because more moisture was trapped between the anvil and the pellet. They, too, were repackaged, but marked for plinking loads against the possibility that there might have been permanent damage. In any event, when they were eventually used, I cannot recall that there were any out-and-out misfires, but neither can I be certain that their performance was entirely up to snuff—at least they went "bang." According to Bruce Hodgdon, of Hodgdon Powder fame, wet propellant should be, ". . . dried for a period of about *6 weeks;* this time frame depends, of course, upon the amount of humidity encountered during drying. At any rate, most propellants will dry out to about 1 percent moisture content; and, the powder will be slightly weakened." If you choose to dry the powder and use it in your normal loading routine, do not assume the above comments are directing you to use "lots more of the dried-out powder." Keep your loading to the *same* propellant-charge figures you were using prior to the drying out phase and be sure to repack the powder in properly identified containers. Bruce further advised that an even better method of drying would be to: "Spread the powder on an elevated bedsheet as it would allow for improved air circulation and, therefore, better, more complete drying."

Let's consider *safety:* If the powder, at the time it became wet, was also exposed to household chemicals, mud, gasoline or what have you, don't take a chance—*flush it down the toilet.* Also, if you feel you *cannot safely* dry the powder out—i.e., leaving it largely unattended for long periods of time, or, in an area which might present an even remote fire hazard—*flush it down the toilet.* In all cases, take a *"SAFETY FIRST"* approach in drying out *all* components; and, do not use the family stove or artificial heat source to speed the process up—you'll regret it.

You're unlikely to ever see any *visible* evidence of primer deterioration. On the otherhand, propellant deterioration is quite evident to the eye, and often to the nose as well. Open a clean, *fresh* can of propellant and familiarize yourself with its aroma and appearance. The ether-like odor comes from the solvents employed during manufacture, a slight trace of which remains, even though the solvent-recovery operations are remarkably efficient. Any significant deviation from that odor *may* indicate some propellant deterioration, especially if of an acrid or sour nature. If upon opening a propellant container it gives off a faint puff of greenish vapor or dust, deterioration is certainly under way. Other evidence of deterioration may be found in "clumping" of the granules where they stick together in lumps; an appearance or feeling of moisture or stickiness; clearly visible variations in color among individual granules; and the appearance of any foreign material such as greenish or reddish powder, mold, etc., indicate deterioration.

As already mentioned, dehydrated or dried-out propellant can produce excessive pressures if used; on the other hand, other forms of deterioration generally reduce its potency, with the result that it will either produce substandard (but safe) performance, or may fail to ignite entirely, or perhaps burn only partially. While dehydrated primers may produce pressures greater than intended, they are so well supported in the case and the amount of compound is so small that gun damage is quite unlikely. Other primer deterioration may produce a misfire, hang-fire, or functioning and ignition that is erratic, though it may not be evident.

Safety becomes an important consideration with primers and propellants. Never remove propellant from its original container except for use, and always make *absolutely certain* that left-over quantities are returned to the *correct* container. If, because of a damaged container, it becomes necessary to repackage propellant, make *absolutely* sure that any previous markings are erased from the new container and that it is indelibly labeled in large characters for its new contents. Spilled propellant (and you'll certainly spill some at one time or another) should be thoroughly cleaned up *immediately,* and preferably destroyed by flushing it down the john. Sweeping it up off the bench top or basement floor is bound to contaminate it with dirt or other debris, and you'll be better off throwing it away than attempting to use it. An accumulation of spilled propellant in the nooks and crannies of your bench or shop is an accident just waiting to happen—a chance spark, excessive heat, or a careless smoker, even bullet-casting, can cause it to ignite and flare up. This *could* ignite other propellant in the area, producing a very violent fire that will destroy your home. Don't think this can't happen—I have been called upon to investigate instances where it *did* happen.

Primers are quite stable when handled with care and left in their original compartmented containers until use. *Never, never,* dump primers from the original pack into a metal, glass, or similar container. Just the impact of the first few primers striking the bottom of such a container can generate sufficient explosive dust that the following primers detonate it and in turn detonate the surrounding primers. The chances of your surviving such an explosion are *very* slim. A handful of primers doesn't look dangerous, but the compound they contain is violently explosive. More than a few primer explosions have occurred in home loading shops due to carelessness. I recall one particular instance to mind: A gentleman had emptied 1,000 primers from their original containers into a coffee can and had further left a couple dozen laying on the bench top nearby. Somehow a heavy object, perhaps a loading can, was dropped or fell on the loose primers, detonating at least one of them which in turn caused the rest to let go by sympathetic detonation. That gentleman was most grievously injured, with multiple torso perforations, and only by dint of great effort and expense were the doctors able to preserve his life and normal functions. Incidentally, a single primer dropped on the floor and left unnoticed may be detonated by a combination of pressure and friction if you step on it—it may punch right through the sole of your boot to cause a painful injury.

For the foregoing reasons, and several others that should be obvious by now, never leave primers or propellants out in your loading shop or area where other people might encounter them. The ideal solution is a securely locked or barricaded room, but lacking that, propellants and primers should always be stored under lock and key when you are not actually engaged in handloading.

Propellants and primers are not particularly fire hazards in your home, but if a fire begins from other causes, they can add much to it. A few pounds of propellant in the middle of a normal fire will eventually ignite and, for only a few seconds, produce a tremendously hot and violent fireball, though *not* an explosion. As primers are heated to the ignition point by surrounding flames, they begin detonating individually. As the package deteriorates or disintegrates they'll go off in bunches, cups, and anvils will be hurled forcefully in all directions. When that happens, the firemen attempting to save your home may depart post-haste in fear of serious injury. That, of course, will allow your house to merrily consume itself, leaving you a somewhat more rigorous fight with the insurance company.

To reduce the hazards just mentioned, propellant and primers should *always* be stored under lock and key in a chest or cabinet which will resist burning for a reasonable period of time and which will insulate the contents from the heat of nearby flames. Pamphlets published by fire-prevention and fire-fighting organizations recommend (and some states require) such a cabinet be constructed on 1-inch thick solid lumber. In most states, such containers are limited to a maximum of 50 pounds of propellant in the original containers. These requirements are based upon the assumption that in a typical residence fire the propellant will not be heated to ignition temperature before it may be reasonably expected that the fire-fighters will arrive and put the fire under control. Incidentally, if you don't feel inclined to build a suitable wood storage cabinet, several firms advertise welded-steel "magazines" which serve the same purpose and meet those requirements.

While recommendations such as those for propellants are seldom encountered for primers, it is my personal belief that they should be stored either with propellants in such a cabinet or chest, or in a similar container alone. It should go without saying that such a storage container for propellant and/or primers should be located away from those areas and conditions where home fires are most likely to start, and further that they should be situated in a cool and dry location.

BULLETS

Now, we are down to bullets. The copper alloy of jacketed bullets is subject to the same type of corrosion already described on brass cases. Consequently, if bullets are stored in damp, muggy areas, their surfaces will surely darken, and they may well develop green verdigris as well. Lead oxidizes very rapidly in a humid atmosphere, so cast bullets should be stored in a dry place. Swaged bullets should be able to resist corrosion better and longer simply because they are normally covered with a thin film of lubricant which retards but does not entirely prevent development of corrosion. Cast bullets don't usually have this protection, so will corrode quickly. I can recall a time in the dim and distant past when I lived in the banana belt and had spent many hours casting a reserve supply of several thousand bullets which were stowed away against pleasanter weather and more free time which would allow me to load my limited supply of cases and get in all the shooting I wanted. Imagine, then, the disappointment and chagrin when, a few months later, those bullets were discovered to be heavily coated with gritty, whitish corrosion. They were salvaged eventually, but not without a lot of fuss and bother—if I'd known then what I know now, there would have been no such problem.

Contrary to what we might like to believe, nothing really takes care of itself. If you would have your handloading components be always ready to give their full measure of service and performance, well then, the responsibility is upon you to store and care for them properly.

Case Capacities—Handgun Cartridges

	Weight of Empty Brass (grs)				Net Capacity (grs. water)				
Case	Rem.	Win.	Other	Other	Rem.	Win.	Other	Other	Average
22 Rem. Jet	74.4				16.9				16.9
221 Remington	82.0				20.8				20.8
25 Automatic	27.8	28.6			2.9	3.0			3.0
256 Winchester			84.5-W				19.0-W		19.0
30 Herrett (Note p)	122.2				29.5				29.5
30 Carbine	73.4	74.0	74.3-G	75.5-F	14.3	14.0	13.9-G	13.5-F	13.9
30 Luger		58.9				11.4			11.4
30 Mauser			70.5-C				14.4-C		14.4
32 S&W	30.8	30.7			4.2	4.1			4.1
32 S&W Long	48.4		*38.5-W	54.4-N	7.6		*9.5-W	7.9-N	8.3
32 Automatic	38.2	45.5			4.4	4.2			4.3
32-20 Win.	68.7	71.0			15.5	15.0			15.3
38 S&W (Old)	*43.9		*44.4-P		*8.5		*8.1-P		*8.3
38 S&W (Late)	59.0				7.8				7.8
380 Automatic	47.3	46.8	48.3-F	45.8-SV	6.7	6.8	6.6-F	6.7-SV	6.7
9mm Luger	55.9	59.0	63.5-F	62.3-SV	7.8	7.7	7.3-F	7.4-SV	7.6
9mm Luger			55.0-S&W	61.3-N			7.9-S&W	7.4-N	
9mm Ultra			55.6-RWS				8.36-RWS		8.36
38 Super Auto	63.9	68.0	69.9-W		9.2	9.0	9.0-W		9.1
38 Special	67.7	66.0	66.5-F	68.9-SV	13.0	13.4	13.2-F	12.8-SV	
38 Special			69.2-Sp	74.4-N			12.7-Sp	12.2-N	12.9
38 Special			66.7-S&W				13.2-S&W		
357 Magnum	77.3	84.4	77.4-F	79.1-SV	16.5	15.7	16.5-F	16.2-SV	16.23
357 Magnum			85.1-S&W	78.5-N			15.4-S&W	16.3-N	15.9
357 AMP (Auto-Mag) (Note q)			127.7-CDM				26.8-CDM		26.8
38-45 Auto (Note r)			86.2-W	86.4-G			14.0-W	13.8-G	13.9
38-40 Win. (Old)	78.4		81.2-P		31.4		31.0-P		31.2
38-40 Win. (Late)	96.5	107.9			29.0	28.4			28.7
41 Rem. Mag.	106.4				21.0				21.0
44 AMP (Auto-Mag)			132.5-CDM				26.2-CDM		26.2
44 S&W Special		111.4				19.8			19.8
44 Rem. Mag.	113.4		113.8-W		24.6		25.2-W		24.9
44-40 Win. (Old)	85.7	79.4	82.9-P		32.0	32.4	32.2-P		32.2
44-40 Win. (Late)		104.5				29.9			29.9
45 Auto Rim	87.5				15.0				15.0
45 ACP	89.5	83.1	86.7-F	87.3-G	14.6	14.9	14.0-F	14.3-G	14.5
45 ACP				88.7-N				14.6-N	14.5
45 Colt (Old)	*86.3	*86.9	*85.9-W		*32.9	*32.6	*32.2-W		*32.5
45 Colt (Late)	107.5				27.7				27.7
455 Colt			76.4-D				20.1-D		20.1

Note that case capacity (interior volume) will vary somewhat depending on the dimensions of the rifle/handgun in which it was fired, cases expanding to fit the chamber(s). Thus, a large or oversize chamber would allow cases to swell, thereby increasing capacity; one more reason why these figures should not be considered as exact. Also, some cases such as the 44 Magnum may have been fired in either rifle or revolver, all of which may help to account for some seeming inconsistencies.

Note p: 30 Herrett cases are formed from 30-30 brass cut to 1.605″.
Note q: 357 AMP (Auto-Mag) cases are formed from 44 Auto-Mag cases necked down.
Note r: 38-45 Automatic cases formed from 45 ACP brass necked down.

*—balloon-head cases.

— KEY —

A=Australian
B=Browning
C=Czech comm.
CDM=Cartuchos
D=Dominion
F=Federal
G=Government
GS=Golden State
H=Herters
K=Kynoch
N=Norma
P=Peters
R=Rigby
RWS=Rhein, Westf. Sprengstoff
S=Sako
Sp=Speer
SV=Super Vel
S&W=Smith & Wesson
W=Western
Wb=Weatherby

Handgun Case Dimension Drawings

.22 "K" Hornet

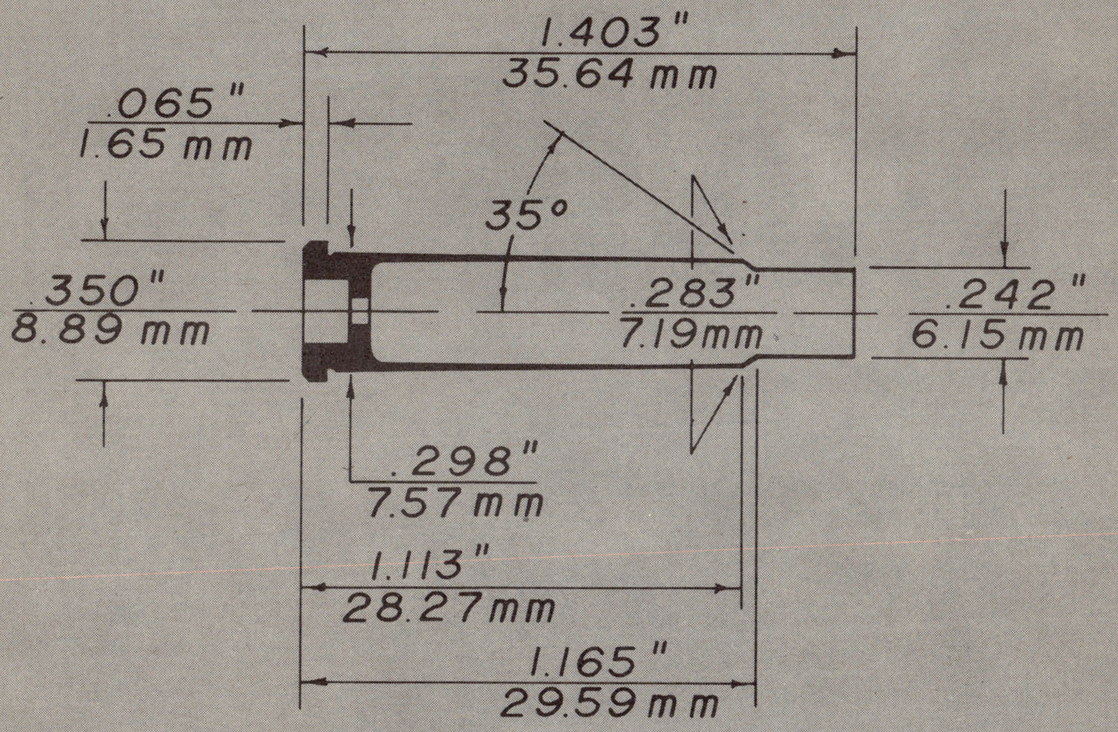

.221 Remington Fireball

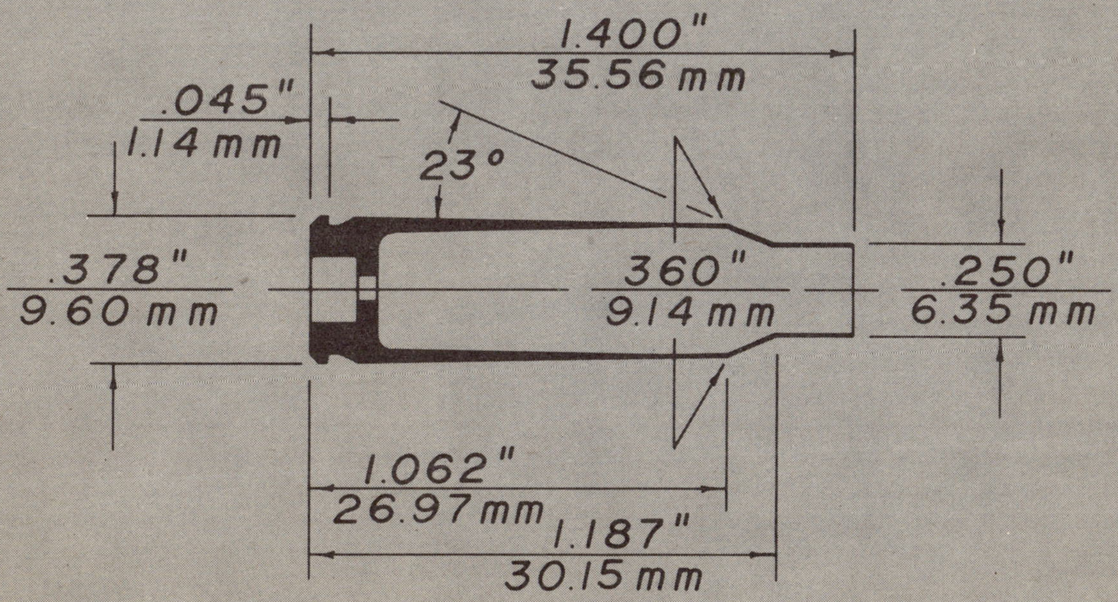

.30 U.S. Carbine

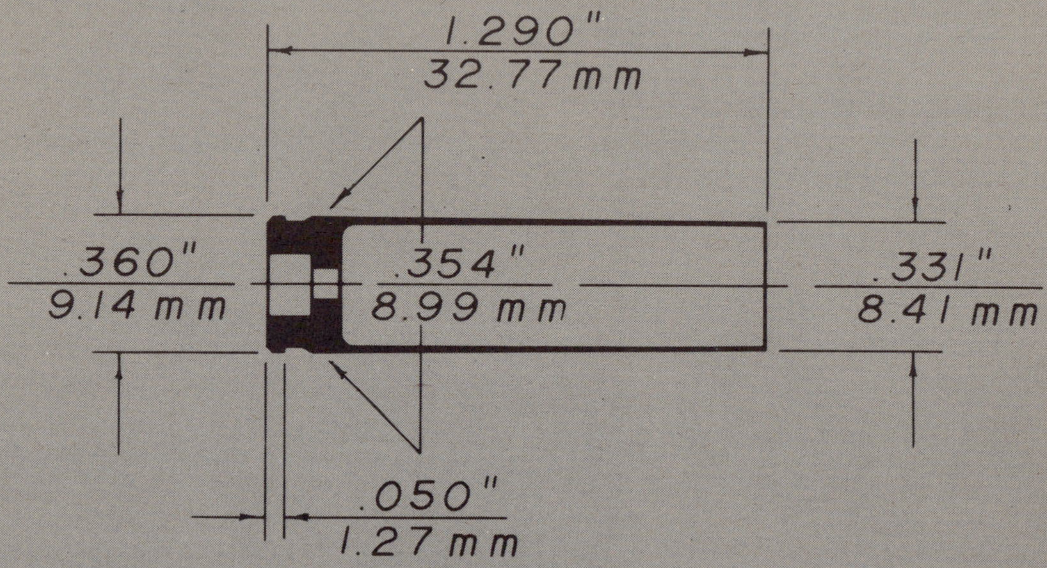

.30 Herrett

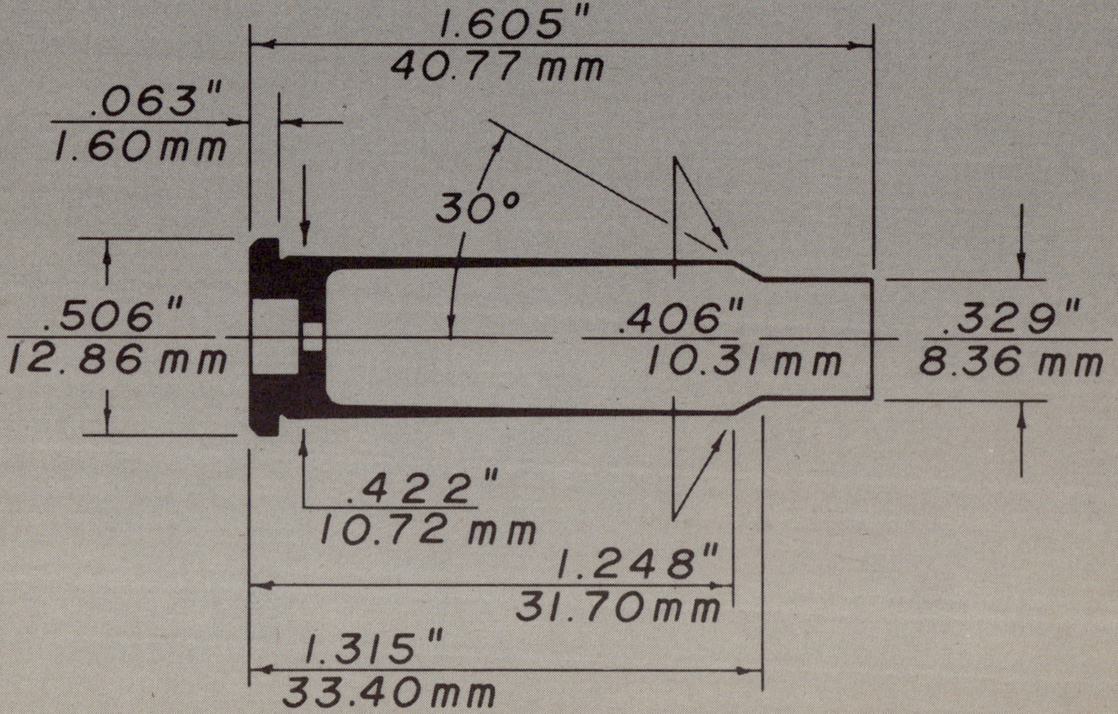

.380 ACP

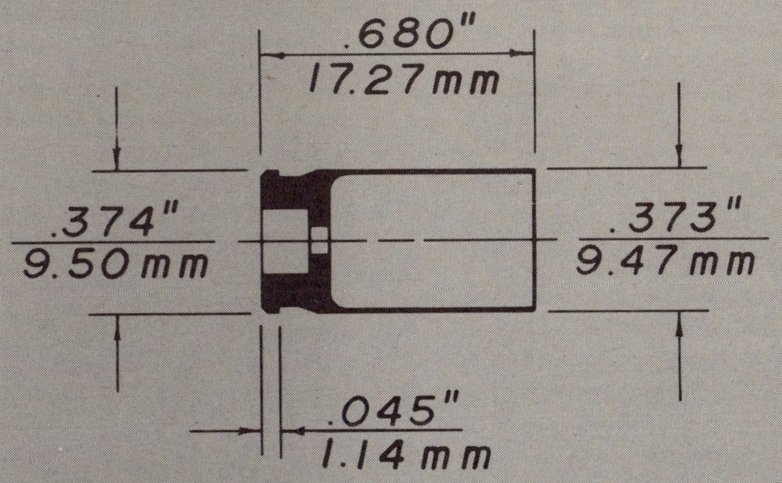

9mm Parabellum (Luger)

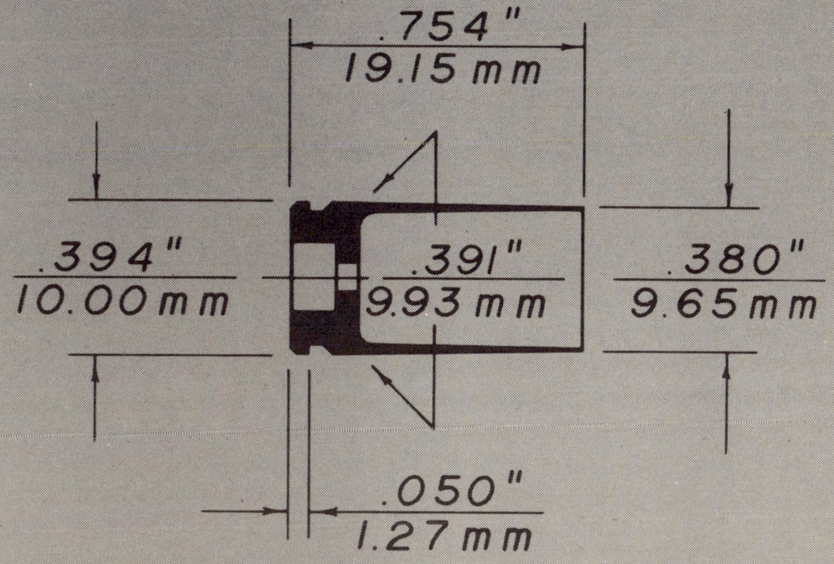

.38 Super

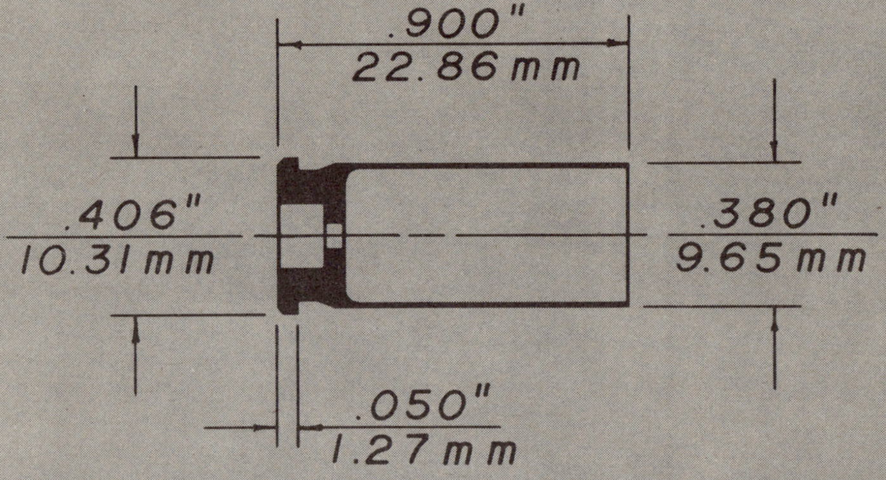

.38 S&W

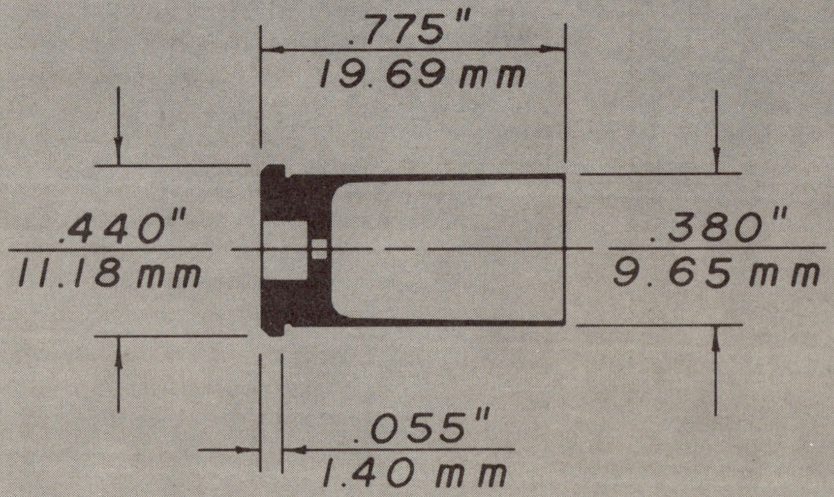

.38 Special

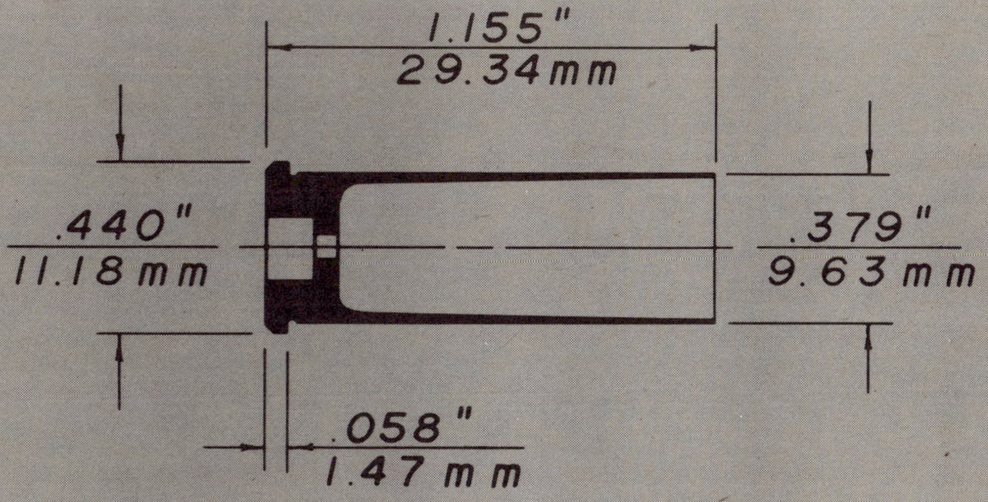

.357 Magnum

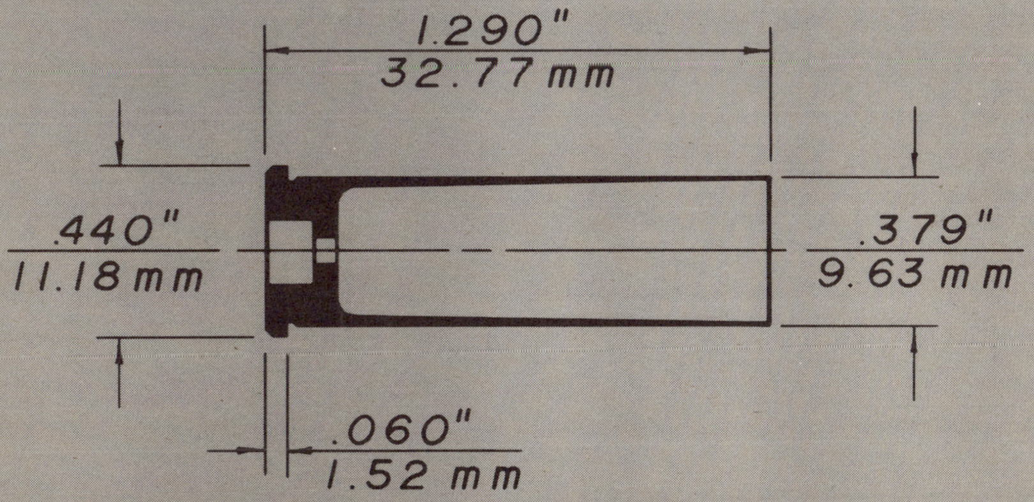

.357 AMP (Auto Mag Pistol)

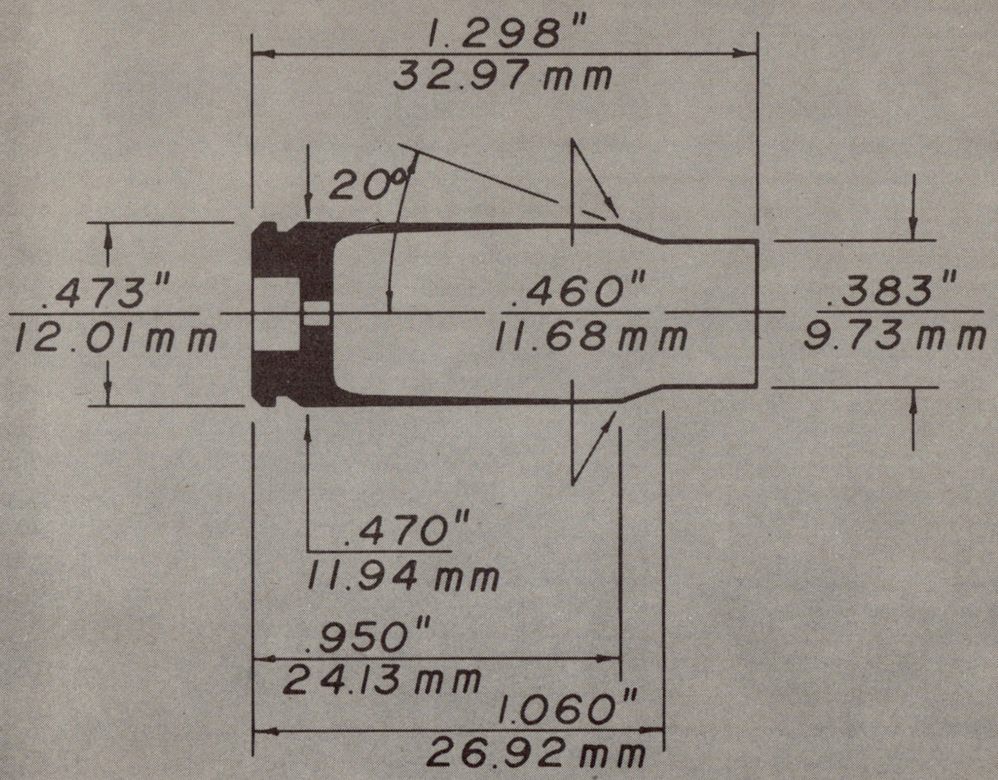

.41 Magnum

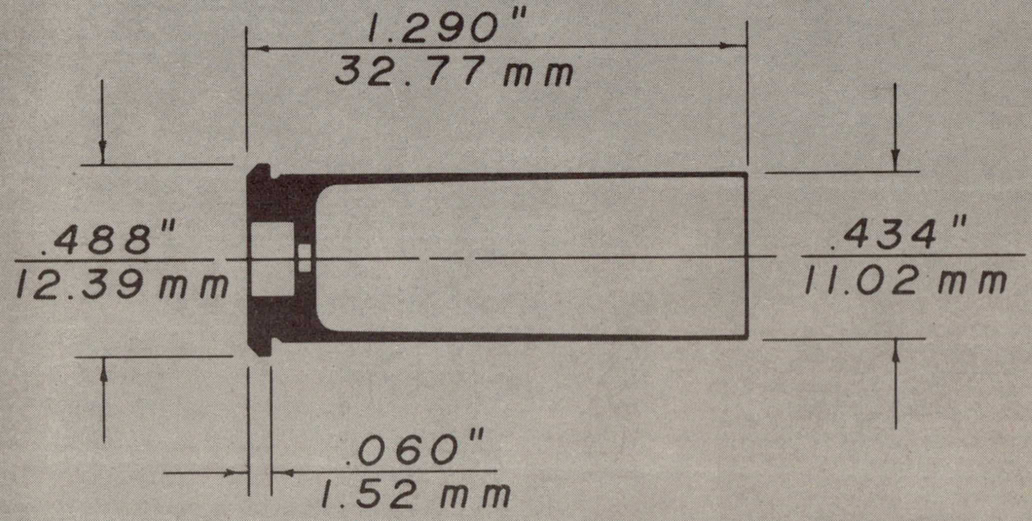

.44 Special

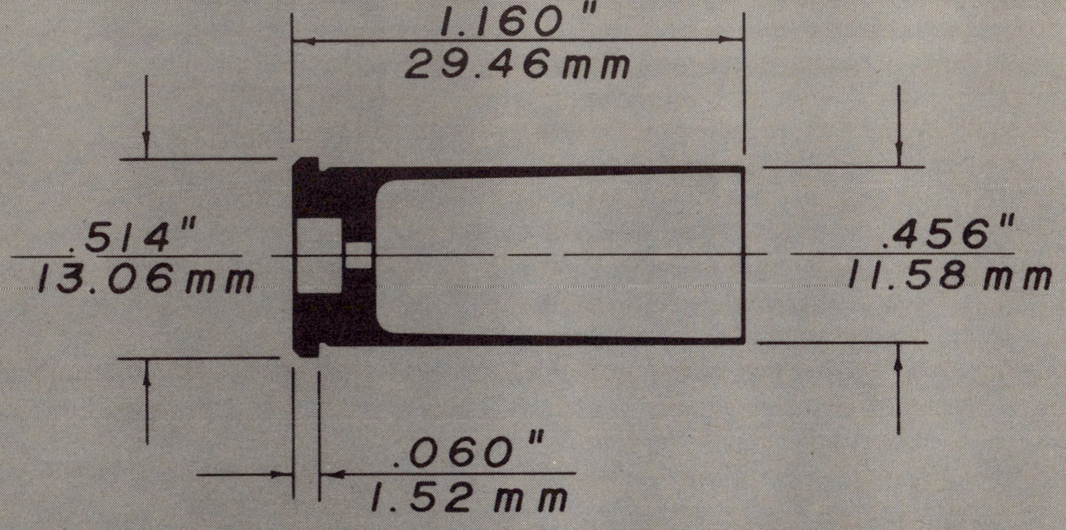

.44 Magnum

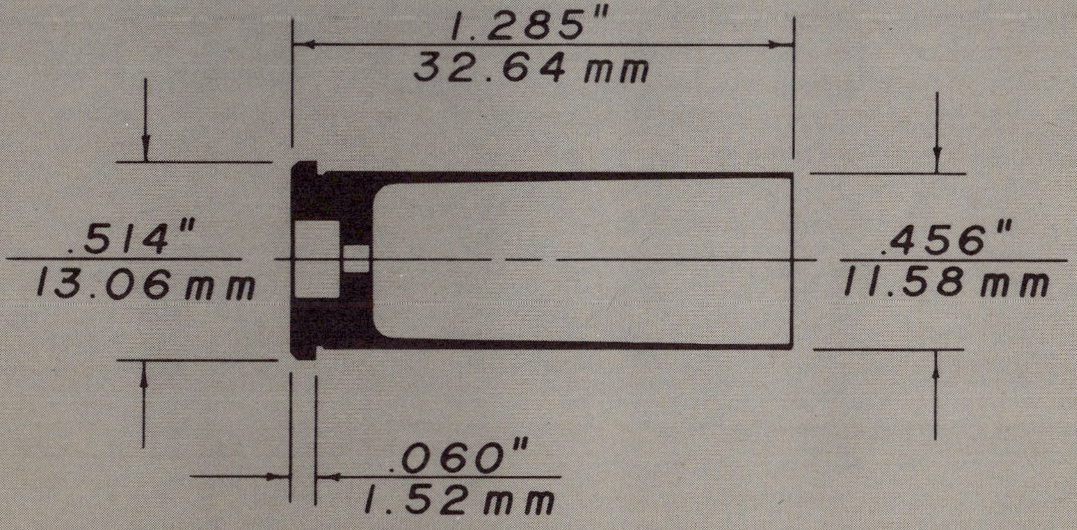

.44 AMP (Auto Mag Pistol)

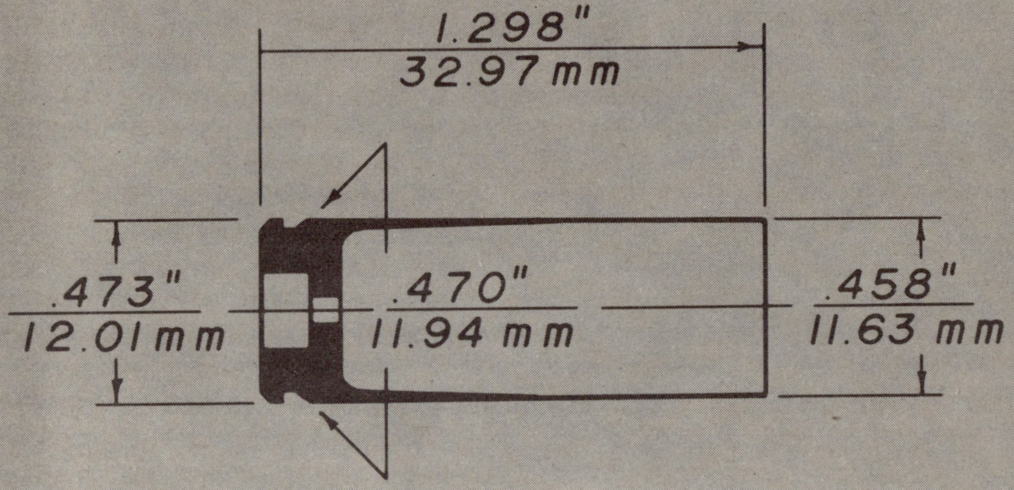

.45 Auto Rim

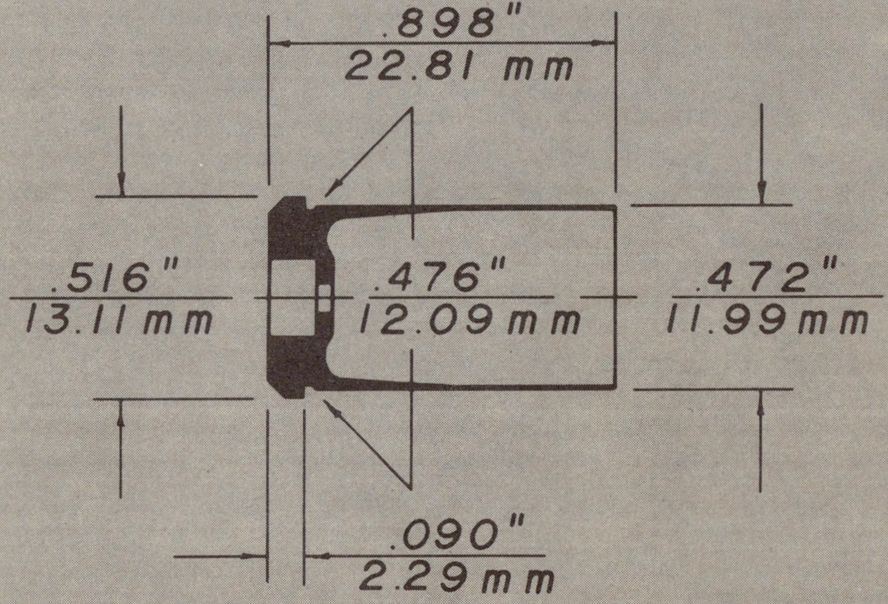

.45 ACP

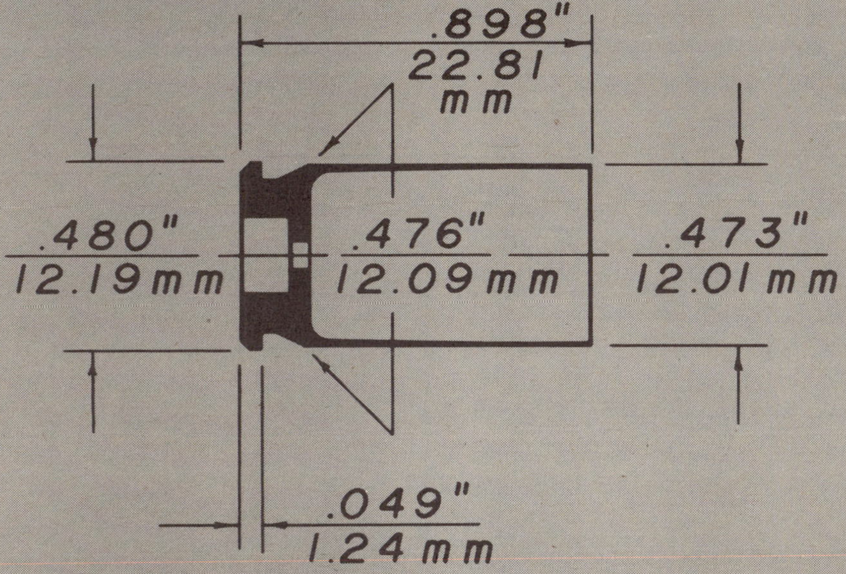

.45 Colt

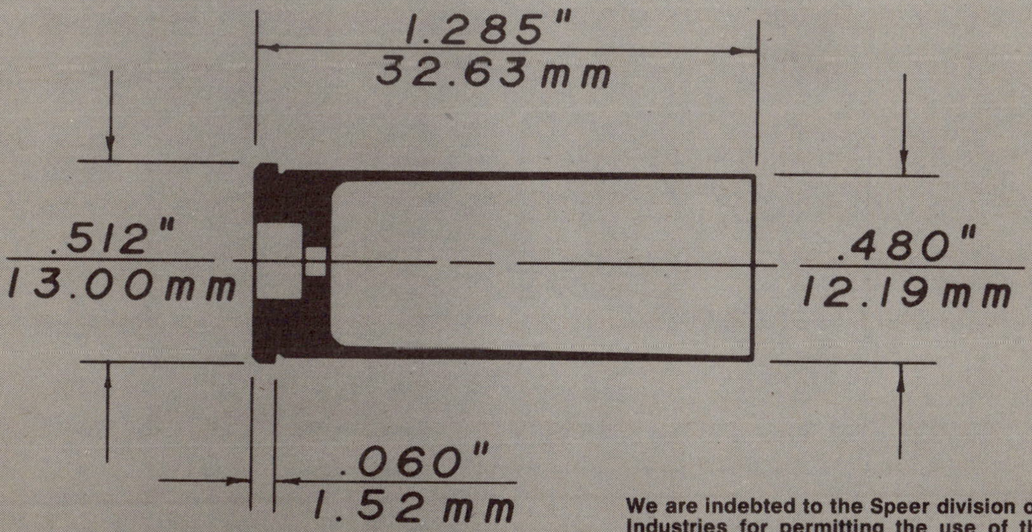

We are indebted to the Speer division of Omark Industries for permitting the use of their cartridge case drawings presented herewith.

Table of Bullet Energies

This table of energies has been worked out by application of the existing formula for computing energy and gives the foot-pounds of striking energy for one grain of bullet weight. The formula for using this table is simple: multiply the foot-pounds opposite the desired velocity by the weight of your bullet. Velocities have been carefully worked out for each increasing ten foot-seconds. For example, to obtain the energy of a 145-grain bullet at 1415 f.s. locate 1410 f.s. in the proper column and you find the energy to be 4.41 foot-pounds. The next figure is for 1420 f.s. and runs 4.47. Difference, .06 foot-pounds. Halve this and get .03, which, added to 4.41, gives 4.44. To get bullet energy, multiply 4.44 by 145 grains, and the figure is 643.8 or 644 foot-pounds. Use of this table saves much time in figuring muzzle or remaining energy of bullets.

Velocity in fps	Energy in ft. lbs.	Velocity in fps	Energy in ft. lbs.	Velocity in fps	Energy in ft. lbs.	Velocity in fps	Energy in ft. lbs.
600	.80	950	2.00	1300	3.75	1650	6.04
610	.82	960	2.04	1310	3.81	1660	6.12
620	.85	970	2.08	1320	3.86	1670	6.19
630	.88	980	2.13	1330	3.92	1680	6.26
640	.91	990	2.17	1340	3.98	1690	6.34
650	.94	1000	2.22	1350	4.04	1700	6.41
660	.96	1010	2.26	1360	4.10	1710	6.49
670	.99	1020	2.31	1370	4.16	1720	6.57
680	1.02	1030	2.35	1380	4.22	1730	6.64
690	1.05	1040	2.40	1390	4.29	1740	6.72
700	1.08	1050	2.45	1400	4.35	1750	6.80
710	1.11	1060	2.49	1410	4.41	1760	6.88
720	1.15	1070	2.54	1420	4.47	1770	6.95
730	1.18	1080	2.59	1430	4.54	1780	7.03
740	1.21	1090	2.63	1440	4.60	1790	7.11
750	1.24	1100	2.68	1450	4.66	1800	7.19
760	1.28	1110	2.73	1460	4.73	1810	7.27
770	1.31	1120	2.78	1470	4.79	1820	7.35
780	1.34	1130	2.83	1480	4.86	1830	7.43
790	1.38	1140	2.88	1490	4.93	1840	7.51
800	1.42	1150	2.93	1500	5.00	1850	7.60
810	1.45	1160	2.99	1510	5.06	1860	7.68
820	1.49	1170	3.04	1520	5.13	1870	7.76
830	1.53	1180	3.09	1530	5.19	1880	7.84
840	1.56	1190	3.14	1540	5.26	1890	7.94
850	1.60	1200	3.19	1550	5.33	1900	8.01
860	1.64	1210	3.25	1560	5.40	1910	8.10
870	1.68	1220	3.30	1570	5.47	1920	8.18
880	1.72	1230	3.36	1580	5.54	1930	8.27
890	1.76	1240	3.41	1590	5.61	1940	8.35
900	1.79	1250	3.47	1600	5.68	1950	8.44
910	1.83	1260	3.52	1610	5.75	1960	8.53
920	1.87	1270	3.58	1620	5.82	1970	8.61
930	1.92	1280	3.63	1630	5.90	1980	8.70
940	1.96	1290	3.69	1640	5.97	1990	8.79
						2000	8.88